The Neuroscience of Intelligence

This new edition provides an accessible guide to advances in neuroscience research and what they reveal about intelligence. Compelling evidence shows that genetics plays a major role as intelligence develops from childhood, and that intelligence test scores correspond strongly to specific features of the brain assessed with neuroimaging. In detailed yet understandable language, Richard J. Haier explains cutting-edge techniques based on DNA and the imaging of brain connectivity and function. He dispels common misconceptions, such as the belief that IQ tests are biased or meaningless. Readers will learn about the real possibility of dramatically enhancing intelligence and the positive implications this could have for education and social policy. The text also explores potential controversies surrounding neuro-poverty, neuro-socioeconomic status, and the morality of enhancing intelligence for everyone.

Richard J. Haier is Professor Emeritus at the University of California, Irvine, USA. He received a PhD from Johns Hopkins University. He pioneered the use of neuroimaging to study intelligence in 1988 and has given invited lectures at meetings sponsored by the National Science Foundation, the National Academy of Sciences, the Defense Advanced Research Projects Agency, the European Molecular Biology Organization, and Cold Spring Harbor Laboratory. In 2013, he created video lectures, *The Intelligent Brain*, for The Great Courses. In 2016, he served as President of the International Society for Intelligence Research and became Editor-in-Chief of *Intelligence*. In 2020, he received the Life Time Achievement Award from the International Society of Intelligence Research. He is the co-editor of *The Cambridge Handbook of Intelligence and Cognitive Neuroscience* (2021) and the co-author of *The Science of Human Intelligence* (due 2023). Dr. Haier has discussed intelligence research on many podcasts including with Jordan Peterson, Scott Barry Kaufman, and Lex Fridman.

Cambridge Fundamentals of Neuroscience in Psychology

Developed in response to a growing need to make neuroscience accessible to students and other non-specialist readers, the *Cambridge Fundamentals of Neuroscience in Psychology* series provides brief introductions to key areas of neuroscience research across major domains of psychology. Written by experts in cognitive, social, affective, developmental, clinical, and applied neuroscience, these books will serve as ideal primers for students and other readers seeking an entry point to the challenging world of neuroscience.

Books in the Series

The Neuroscience of Expertise by Merim Bilalić
The Neuroscience of Intelligence by Richard J. Haier
Cognitive Neuroscience of Memory by Scott D. Slotnick
The Neuroscience of Adolescence by Adriana Galván
The Neuroscience of Suicidal Behavior by Kees van Heeringen
The Neuroscience of Creativity by Anna Abraham
Cognitive and Social Neuroscience of Aging by Angela Gutchess
The Neuroscience of Sleep and Dreams by Patrick McNamara
The Neuroscience of Addiction by Francesca Mapua Filbey
The Neuroscience of Sleep and Dreams, 2e, by Patrick McNamara
The Neuroscience of Intelligence, 2e, by Richard J. Haier

The Neuroscience of Intelligence

SECOND EDITION

Richard J. Haier
University of California, Irvine

CAMBRIDGE
UNIVERSITY PRESS

Shaftesbury Road, Cambridge CB2 8EA, United Kingdom

One Liberty Plaza, 20th Floor, New York, NY 10006, USA

477 Williamstown Road, Port Melbourne, VIC 3207, Australia

314–321, 3rd Floor, Plot 3, Splendor Forum, Jasola District Centre,
New Delhi – 110025, India

103 Penang Road, #05–06/07, Visioncrest Commercial, Singapore 238467

Cambridge University Press is part of Cambridge University Press & Assessment,
a department of the University of Cambridge.

We share the University's mission to contribute to society through the pursuit of
education, learning and research at the highest international levels of excellence.

www.cambridge.org
Information on this title: www.cambridge.org/9781009295062

DOI: 10.1017/9781009295055

© Richard J. Haier 2023

This publication is in copyright. Subject to statutory exception and to the provisions
of relevant collective licensing agreements, no reproduction of any part may take
place without the written permission of Cambridge University Press & Assessment.

First published 2017
8th printing 2022
Second edition 2023

Printed in the United Kingdom by TJ Books Limited, Padstow Cornwall

A catalogue record for this publication is available from the British Library.

Library of Congress Cataloging-in-Publication Data
Names: Haier, Richard J., author.
Title: The neuroscience of intelligence / Richard J. Haier, UC Irvine.
Description: Cambridge, United Kingdom ; New York, NY, USA : Cambridge
University Press, 2023. | Series: Cambridge fundamentals of neuroscience
in psychology | Includes bibliographical references and index.
Identifiers: LCCN 2022062057 (print) | LCCN 2022062058 (ebook) |
ISBN 9781009295062 (paperback) | ISBN 9781009295055 (ebook)
Subjects: LCSH: Intellect. | Cognitive neuroscience.
Classification: LCC BF431 .H255 2023 (print) | LCC BF431 (ebook) |
DDC 153.9–dc23/eng/20230303
LC record available at https://lccn.loc.gov/2022062057
LC ebook record available at https://lccn.loc.gov/2022062058

ISBN 978-1-009-29506-2 Paperback

Cambridge University Press & Assessment has no responsibility for the persistence
or accuracy of URLs for external or third-party internet websites referred to in this
publication and does not guarantee that any content on such websites is, or will
remain, accurate or appropriate.

To my family, who changed the orbit of my life, and

To the memory of my father who died too young, and

To the memory of my mother who carried on, and

To the memory of my grandparents who sacrificed for a future they could scarcely imagine.

"Forty years of Haier's research and thinking about the neuroscience of intelligence have been condensed into this captivating book. He consistently gets it right, even with tricky issues like genetics. It is an intelligent and honest book."

Robert Plomin, *Institute of Psychiatry, Professor of Psychology and Neuroscience, King's College London.*

"An original, thought-provoking review of modern research on human intelligence from one of its pioneers."

Aron K. Barbey, *Director, Decision Neuroscience Laboratory, Professor in Psychology, Neuroscience, and Bioengineering, Beckman Institute for Advanced Science and Technology, University of Illinois at Urbana-Champaign*

"Deftly presenting the latest insights from genetics and neuroimaging, Haier provides a brilliant exposition of the recent scientific insights into the biology of intelligence. Highly timely, clearly written, certainly a must-read for anyone interested in the neuroscience of intelligence!"

Danielle Posthuma, *Professor of Complex Trait Genetics, VU University Amsterdam, The Netherlands*

"The trek through the maze of recent work using the modern tools of neuroscience and molecular genetics will whet the appetite of aspiring young researchers. The author's enthusiasm for the discoveries that lie ahead is infectious. Kudos!"

Thomas J. Bouchard, Jr., *Emeritus Professor of Psychology, University of Minnesota*

"Richard Haier invites us to a compelling journey across a century of highs and lows of intelligence research, settling old debates and fueling interesting questions for new generations to solve. From cognitive enhancement to models predicting IQ based on brain scans, the quest to define the neurobiological basis of human intelligence has never been more exciting."

Emiliano Santarnecchi, *Associate Professor, Berenson-Allen Center for Noninvasive Brain Stimulation, Harvard Medical School*

"Loud voices have dismissed and derided the measurement of human intelligence differences, their partial origins in genetics, and their associations with brain structure and function. If they respect data, Haier's book will quieten them. It's interesting to think how slim a book with the title 'The Neuroscience of Intelligence' would have been not long ago, and how big it will be soon; Haier's lively book is a fingerpost showing the directions in which this important area is heading."

Ian J. Deary, *Professor Emeritus of Differential Psychology, University of Edinburgh*

"The biology of few psychological differences is as well understood as that of intelligence. Richard Haier pioneered the field of intelligence neuroscience and he is still at its forefront. This book summarizes the impressive state the field has reached, and foreshadows what it might become."

Lars Penke, *Professor of Biological Personality Psychology, Georg August University Göttingen*

"It increasingly appears that we are within years, not decades, of understanding intelligence at a molecular level – a scientific advance that will change the world. Richard Haier's *The Neuroscience of Intelligence* gives us an overview of the state of knowledge that covers not only his own field, the brain, but also recent developments in genetics, and he does so engagingly and accessibly for the non-specialist. I highly recommend it."

Charles Murray, *WH Brady Scholar Emeritus, American Enterprise Institute*

"This book was overdue: a highly readable and inspiring account of cutting-edge research in neuroscience of human intelligence. Penned by Richard Haier, the eminent founder of this research field, the book is an excellent introduction for beginners and a valuable source of information for experts."

Dr. Aljoscha Neubauer, *University of Graz, Austria, & past president of the International Society for the Study of Individual Differences*

"This book is 'A Personal Voyage through the Neuroscience of Intelligence'. Reading this wonderful volume 'forces thinking,' which can be said only about a very small fraction of books. Here the reader will find reasoned confidence on the exciting advances, waiting next door, regarding the neuroscience of intelligence and based on the author's three basic laws: 1) no story about the brain is simple, 2) no one study is definitive, and 3) it takes many studies and many years to sort things out."

Roberto Colom, *Professor of Differential Psychology,*
Universidad Autonoma de Madrid

"Richard Haier's *The Neuroscience of Intelligence* is an excellent summary of the major progress made in the fields of psychology, genetics and cognitive neuroscience, expanding upon the groundbreaking work of "The Bell Curve." He addresses the many misconceptions and myths that surround this important human capacity with a clear summary of the vast body of research now extending into the human brain and genome."

Rex E. Jung, *Assistant Research Professor of Psychology,*
University of New Mexico

Contents

Preface to the First Edition   *page* xi
Preface to the Second Edition   xv
Acknowledgments (First Edition)   xvii
Acknowledgments (Second Edition)   xix

1 **What We Know about Intelligence from the Weight
   of Studies**   1

2 **Nature More than Nurture: The Impact of Genetics
   on Intelligence**   40

3 **Peeking Inside the Living Brain: Neuroimaging Is a Game
   Changer for Intelligence Research**   75

4 **50 Shades of Gray Matter: A Brain Image of Intelligence
   Is Worth a Thousand Words**   105

5 **The Holy Grail: Can Neuroscience Boost Intelligence?**   150

6 **As Neuroscience Advances, What's Next
   for Intelligence Research?**   184

Glossary   227
References   234
Index   287

The color plate section can be found between pp. 76 and 77.

Preface to the First Edition

Why are some people smarter than others? This book is about what neuroscience tells us about intelligence and the brain. Everyone has a notion about defining intelligence and an opinion about how differences among individuals may contribute to academic success and life achievement. Conflicting and controversial ideas are common about how intelligence develops. You may be surprised to learn that the scientific findings about all these topics are more definitive than you think. The weight of evidence from neuroscience research is rapidly correcting outdated and erroneous beliefs.

I wrote this book for students of psychology and neuroscience, educators, public policy makers, and for anyone else interested in why intelligence matters. On one hand this account is an introduction to the field that presupposes no special background; on the other hand it is more in depth than popularized accounts in the mass/social media. My emphasis is on explaining the science of intelligence in understandable language. The viewpoint that suffuses every chapter is that intelligence is 100 percent a biological phenomenon, genetic or not; influenced by environment or not; and that the relevant biology takes place in the brain. That is why there is a neuroscience of intelligence to write about.

This book is not neutral but I believe it is fair. My writing is based on over 40 years of experience doing research on intelligence using mental ability testing and neuroimaging technology. My judgments about the research to include are based on the existing weight of evidence. If the weight of evidence changes for any of the topics covered, I will change my mind and so should you. No doubt, the way I judge the weight of evidence will not please everyone but that is exactly why a book like this elicits conversation, potentially opens minds, and with luck, fosters a new insight or two.

Be advised, if you already believe that intelligence is due all or mostly to the environment, new neuroscience facts might be difficult to accept. Denial is a common response when new information conflicts with prior beliefs. The older you are, the more impervious your beliefs may be. Santiago Ramon Cajal (1852–1934), the father of neuroscience, once wrote: "Nothing inspires more reverence and awe in me than an old man who knows how to change his mind" (Cajal, 1924). Students have no excuse.

The challenge of neuroscience is to identify the brain processes necessary for intelligence and discover how they develop. Why is this important?
The ultimate purpose of all intelligence research is to enhance intelligence.
Finding ways to maximize a person's use of their intelligence is one goal of
education. It is not yet clear from the weight of evidence how neuroscience
can help teachers or parents do this. Finding ways to increase intelligence
by manipulating brain mechanisms is quite another matter and one where
neuroscience has considerable potential. Surely, most people would agree
that increasing intelligence is a positive goal for helping people in the
lower than normal range who often cannot learn basic self-care routines
or employment skills. What then is the argument against enhancing intelligence so students can learn more, or adults can enjoy increased probability
of greater achievement? If you have a negative reaction to this bold statement of purpose, my hope is that by the end of this book you reconsider.

Three laws govern this book: (1) No story about the brain is simple;
(2) no one study is definitive; (3) it takes many years to sort out conflicting and inconsistent findings and establish a compelling weight of
evidence. With these in mind, Chapter 1 aims to correct popular mis-
information and summarizes how intelligence is defined and measured
for scientific research. Some of the validity data will surprise you. For
example, childhood IQ scores predict adult mortality. Chapter 2 reviews
the overwhelming evidence that there are major genetic effects on intelligence and its development. Conclusive studies from quantitative and
molecular genetics leave no doubt about this. Since genes always work
through biological mechanisms, there must be a neurobiological basis
for intelligence, even when there are environmental influences on those
mechanisms. Genes do not work in a vacuum; they are expressed and
function in an environment. This interaction is a theme of "epigenetics"
and we will discuss its role in intelligence research.

Chapters 3 and 4 delve into neuroimaging and how these revolution-
ary technologies visualize intelligence in the brain, and indicate the neurobiological mechanisms involved. New twin studies of intelligence, for
example, combine neuroimaging and DNA analyses. Key results show
common genes for brain structure and intelligence. Chapter 5 focuses
on enhancement. It begins with critiques of three widely publicized
but incorrect claims about increasing IQ and ends with electrical brain
stimulation. So far, there is no proven way to enhance intelligence but I
explain why there is a strong possibility that manipulation of some genes
and their biological processes may achieve dramatic increases. Imagine
a moonshot-like national research effort to reach this goal; guess which
nation apparently is making this commitment (it is not the United States).

Chapter 6 introduces several astonishing neuroscience methods for studying synapses, neurons, circuits, and networks that move intelligence research even deeper into the brain. Soon we might measure intelligence based on brain speed, and build intelligent machines based on how the brain actually works. Large collaborative efforts around the world are hunting intelligence genes, creating virtual brains, and mapping brain fingerprints unique to individuals – fingerprints that predict intelligence. Overlapping neuro-circuits for intelligence, consciousness, and creativity are explored. Finally, I introduce the terms "neuro-poverty" and "neuro-SES" (social economic status) and explain why neuroscience advances in intelligence research may inform education policies.

Personally, I believe we are entering a Golden Age of intelligence research that goes far beyond nearly extinct controversies about whether intelligence can be defined or measured and whether genes are involved. My enthusiasm about this field is intended to permeate every chapter. If you are an educator, policy maker, parent, or student you need to know what twenty-first-century neuroscience says about intelligence. If any of you are drawn to a career in psychology or neuroscience and pursue the challenges of intelligence research, then that is quite a bonus.

Preface to the Second Edition

A lot has happened since I submitted the final manuscript for the first edition of this book in 2015, but not much has changed. As I discovered writing this second edition in 2022, the weight of evidence for the key topics is even stronger now. This is due in no small way to a new generation of researchers who have access to more advanced technology and data analysis methods, and access to incredibly large databases that include DNA, neuroimaging, and cognitive testing. Intelligence research is subject to the same inevitable progress as all other areas of science driven mostly by observations and insights from methodological and technical advances. I am happy to report that the last seven years have sharpened our neuroscience understandings and the formulations of questions still to be answered.

This edition includes new sections on predicting intelligence measures from DNA using polygenic scores. The fact that such prediction is possible further debunks stubborn views that intelligence has little to do with genes and that intelligence cannot be assessed for scientific study. But, more importantly, these findings invigorate efforts to understand how gene expression influences intelligence. They open the door for molecular biology research that could identify salient mechanisms that underlie the cognitive processes necessary for intelligence. It is this kind of research that I believe could someday result in ways to dramatically enhance intelligence for individuals. I still believe that is the ultimate and noble goal of intelligence research.

There are also new sections on neuroimaging and connectivity analyses to identify specific brain networks and circuits relevant to intelligence and to individual differences. This kind of research similarly has potential to identify ways to manipulate network information flow within and across brain areas to enhance intelligence. Related to these advances, many cognitive neuroscientists previously focused on learning and memory have broadened their interests to specifically study intelligence and individual differences. It's not just about psychometrics anymore, but meaningful collaborations still require psychometric sophistication.

The chapters in this edition have been updated with new research findings and references. There were many to choose from and I could not include them all. Mostly, their findings are consistent with those discussed in the first edition. This may be the result of my unconscious cherry-picking or it may reflect the robustness of the underlying phenomena. Time will tell and I will be the first to change my mind if the weight of evidence changes. But, as always in science, new data typically make the explanatory picture more complex, not less. We see this in the stunning images from the James Webb telescope and their impact on cosmology theories. There is poetry and a bit of magic in these images. As neuroscience approaches go deeper and deeper into smaller and smaller brain structures and faster and faster functions, the complexity is both beautifully grand and nightmarishly challenging. That dynamic is the excitement of intelligence research that I hope to convey on every page. The poetry and magic are up to you.

Acknowledgments (First Edition)

Because my academic appointments have been in medical schools, I have never had psychology graduate students working with me on research so I have none to thank. I have had fabulous collaborators over the years and they have made all the difference. Most of my neuro-imaging studies of intelligence are co-authored by friends Rex Jung, Roberto Colom, Kevin Head, Sherif Karama, and Michael Alkire. Many others, too numerous to name, contributed time, effort, and ideas over the last 40 years. I am indebted to all of them. I am especially grateful to Matthew Bennett at Cambridge University Press for inviting me to contribute to their Neuroscience series. It is the first time intelligence has been included. Drafts of this book were read all or in part by Rosalind Arden, Roberto Colom, Doug Detterman, George Goodfellow, Earl Hunt, Rex Jung, Sherif Karama, Marty Nemko, Aljoscha Neubauer, Yulia Kovas, and Lars Penke. Their corrections and insights were invaluable; any remaining errors are mine. Although I have included a substantial number of citations to relevant work, I could not possibly include everything I would have liked to. In fact, the field is moving so quickly, I added newly published papers right up to the last days before my deadline. I apologize to anyone who feels his or her work was left out. Some topics, explanations, and illustrations in this book were included in a set of my video lectures, *The Intelligence Brain* (copyright 2013 The Teaching Company. LLC. www.thegreatcourses.com). Most of all, my wife protected my work time against all intrusions, and that is why this book exists.

Acknowledgments (Second Edition)

Many people reviewed parts of this edition and I thank them all. Of course, any errors are mine alone. I also thank David Repetto and Rowan Groat at Cambridge University Press for bringing this edition to fruition. And, I thank all the readers of the first edition who commented publicly or privately both on the exposition and the content. They made me eager to write this second edition. It turned out to be a harder job than I had imagined because there are so many new interesting papers to consider. I am grateful to have had the time and resources to focus on writing, especially with the unconditional support of my wife, who makes everything possible.

What We Know about Intelligence from the Weight of Studies

[T]he attack on tests is, to a very considerable and very frightening degree, an attack on truth itself by those who deal with unpleasant and unflattering truths by denying them and by attacking and trying to destroy the evidence for them.

Barbara Lerner (1980)

Intelligence is surely not the only important ability, but without a fair share of intelligence, other abilities and talents usually cannot be fully developed and effectively used ... It [intelligence] has been referred to as the "integrative capacity" of the mind.

Arthur Jensen (1981)

The good thing about science is that it's true whether or not you believe in it.

Neil deGrasse Tyson, HBO's *Real Time with Bill Maher*, February 4, 2011

The University of California will no longer consider SAT and ACT scores.

Los Angeles Times, May 15, 2021

Learning Objectives

- How is intelligence defined for most scientific research?
- How does the structure of mental abilities relate to the concept of a general intelligence factor?
- Why do intelligence test scores estimate but not measure intelligence?
- What are four kinds of evidence that intelligence test scores have predictive value?
- Why do myths about intelligence persist?

Introduction

When a computer beats a human champion at games such as chess or Go that require strategy, or a verbal knowledge game such as *Jeopardy*, is the computer smarter than the person? Why can some people memorize exceptionally long strings of random numbers or tell the day of the week for any date in the past, present, or future? What is artistic genius and is it related to intelligence? These are some of the challenges to defining intelligence for research. It is obvious that no matter how you define it, intelligence must have something to do with the brain, and that is why this book is about neuroscience research.

Among the many myths about intelligence, perhaps the most pernicious is that intelligence is a concept too amorphous and ill-defined for scientific study. In fact, the definitions and measures used for research are sufficiently developed for empirical investigations and have been so for over 100 years. This long research tradition used various kinds of mental ability tests and sophisticated statistical methods known collectively as psychometrics. The new science of intelligence builds on that database and melds it with new technologies of the last two decades or so, especially genetic and neuroimaging methods. These advances, the main focus of this book, are helping to evolve a more neuroscience-oriented approach to intelligence research. The trajectory of this research is similar to that in other scientific fields, which has led from better measurement tools to more sophisticated definitions and understandings of, for example, an "atom" and a "gene." Before we address the brain in subsequent chapters, this chapter reviews the current state of basic research issues regarding the definition of intelligence as a general mental ability, the measurement of intelligence relative to other people, and the validity of intelligence test scores for predicting real-world variables.

1.1 What Is Intelligence? Do You Know It When You See It?

It may seem odd, but let's start our discussion of intelligence with the value of pi, the circumference of a circle divided by its diameter. As you know, the value of pi is always the same: 3.14 … carried out to an infinite, nonrepeating sequence of decimals. For our purpose here, it's just a very long string of numbers in seemingly random order that is always the same. This string of numbers has been used as a simple test of memory. Some people can memorize a longer string of the pi sequence than others. And a few people can memorize a very long string.

Daniel Tammet, a young British man, studied a computer printout of the pi sequence for a month. Then, for a demonstration organized by the BBC, Daniel repeated the sequence from memory publicly while checkers with the computer printout followed along. Daniel stopped over five hours later after correctly repeating 22,514 digits in the sequence. He stopped because he was tired and feared making a mistake (Tammet, 2007).

In addition to his ability to memorize long strings of numbers, Daniel also has a facility to learn difficult languages. The BBC arranged a demonstration of his language ability when they moved him to Iceland to learn the local language with a tutor. Two weeks later, he conversed on Icelandic TV in the native tongue. Do these abilities indicate that Daniel is a genius or, at least, more intelligent than people who do not have these mental abilities?

Daniel has a diagnosis of autism and he may have a brain condition called synesthesia. Synesthesia is a mysterious disorder of sensory perception where numbers, for example, may be perceived as colors, shapes, or even odors. Something about brain wiring seems to be amiss, but it is so rare a condition that research is quite limited. In Daniel's case, he reports that he sees each digit as a different color and shape, and when he recalls the pi sequence, he sees a changing "landscape" of colors and shapes rather than numerical digits. Daniel is also atypical among people with autism because he has a higher-than-average intelligence quotient (IQ) score.

Recalling 22,514 digits of pi from memory is a fascinating achievement no matter how it is accomplished (the official record is an astonishing 70,000 digits – see Chapter 6.2). So is learning to converse in the Icelandic language in two weeks. There are people with extraordinary, specific mental abilities. The term savant is typically used to describe these rare individuals. Sometimes the savant ability is an astonishing memory or the ability to rapidly calculate large numbers mentally or the ability to play any piece of music after only hearing it once or the ability to rapidly create sophisticated artistic drawings or sculptures.

Kim Peek (1951–2009), for example, was able to remember an extraordinary range of facts and figures. He read thousands of books, especially almanacs, and he read each one by quickly scanning page after page. He could then recall this information at will as he demonstrated many times in public forums in response to audience questions: Who was the 10th king of England? When and where was he born? Who were his wives? And so on. Kim's IQ was quite low and he could not care for himself. His father managed all aspects of his life except when he answered questions from memory.

Stephen Wiltshire has a different savant ability. Stephen draws accurate, detailed pictures of city skylines and he does so from memory after a short helicopter tour. He even gets the number of windows in buildings correct. You can buy one of his many city skyline drawings at a gallery in London or online. Alonzo Clemons is a sculptor. He also has a low IQ. His mother claims he was dropped on his head as a baby. Alonzo creates animal sculptures in precise detail, typically after only a brief look at his subject. The artistry is amazing. Derek Paravicini has a low IQ and cannot care for himself. Blind from birth, Derek is a virtuoso piano player. He amazes audiences by playing any piece of music after hearing it only once, and can play it in any musical style. It is worth noting that Albert Einstein and Isaac Newton did not have any of these memory, drawing, sculpting, or musical abilities.

Savants raise two obvious questions: How do they do it, and why can't I? We don't really know the answer to either question. These individuals also raise a core question about the definition of intelligence. They are important examples of the existence of specific mental abilities. But is extraordinary specific mental ability evidence of intelligence? Most savants are not intelligent. In fact, they typically have low IQ and often cannot care for themselves. Clearly extraordinary but narrow mental ability is not what we usually mean by intelligence.

One more example is Watson, the IBM computer that beat two all-time *Jeopardy* champions. *Jeopardy* is a game where answers are provided and players must deduce the question. The rules were that Watson could not search the web and all information had to be stored inside Watson's 15 petabytes of memory, which was about the size of 10 refrigerators. Here's an example. In the category "Chicks Dig Me," the answer is: "This mystery writer and her archeologist husband dug to find the lost Syrian City of Arkash." This sentence is actually quite complex for a computer to understand, let alone formulate the answer in the form of a question. In case you're still thinking, the answer, in the form of a question is: "Who was Agatha Christie?" Watson answered this faster than the humans, and in the actual match, Watson trounced the two human champions. Does Watson have the same kind of intelligence as humans, or better? Let's look at some definitions to consider if Watson is more like a savant or Albert Einstein.

1.2 Defining Intelligence for Empirical Research

No matter how you define intelligence, you know someone who is not as smart as you are. It would be unusual if you have never called someone an "idiot" or a "moron" or just plain dumb, and meant it literally. And, in all honesty, you know someone who is smarter than you are. Perhaps you refer to such a person in equally pejorative terms such as "nerd" or "egghead," even if in your innermost self you wish you had more "brains." Given their rarity, it is less likely you know a true genius, even if many mothers and fathers say they know at least one.

There are everyday definitions of intelligence that do not lend themselves to scientific inquiry: Intelligence is being smart. Intelligence is what you use when you don't know what to do. Intelligence is the opposite of stupidity (and we all know stupidity when we see it). Intelligence is what we call individual differences in learning, memory, and attention. Researchers, however, have proposed a number of definitions, and mostly they all share a single attribute. Intelligence is a *general* mental ability. Here are two examples:

1. From the American Psychological Association Task Force on Intelligence:

Individuals differ from one another in their ability to understand complex ideas, to adapt effectively to the environment, to learn from experience, to engage in various forms of reasoning, to overcome obstacles by taking thought. (Neisser et al., 1996)

2. Here's a widely accepted definition among researchers:

[Intelligence is] a very general mental capability that, among other things, involves the ability to reason, plan, solve problems, think abstractly, comprehend complex ideas, learn quickly and learn from experience ... It is not merely book learning, a narrow academic skill, or test-taking smarts. Rather it reflects a broader and deeper capability for comprehending our surroundings – "catching on," "making sense" of things, or "figuring out" what to do. (Gottfredson, 1997)

The concept of intelligence as a general mental ability is widely accepted among many researchers but it is not the only concept. What evidence supports the concept of intelligence as a general mental ability, and what other mental abilities are relevant for defining intelligence? How do we reconcile intelligence as a general ability with the specific abilities of savants?

1.3 The Structure of Mental Abilities and the *g*-factor

We all know from our experience that there are many mental abilities. Some are specific, such as spelling or the ability to mentally rotate 3D objects or to rapidly calculate winning probabilities of various poker hands. There are many tests of specific mental abilities. We have over 100 years of research about how such tests relate to each other. Here's what we know: Different mental abilities are not independent. They are all related to each other and the correlations among mental tests are always positive. That means that if you do well in one kind of mental

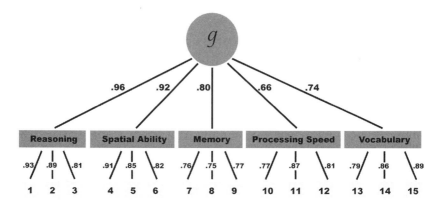

Figure 1.1 The structure of mental abilities. The *g*-factor is common to all mental tests. Numbers are correlations that show the strength of relationship between tests, factors, and *g*. Note all correlations are positive; these are simulated data. (Courtesy Richard Haier)

ability test, you tend to do well in other tests. This may not be the case for any specific person but it is true statistically for populations.

This is the core finding about intelligence assessment and, as we'll see throughout this book, it is the basis for most modern research. Please note this important point: *tend* means there is a higher probability, not a perfect prediction. Whenever we say that one score *predicts* something, we always mean that the score predicts a higher probability for the something.

The relationship among mental tests is called the structure of mental abilities. To picture one possible structure, imagine a three-level pyramid, as shown in Figure 1.1.

At the bottom of Figure 1.1, we have a row of 15 different tests of specific abilities. At the next level up, tests of similar abilities are grouped into more specific factors: reasoning, spatial ability, memory, speed of information processing, and vocabulary. In the illustration, tests 1, 2, and 3, for example, are all reasoning tests and tests 7, 8, and 9 are all memory tests. But all these more specific factors are also related to each other. Basically, people who score high on one test or factor tend to score high on the others (the numbers in the figure are illustrative correlations that show the strength of relationship between tests and factors; see more about correlations in Textbox 1.1). This is a key finding that is demonstrated over and over again. It strongly implies that all the factors derived from individual tests have something in common, and this common factor is called the general factor of intelligence or *g* for short: *g* sits at the highest point on the pyramid in Figure 1.1. The *g*-factor provides a bridge between the definitions of intelligence that emphasize a general mental ability and individual tests that measure (or, more accurately, estimate) specific abilities.

Most theories about factors of intelligence start with the empirical observation that all tests of mental abilities are positively correlated with each other. This is called the "positive manifold," and Charles Spearman first described it more than 115 years ago (Spearman, 1904). Spearman worked out statistical procedures for identifying the relationships among tests based on their correlations with one another. The basic method is called factor analysis. It works essentially by analyzing correlations among tests. You probably already know about correlations, but see the brief review in Textbox 1.1.

Textbox 1.1: Correlations

Many of you know about correlations. Since they are ubiquitous throughout this book, here is a brief explanation so everyone starts with an understanding of the concept. Let's say we measure height and weight in many people. We can graph each person by locating the height and weight as a single point with height ranges on the y-axis and weight ranges on the x-axis. When we add points on the graph for each person, we begin to see an association. Taller people tend to weigh more. You can see this in Figure 1.2. This association is obvious without needing to plot the points, but associations between other variables are not so obvious. Moreover, correlations quantify the strength of association.

If height and weight were perfectly related, the points would all fall on a straight line and we could predict one from the other without error. A correlation has a value of +1 if a high value on one variable goes perfectly with a high value on the other variable. A strong but not perfect positive correlation is shown in Figure 1.2. A perfect negative correlation is where a high value on one variable predicts a low value on the other without error. A strong but not perfect negative correlation (also called an inverse correlation) is also shown in Figure 1.2. A perfect negative correlation has a value of minus 1. In the Figure 1.2 example, the higher the family income, the lower the rate of infant mortality. Finally, in Figure 1.2 the bottom panel shows no relationship at all (zero correlation) between height and hours of video game playing.

Correlations between two variables are calculated based on how much each point deviates from the perfect line. The higher the correlation, positive or negative, the stronger the relationship and the better one variable predicts the other. Correlations always fall between plus and minus 1. Here is a critical point: A correlation between two variables does not mean one causes the

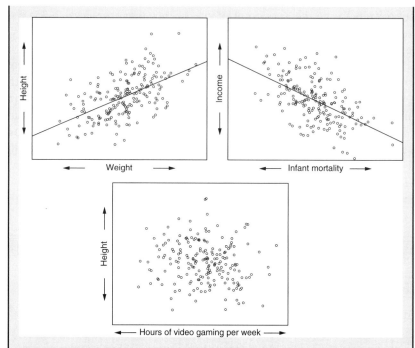

Figure 1.2 An example of a positive correlation is on the left, showing that as height increases weight also increases. A negative correlation is on the right, showing that as family income goes up infant mortality goes down (simulated data). No correlation between height and hours spent playing video games is shown on the bottom. For all of these scatterplots, each circle is a data point. The solid line shows a perfect correlation; the amount that points scatter above and below this line is used to calculate the correlation. (Courtesy Richard Haier)

other. The correlation only means there is a relationship such that as one goes up or down so does the other. To repeat, correlation does not mean causality. Two variables may be correlated to each other but neither causes the other. For example, salt consumption and cholesterol level in the blood may be somewhat correlated but that does not mean one causes the other. The correlation could be caused by a third factor common to both, such as poor diet.

Factor analysis is based on the pattern of correlations among multiple variables. In our case we are interested in the correlations among different tests of mental abilities. So the point of factor analysis is to identify what tests go with other tests, based not on content but rather on correlations of scores irrespective of content. The set of tests that go with each other define a factor because they have something in common that causes the correlation. Studies in this field typically apply factor analysis to data sets where hundreds or thousands of people have completed dozens of tests.

There are many forms of factor analysis but this is the basic concept, the basis for models of the structure of mental abilities such as the pyramid described in Figure 1.1. Going back to that, note that the correlation values show how strong the associations are among tests, factors, and *g*. Note that all the correlations are positive and illustrative of Spearman's positive manifold.

Let's look at some details of this example in Figure 1.1. The reasoning factor is related to *g* with the strongest correlation of 0.96. This indicates that the reasoning factor is the strongest factor related to *g*, so tests of reasoning are regarded as among the best estimates of *g*. Another way of saying this is that reasoning tests have high *g*-loadings. Note that test 1 has the single highest loading of 0.93 on the reasoning factor so it might provide the single best estimate of *g* if only one test is used rather than a battery of tests. The second strongest correlation is between the spatial ability factor and *g*. It turns out that spatial ability tests are also good estimates of *g*. The vocabulary factor is fairly strong at 0.74, followed by the other factors including memory. In this example, memory tests are good but not the best estimators of *g* with a correlation of 0.80, although other research shows much stronger correlations between working memory and *g* (see Section 6.2).

1.4 Alternative Models

Other statisticians and researchers worked out alternative factor analysis methods. The details don't concern us, but different factor analysis models of intelligence were derived using these various methods. Each identified a different factor structure for intelligence. These various factors emphasize that the *g*-factor alone is not the whole story about intelligence; no intelligence researcher ever asserted otherwise or claimed that a single score captures all aspects of intelligence. The other broad factors and specific mental abilities are important. Depending on how researchers derive factors from a battery of tests, a different number of factors secondary to *g* emerge. In the pyramid structure diagram example, there are five broad factors. Another widely used model is based on only two core factors: crystalized intelligence and fluid intelligence (Cattell, 1971, 1987). Crystalized intelligence refers to the ability to learn facts and absorb information based on knowledge and experience. This is the kind of intelligence shown by some savants. Fluid intelligence refers to inductive and deductive reasoning for novel problem solving. This is the kind of intelligence we associate with Einstein or Newton. Measures of fluid intelligence are typically highly correlated to measures of g, and the two are often used synonymously. Crystalized intelligence is relatively stable over the life span with little deterioration with age, whereas fluid intelligence decreases slowly with age (Schaie, 1993). The distinction between fluid and crystalized intelligence is widely recognized as an important evolution in the definition of intelligence. Both are related so they are not in conflict with the g-factor. They represent factors just below g in the pyramid structure of mental abilities.

Another factor analysis model focuses on three core factors – verbal, perceptual, and spatial rotation – in addition to g (Johnson & Bouchard, 2005). There are also models with less empirical evidence such as those of Robert Sternberg (Brody, 2003; Gottfredson, 2003; Sternberg, 2000, 2003, 2014) that deemphasize g, and Howard Gardner (Ferrero, Vadillo, & León, 2021; Gardner, 1987; Gardner & Moran, 2006; Waterhouse, 2006) that ignore the g-factor. Virtually all of the neuroscience studies of intelligence, however, use various measures with high g-loadings. We will focus on these, but also include several neuroscience studies that investigate factors and specific abilities other than g.

1.5 Focus on the *g*-factor

The g-factor is the basis of most intelligence assessment used in research today because it alone accounts for about half of the intelligence test score variability among people. It is not the same as IQ, but IQ scores are good estimates of g because most IQ tests are based on a battery of tests that sample many mental factors, an important aspect of g. Many of the controversies about intelligence have their origins in confusion about how we use words such as mental abilities, intelligence, the g-factor, and IQ. Figure 1.3 shows a diagram that will help clarify how I use these words throughout this book.

We have many mental abilities – all the things you can think of from multiplying in your head to picking stocks to naming state capitals. The large circle in Figure 1.3 represents all mental abilities. Intelligence is a catchall word that means the mental abilities most related to responding to everyday problems and navigating the environment, as per the American Psychological Association and the Gottfredson definitions. The circle labeled intelligence is smaller than all mental abilities. IQ is a

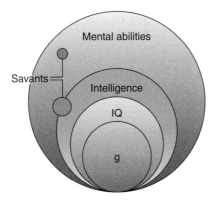

Figure 1.3 Conceptual relationships among mental abilities, intelligence, IQ, and the *g*-factor. (The Intelligent Brain, copyright 2013 The Teaching Company. LLC. Reproduced with permission of The Teaching Company, LLC, www.thegreatcourses.com)

test score based on a subset of the mental abilities that relate to everyday intelligence. The IQ circle is a fairly large part of the intelligence circle because IQ is a good predictor of everyday intelligence. This circle also includes broad factors such as those shown in the diagram of the pyramid structure in Figure 1.1. We describe IQ in more detail in Section 1.6. Finally, the *g*-factor is what is common to all mental abilities. The *g*-factor is a fairly large part of IQ. Whereas everyday intelligence and IQ test scores can be influenced by many factors, including social and cultural ones, the *g*-factor is thought to be relatively more biological and genetic, as we discuss in Chapters 2 and 6.

The savant examples described earlier speak to the level of very specific abilities with little if any *g* in many cases, such as Kim and Derek. They show that powerful independent abilities can exist, but they also show the problems when *g* is lacking. The IBM computer Watson demonstrates a specific ability to analyze verbal information and solve problems based on the meanings of words. This is an amazing accomplishment, but, in my view, Watson does not show the *g*-factor. Watson is more like Kim Peek than Albert Einstein – at least for now. There is a concerted effort among artificial intelligence (AI) researchers to develop general AI, but it is a daunting challenge (Chollet, 2019). Perhaps psychometric or neuroscience insights from the human *g*-factor will be helpful (see Section 6.4).

The savant examples are exceedingly rare cases. Most people have *g* and independent factors to varying degrees, and two people with the same level of *g* can have different patterns of mental strengths and weaknesses across different mental abilities. Can we ever hope to learn how savants do amazing mental feats, and why we can't? Is it possible that we all have the potential to memorize 22,514 digits or the potential for musical or artistic genius? And why are some people just smarter than others? Does everyone have equal potential for learning all subjects? There are many questions and, as in every scientific field, the answers depend entirely on measurement.

1.6 Measuring Intelligence and IQ

IQ is what most people associate with measuring intelligence. Criticism of IQ and all mental tests is widespread, and has been so for decades (Lerner, 1980). It is worth remembering that the concept of testing mental ability arose to help children get special education. It is also worth stating that intelligence tests are regarded as one of the great achievements of psychology despite many concerns. Let's briefly discuss both these points. Informative, detailed discussions about IQ testing are also found in two classic textbooks (Hunt, 2011; Mackintosh, 2011) and a recent one (Haier & Colom, 2023; see also Coyle, 2021).

In the early part of the twentieth century, the minister of education in France was concerned about identifying children with low school achievement who needed special attention. The problem was how to distinguish children who were "mentally defective" from other children who were low achievers owing to behavioral or other reasons. They wanted the distinction to be made objectively by means of testing so a teacher could not assign a child with discipline issues to a special school as a punishment, as was apparently common at the time.

In this context, Alfred Binet and his collaborator, Theodore Simon, devised the first IQ test to identify children who could not benefit mentally from ordinary school instruction. So the IQ test was born as an objective means for identifying low mental ability in children so they could get special attention, and also to identify children erroneously sent to special schools not because of low mental ability but as a punishment for bad behavior. Both goals were admirable.

The test constructed by Binet and Simon consisted of several subtests that sampled different mental abilities with an emphasis on tests of judgment because Binet felt that judgment was a key aspect of intelligence. He gave each test to many children and developed average scores for each age and sex. He was then able to say at what age level any individual child scored. This was called the child's mental age. A German psychologist named William Stern took the concept of mental age another step, by dividing mental age by chronological age. This resulted in an IQ score that was the ratio of a child's mental age (averaged across all the subtests) divided by the child's chronological age. Multiplying this ratio by 100 avoided fractions.

For example, if a child was reading at the level of an average 9-year-old, the child's mental age was nine. If this child actually had a chronological age of 9, the IQ would be 9 divided by 9 = 1 × 100, or an IQ of 100. If a child had a mental age of 10, but was only 9 years old, the IQ would be 10 divided by 9 = 1.11 × 100, or 111. A 9-year-old with a mental age of 8 would have an IQ of 8 divided by 9 = 0.89 × 100, IQ= 89.

The point of these early tests was to find children who were not doing so well in school *relative to their peers*, and get them special attention. The Binet–Simon test actually worked reasonably well for this purpose. However, one problem with the concept of mental age is that it is hard to assess after about age 16. Can we really see a mental age difference between a 19-year-old and a 21-year-old? We're not talking about maturity here. The mental age of a 30-year-old really isn't much different than a 40-year-old, so the Binet–Simon test was not really useful or intended for adults.

But there is a much more important measurement problem to keep in mind. Note that the IQ score is a measure of a child *relative to his or her peers*. Even today, newer IQ tests based on a different calculation, discussed later, show how an individual scores *relative to his or her peers*. IQ scores are not absolute measures of a quantity, such as pints of water or feet of distance. *IQ scores are meaningful only relative to other people.* Note that intelligence differences among people are quite real, but our methods of measuring these differences depend on test scores that are interpretable only in a relative way. We elaborate this key point shortly and return to it throughout this book.

Nonetheless, the Binet–Simon test was an important advance for assessing the abilities of children in an objective way. The Binet–Simon test was translated to English and redone at Stanford University in the 1920s by Professor Louis Terman and the test is now known as the Stanford–Binet test. Professor Terman used very high IQ scores from this test to identify a sample for a longitudinal study of "genius," which we discuss in Section 1.10.4.

The Wechsler Adult Intelligence Scale (WAIS) was designed with subtests such as the Stanford–Binet, but as its name states, it was designed for adults. It is the most widely used intelligence test today. The current version consists of a battery of ten core subtests and another five supplemental subtests. Together, they sample a broad range of mental abilities.

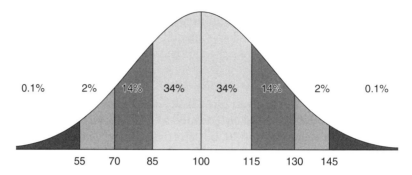

Figure 1.4 The normal distribution of IQ scores and the percentage of people within each level. (Courtesy Richard Haier)

One key change is in the way IQ is calculated in both the WAIS and the Stanford–Binet tests. Mental age is no longer used. IQ is now based on the statistical properties of the normal distribution and deviation scores. The concept is simple: How far from the norm does an individual's score deviate?

Here's how deviation scores work. Let's start with the properties of a normal distribution (also called a bell curve because of its shape), as shown in Figure 1.4.

Many variables and characteristics such as height or income or IQ scores are normally distributed in large populations of randomly selected individuals. Most people have middle values, and the number of individuals decreases toward the low and high extremes of the distribution. Any normal distribution has specific statistical properties in that any individual score can be expressed as a percentile relative to other people. This is shown in the illustration of IQ scores where the mean score is 100 and the standard deviation is 15 points. Standard deviations show the degree of spread around the mean and are calculated as a function of how much each person deviates from the group mean. In a normal distribution, 50 percent of people score below 100, while 68 percent of individuals fall between plus one and minus one standard deviations, so scores between 85 and 115 are regarded as the range of average IQ. A score of 130, two standard deviations above the mean, would be at about the 98th percentile, which is the top 2 percent. A score of 70 would be two standard deviations below the mean and represent about the second percentile. A score of 145 represents the top 10th of 1 percent. Scores over 145 are often considered to be in the genius range, although few tests are accurate at this extreme high end of the distribution.

IQ tests were developed so scores would be normally distributed. Each subtest has been taken by a large number of males and females of different ages. These are the norm groups. Each norm has an average score called the mean, and the spread of scores around the mean is measured by a statistic called the standard deviation.

Let's say a subtest has a perfect possible score of 20 points. Each norm group may have a different average score on this test depending, say, on age. Younger test takers may average 8 points if they are 10 years old, and older children taking the same test, say at age 12, may average a score of 14 points. This is why it's important to have norm groups for each age. If a new 12-year-old takes the subtest and scores 14, he is scoring at the average for his age. If he scores above or below 14, the deviation from the norm average can be calculated and his score can be expressed by how much it deviates from the mean. The average deviation across all the subtests is used to calculate the deviation IQ for the full battery. As illustrated, deviation scores are easily convertible into percentiles.

Each deviation point is equal, but these scores only have meaning relative to other people. In technical terms, these scores are not a ratio scale because there is no actual zero point. This is unlike quantitative units of weight or distance or liquid, which are ratio scales. IQ scores and their interpretation depend on having good normative groups. This is one reason that new norms are generated periodically for these tests. It is also why there is a separate version of the test for children called the Wechsler Intelligence Scale for Children.

The WAIS can be divided into specific factors other than the full scale IQ (FSIQ) score that closely resemble the pyramid structure of mental abilities shown in Figure 1.1. The individual subtests are grouped at the next highest level into factors of verbal comprehension, working memory, perceptual organization, and processing speed. These four specific factors are grouped into more general factors of verbal IQ and performance IQ, and these two broad factors have a common general factor defined by the total IQ score, or FSIQ. This is based on several tests that sample a range of different mental abilities, and is therefore a good estimate of the g-factor. Each of the factor scores can be used for other predictions, but FSIQ is the most widely used score in research.

1.7 Some Other Intelligence Tests

So far, the IQ tests we have described are administered by a trained test-giver interacting with one individual at a time until the test is completed, often taking 90 minutes or more. Other kinds of psychometric intelligence

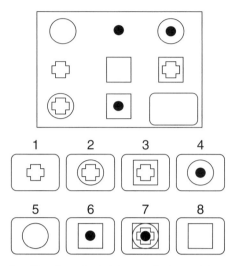

Figure 1.5 Simulated problem from the RAPM test. The lower right corner is missing from the matrix of symbols. Only one of the eight choices fits that spot once you infer the pattern or rule. In this case the answer is seven (add one row or column to the next). (Courtesy of Rex Jung)

tests can be given in a group setting or without direct interaction with the test-giver. Some tests are designed to assess specific mental abilities and others are designed to assess general intelligence. Typically, the more a test requires complex reasoning, the better it estimates the g-factor. Such tests have a "high g-loading." Here, briefly, are three important high-g tests used in neuroscience studies in addition to IQ.

1. The Raven's Advanced Progressive Matrices (RAPM) test (named for its developer, Dr. Raven) can be given in a group format and usually has a time limit of 40 minutes. It's regarded as a good estimate of the g-factor, especially because of the time constraint. Tests with a time limit tend to separate individuals better. It's a nonverbal test of abstract reasoning. Figure 1.5 is an example of one item. In the large rectangle, you see a matrix of eight symbols and a blank spot in the lower right corner. The eight symbols are not arranged randomly. There is a pattern or a rule linking them. Once you deduce the pattern or rule, you can decide which of the eight choices below the matrix completes the pattern or rule and goes in the lower right corner.

In this example, the answer is seven. If you add the left column to the middle column in the matrix, you get the symbols in the right column. If you add the top row to the middle row, you get the bottom row. The actual test items get progressively more difficult. The underlying pattern or rule can be quite hard to infer and there are different versions of the test that vary in difficulty. But because of its simple administration, this test has been used in many research studies. Performance on a test like this is fairly independent of education or culture. Scores are reasonable estimates of the g-factor, but they should not be mistaken as the g-factor (Gignac, 2015).

2. Analogy tests also are very good estimators of g. For example, wing is to bird as window is to _____ (answer is house). Or, helium is to balloon as yeast is to _____ (answer is dough). Or how about Monet is to art as Mozart is to _____ (answer is music). Analogy tests look as if they could be easily influenced by education and culture, so they have been dropped from many assessment test batteries despite the fact that empirically they are good estimates of g.

3. The Scholastic Assessment Test (SAT) is an interesting example. Until recently, it was widely used for college admission (see Textbox 1.2). Is it an achievement test, an aptitude test, or an intelligence test? The SAT was originally called the Scholastic Aptitude Test, then it was renamed the Scholastic Achievement Test, and now it's called the Scholastic Assessment Test. Achievement tests measure what you have learned. Aptitude tests measure what you might learn, especially in a specific area, such as music or a foreign language. It turns out that the SAT, especially the overall total score, is a good estimator of g because the problems require reasoning (Coyle, 2015; Frey & Detterman, 2004); see also (Beaujean et al., 2006; Coyle, 2021; Koenig, Frey, & Detterman, 2008). Like IQ scores, SAT scores are normally distributed and interpreted best as percentiles. For example, people in the top 2 percent of the SAT distribution tend to be in at least the top 2 percent of the IQ distribution. Sometimes this surprises people, but why should intelligence not be related to how much someone learns?

Achievement, aptitude, and intelligence test scores are all related to each other. They are not independent. Remember, the g-factor is common to all tests of mental ability. It would be unusual if learning and intelligence were unrelated. So your performance on achievement tests is related to the general factor, just as IQ scores and aptitude test scores are related to g. It can be confusing because we all know examples of bright students who are underachievers, and not-so-bright students who are overachievers. However, such examples are exceptions. In reality, there are some valid distinctions among achievement, aptitude, and intelligence testing. Each kind of test is useful in different settings, but they are also all related to g.

> **Textbox 1.2: Is this a case for or against using standardized tests for college admission?**
> The University of California (UC) decided in 2021 to end the use of standardized tests but this was not consistent with findings from a UC task force charged with evaluating the issues. Part of the task force conclusion was that they found "that standardized test scores aid in predicting important aspects of student success, including undergraduate grade point average (UGPA), retention, and completion. At UC, test scores are currently better predictors of first-year grade point average (GPA) than high school grade point average (HSGPA), and about as good at predicting first-year retention, UGPA, and graduation. For students within any given (HSGPA) band, higher standardized test scores correlate with a higher freshman UGPA, a higher graduation UGPA, and higher likelihood of graduating within either four years (for transfers) or seven years (for freshmen). Further, the amount of variance in student outcomes explained by test scores has increased since 2007, while variance explained by high school grades has decreased, although altogether does not exceed 26 percent. Test scores are predictive for all demographic groups and disciplines, even after controlling for HSGPA. In fact, test scores are better predictors of success for students who are Underrepresented Minority students (URMs), who are first-generation, or whose families are low-income: that is, test scores explain more of the variance in UGPA and completion rates for students in these groups. One consequence of dropping test scores would be increased reliance on HSGPA in admissions. The [task force] found that California high schools vary greatly in grading standards, and that grade inflation is part of why the predictive power of HSGPA has decreased since the last UC study."
> They also "noted the average differences in test scores among groups and expected to find that test score differences explain differences in admission rates. That is not what we found. Instead, the [task force] found that UC admissions practices compensated well for the observed differences in average test scores among demographic groups. This likely reflects UC's use of comprehensive review, as well as UC's practice of referencing each student's performance to the context of their school" (https://senate.universityofcalifornia.edu/_files/underreview/sttf-report.pdf). These empirical findings appear to undermine the administrative decision that ended the use of these tests.

1.8 Myth: Intelligence Tests Are Biased or Meaningless

Are intelligence test questions fair or do correct answers depend on an individual's education, social class, or factors other than intelligence? A professor I had in graduate school used to say that most people define a fair question as one they can answer correctly. Is a question unfair or biased because you don't know the answer?

Just what do intelligence test scores actually mean? Low test scores result because a person doesn't know the answers to many questions. There are numerous possible reasons for not knowing the answer to a question: You were never were taught it, never learned it on your own, learned it but forgot it a long time ago, learned it but forgot it during the test, were taught it but couldn't learn it, didn't know how to reason it out, or couldn't reason it out. Most, but not all, of these reasons seem related to general intelligence in some way. High test scores, on the other hand, mean the person knows the answers. Does it matter how you came to know the answer? Is it better education, just good memory, good test-taking skills, or good learning? The definitions of general intelligence combine all these things.

Test bias has a specific meaning. If scores on a test consistently over or underpredict actual performance, the test is biased. For example, if people in a particular group with high SAT scores consistently fail college courses, the test is overpredicting success and it is a biased test. Similarly, if people with low SAT scores consistently excel in college courses, the test is underpredicting success and it is biased. A test is not inherently biased just because it may show an average difference between two groups. A spatial ability test, for example, may have a different mean score for men and women, but that does not make the test biased. If scores for men and for women predict spatial ability equally well, the test is not biased even if there is a mean difference. Note that a few cases of incorrect prediction do not constitute bias. For a test to be biased, there needs be a consistent failure of prediction in the wrong direction. The lack of any prediction is not bias; it means the test is not valid.

Considerable research on test bias for decades shows this is not the case for IQ and other intelligence test scores (A. Jensen, 1980). Test scores do predict academic success irrespective of social economic status (SES), age, sex, race, and other variables. Scores also predict many other important variables, including brain characteristics such as regional cortical thickness or cerebral glucose metabolic rate, as we detail in Chapters 3 and 4. If intelligence test scores were meaningless, they would not predict any other measures, especially quantifiable brain characteristics. In this context, "predict" also has a specific meaning. To say a test score predicts something only refers to a higher probability of the something occurring. No test is 100 percent accurate in its predictions, but the reason intelligence tests are considered by many psychologists to be a great achievement is that the scores are good predictors for success in many areas, and in some areas test scores are excellent predictors. Before we

review key research that is the basis for this conclusion, there is a fundamental problem to discuss.

1.9 The Key Problem for "Measuring" Intelligence

As briefly noted earlier in this chapter, the main problem with all intelligence test scores is that they are not on a ratio scale. This means there
is no true zero, unlike measures for height and weight. For example, a
person who weighs 200 pounds is literally twice the weight of a person
who weighs 100 pounds because a pound is a standard unit on a scale
with an actual zero point. Ten miles is twice the distance of five miles.
This is not the case for IQ scores. A person with an IQ score of 140 is not
literally twice as smart as a person with a score of 70. Even if you believe
you have encountered at least one person with zero intelligence, zero is
certainly not the case. For IQ, it's the percentile that counts. Someone
with an IQ of 140 is in the top 1 percent and someone with a score of 70
is in the bottom 2 percent. A person with an IQ of 130 is not 30 percent
smarter than a person whose score is 100. The person with an IQ of 100
is at the 50th percentile and the person with an IQ of 130 is at the 98th
percentile. No psychometric test score is based on a ratio scale. All IQ
test scores have meaning only relative to other people.

Here's the key point about this limitation of all intelligence test scores:
They only estimate intelligence because we don't yet know how to measure intelligence as a quantity like we measure liquid in liters or distance
in feet (Haier, 2014). If you take an intelligence test when you are sick
and unable to concentrate, your score may be a bad estimate of your
intelligence. If you retake the test when you are well, your score is a
better estimate. However, just because your score goes up, it does not
mean your intelligence increased in the interval between the two tests.
This becomes an issue in Chapter 5 when we talk about why claims of
increasing intelligence are not yet meaningful.

Despite this fundamental problem, researchers have made considerable
progress. The main point is that measurement is required to do scientific
research on intelligence. No one test may be a perfect measure of a single
definition, but as research findings accumulate, both definition and measurement evolve and our understanding of the complexities increases. The
empirical robustness of research on the g-factor essentially negates the myth
that intelligence cannot be defined or measured for scientific study. It is
this research base that allows neuroscience approaches to take intelligence
research to the next level, as detailed in subsequent chapters. But first, we
will summarize some compelling studies of intelligence test validity.

1.10 Four Kinds of Predictive Validity for Intelligence Tests

1.10.1 Learning Ability

IQ scores predict general learning ability, which is central to academic and vocational success and to navigating the complexities of everyday life (Gottfredson, 2003). For people with lower IQs of around 70, simple learning typically is slow and requires concrete, step-by-step teaching with individual instruction. Learning complex material is quite difficult or not possible. IQs around 80–90 still require very explicit, structured individual instruction. When it comes to learning by written materials, IQs of at least 100 are usually required, and college-level learning usually works best at 115 and over. Higher IQs over 130 usually mean that more abstract material can be learned relatively quickly, and often independently. There are exceptions, and there is good evidence that lower-IQ (< 90) students who complete college also benefit in later life success as much as higher-IQ students, possibly because of strong compensatory factors such as personality and family support (McGue et al., 2022).

The US military uses their own test but the rough equivalent IQ score minimum cutoff is about 85–90 for recruits, although this moves down a bit when recruitment is strained; sometimes with tragic results (Gregory, 2015). Most graduate programs in the United States require tests such as the Graduate Record Exam or the Medical College Admission Test for medical school or the Law School Admission Test for law school. Cutoffs for these tests usually ensure that individuals with IQs over 120 are most likely to be accepted, and the top programs have higher cutoffs to maximize accepting applicants in the top 1 or 2 percent of the normal distribution. This doesn't mean that people with lower scores cannot complete these programs, but the higher-scoring students are usually more efficient, faster learners and more likely to successfully finish the program.

Keep in mind, these are not perfect relationships and there are exceptions, as noted (McGue et al., 2022). The relationship between IQ scores and learning ability, however, is strong. Many people find this disturbing because it indicates a limitation on personal achievement that runs counter to a prevalent notion expressed in the phrase: "You can be anything you want to be if you work hard." This is a restatement of another notion: "If you work hard, you can be successful." The latter may often be true because success comes in many forms and for many reasons, but the former is seldom true unless a caveat is added: "You can be anything you want to be if you work hard and have the ability." Not everyone has the ability to do everything successfully, although, surprisingly, many students arrive at college determined to succeed but naïve about the role ability plays. Few students with low SAT-Math scores, for example, are successful majors in the physical sciences even if they are highly motivated and work hard.

Given the powerful influence of g on educational success, it is surprising that intelligence is rarely considered explicitly in vigorous debates about why pre-college education appears to be failing many students. Doug Detterman has noted, "[a]s long as educational research fails to focus on students' characteristics, we will never understand education or be able to improve it" (Detterman, 2016: 1). The best teachers cannot be expected to attain educational objectives beyond the capabilities of students. The best teachers can maximize a student's learning, but the intelligence level of the student creates some limitations, although it is fashionable to assert that no student has inherent limitations. Many factors that limit educational achievement can be addressed, including poverty, poor motivation, lack of role models, family dysfunction, and so on, but so far there is no evidence that alleviation of these factors increases g. As we will see in Chapter 2, early childhood education has a number of beneficial effects but increasing intelligence is not one of them. Imagine a pie chart with all the factors that influence a student's school achievement. Surely the g-factor would deserve representation as a slice greater than zero. The strong correlations between intelligence test scores and academic achievement indicate that the slice could represent a sizable portion of the whole. In my view, this alone should justify more research on intelligence and how it develops.

1.10.2 Job Performance

In addition to academic success, IQ scores also predict job performance (F. Schmidt, 2016; F. L. Schmidt & Hunter, 2004; F. L. Schmidt & Hunter, 1998), especially when jobs require complex skills. In fact, for complex jobs the g-factor predicts success more than any other cognitive ability (Gottfredson, 2003). A large study conducted by the US Air Force, for example, found that g predicted virtually all the variance in pilot performance (M. J. Ree & Carretta, 1996; Malcolm James Ree & Carretta, 2022; M. J. Ree & Earles, 1991). Most of us are not pilots, but in general, lower IQ is sufficient for jobs that require a minimum of complex, independent reasoning. These jobs tend to follow specific routines such as assembling a simple product, food service, or nurse's aide. IQs around a 100 are minimally necessary for more complex jobs such as bank teller and police officer. Successful managers, teachers, accountants, and others in similar professions usually have IQs of at least 115. Professions such as attorney, chemist, doctor, engineer, and business executive usually require higher IQs to finish the advanced schooling that is required and to perform at a high level of complexity.

Complex job performance is largely *g*-dependent, but of course there are other factors, including how well one deals with other people. This is the concept of emotional intelligence. Emotional intelligence, that is, the personality and social skills one has, may contribute to greater success compared to a person of equal *g* but lacking people skills. This does not diminish the importance of the *g*-factor. In some circumstances emotional intelligence might compensate for a lack of job-appropriate *g* but only for so long, if at all. Good evidence shows that IQ scores are more predictive of educational attainment (years of education) than personality measures (Zisman & Ganzach, 2022).

As with academic success, intelligence–job performance relationships are general trends, and there are always exceptions. But, from a practical point of view, a person with an IQ under 100 is not very likely to complete medical school or engineering school. Of course, it's possible, especially if the IQ score is not a good estimate of intelligence for that person, or if that person has a very specific ability such as memorization to compensate for low or average general intelligence. Similarly, a high score does not guarantee success. This is why an IQ score by itself is not usually used to make education or employment decisions. IQ is usually considered in the context of other information but a low score is typically a red flag in many areas that require complex, independent reasoning.

Here's another point about predicting job success. Some researchers suggest that expertise in any area requires at least 10,000 hours of practice (Ericsson, 2014; Ericsson & Towne, 2010). That's 1,250 8-hour days, or about 3.4 years of constant effort. This implies that expertise can be achieved in any field with this level of practice irrespective of intelligence or talent. Studies of chess grand masters, for example, report that the group average IQ is about 100. This suggests that becoming a grand master may depend more on practice of a specific ability such as spatial memory than on general intelligence. Grand masters may actually have a savant-like spatial memory, but the idea of a chess grand master being a super all-purpose giant intellect is not necessarily correct. Many studies refute the idea that 10,000 hours of practice can lead to expertise if there is no pre-existing talent to build on (Detterman, 2014; Grabner, 2014; Grabner, Stern, & Neubauer, 2007; Gullich, Macnamara, & Hambrick, 2022; Hambrick, Macnamara, & Oswald, 2020; Macnamara & Maitra, 2019; Plomin et al., 2014a, 2014b).

1.10.3 Everyday Life

The importance of general intelligence in everyday life often is not obvious but it is profound. As Professor Earl Hunt has pointed out, if you are a college-educated person, it is highly likely that most of your friends and acquaintances are as well. When is the last time you invited someone to your home for dinner that was not college-educated? Professor Hunt calls this cognitive segregation and it is powerful in fostering the erroneous belief that everyone has a similar capacity or potential for reasoning about daily problems and issues. Most people with high *g* cannot easily imagine what daily life is like for a person with low *g*.

The complexity of everyday life is often quite challenging, especially when a nonroutine or a novel problem presents itself. Professor Robert Gordon summarizes this with a simple statement: "Life is a long mental test battery" (Gordon, 1997). This was true as early humans navigated unforgiving natural environments and solved continuous problems of finding food, water, shelter, and safety. It was true as early civilizations developed and great thinkers (likely with high *g*) solved even more complex problems (e.g., just how does one build a seaworthy ship or a pyramid?). And it is still true today as we grapple with connecting our new television sets and audio systems with HDMI cables or using all the functions in our word processor or on our "smart" phones or digital cameras beyond the auto mode. Do you know how to use the scanners in the self-checkout lines at the supermarket or do you wait in a long slow line for a human cashier? How much do you understand about money management and investing in stocks, bonds, and mutual funds? Do you do your own taxes? Many people grapple daily with the challenges of navigating nearly impenetrable systems for healthcare, social support, or justice. Poverty presents myriad daily problems to solve. It could be said that in the modern world, nothing is simple for anyone all the time.

Consider some statistics comparing low- and high-IQ groups (low = 75–90; high = 110–125) on relative risk of several life events. For example, the odds of being a high school dropout are 133 times more likely if you're in the low group. People in the low group are 10 times more at risk for being a chronic welfare recipient. The risk is 7.5 times greater in the low group for incarceration, and 6.2 times more for living in poverty. Unemployment and even divorce are a bit more likely in the low group. IQ even predicts traffic accidents. In the high-IQ group, the death rate from traffic accidents is about 51 per 10,000 drivers, but in the low-IQ group this almost *triples* to about 147. This may be telling us that people with lower IQ, on average, have a poorer ability to assess risk and may take more chances when driving or performing other activities (Gottfredson, 2002, 2003).

Textbox 1.3: Functional literacy

Another way to look at the role of thinking skills and everyday life is based on functional literacy data. Functional literacy is assessed by the complexity of everyday tasks that a person can complete. Like IQ scores, functional literacy scores are meaningful relative to other people, but they provide more concrete examples of ability. The last comprehensive US national survey of functional literacy was done in 1992.

The chart in Table 1.1 is from that survey. On the left side, we have five levels of functional literacy: 1 is the lowest, 5 is the highest. In the middle we have the percentage of people who are in each category, and on the left we have some sample tasks that people in each category can complete successfully. Let's look at the top row. If you're like me, you will be quite surprised to see that only 4 percent of the white population is in the top category and can complete tasks such as using a calculator to figure out the cost of carpeting a room. This requires determining the area, converting to square yards, and multiplying by the price. In the next row down, 21 percent of people are at level four of functional literacy. They can calculate social security benefits from a table and understand basic issues of how employee benefits work. Then 36 percent are in the middle category. They can calculate miles per gallon from a chart, and they can write a letter explaining a credit card

Table 1.1 Everyday literacy levels from the National Adult Literacy Survey along with sample problems from each level

Everyday literacy (NALS)

NALS level	% pop. (white)	Simulated everyday tasks
5	4	• Use calculator to determine cost of carpet for a room • Use table of information to compare two credit cards
4	21	• Use eligibility pamphlet to calculate Supplemental Security Income benefits • Explain difference between two types of employee benefits
3	36	• Calculate miles per gallon from mileage record chart • Write brief letter explaining error on credit card bill
2	25	• Determine difference in price between two show tickets • Locate intersection on street map
1	14	• Total bank deposit entry • Locate expiration date on driver's license

*error. Twenty-five percent are in category two. They can determine price dif-
ferences between two tickets, and they can locate an intersection on a map.
Fourteen percent are in the lowest category. They can accomplish tasks such
as filling out a bank deposit slip, but more complex tasks, such as locating
an intersection on a map, would present difficulty. Note that these data are
more than 30 years old and it may be that the percentages of people in each
category differ now, but the main point is still the same: intelligence matters
in everyday life.*

The examples in Textbox 1.3 and Table 1.1 demonstrate that intelligence helps us navigate the problems of everyday life. It's not a shocking idea. But this is easy to take for granted, especially if you are navigating reasonably well and most of the people you spend time with are like you. The key point here is that functional literacy is another indicator of intelligence, and you can see from the functional literacy data that intelligence matters for daily tasks. But, of course, the *g*-factor does not predict many other important things such as being a kind or likable or honest person. No intelligence researcher has ever asserted otherwise.

Let's talk for a moment about a controversial book in 1994 that explored the role of intelligence in social policy, *The Bell Curve* by Richard Herrnstein and Charles Murray (Herrnstein & Murray, 1994). The main theme was that modern society increasingly requires and rewards people with the best reasoning skills. This is to say people with high intelligence. Therefore, people in the bottom part of the normal distribution of IQ (a normal distribution is also called a bell curve because of its shape) will be at a serious disadvantage for succeeding, especially in school and some vocations. Herrnstein had introduced this theme in an earlier book, *IQ in the Meritocracy* (Herrnstein, 1973), that also generated considerable acrimony (see the detailed description of hostility on the Harvard campus recounted in the preface to get a sense of the times); a few years later another Harvard professor, Edward O. Wilson, encountered similar outrage when he proposed the concept of sociobiology (Wilson, 1975). *The Bell Curve* continued the argument with over 900 pages of data and statistical analyses mostly comparing high and low intelligence groups, but the one chapter that discussed black/white IQ differences aroused the fiercest controversy (please note that the terms black and white are used here because most of the research, from America and other countries, uses these terms). This issue of group differences haunts all intelligence research and I refer the reader to in-depth accounts of the complexities involved (see Further Reading).

My point about *The Bell Curve* is whether public policy discussions benefit by recognizing that people with low *g* may need help in navigating life, irrespective of race, background, or why they might have low *g*. This is a fundamental issue today in politics, although the role of intelligence is hardly mentioned as explicitly as it was in *The Bell Curve*, or in a later book by Murray expanding the theme of societal implications of cognitive segregation (C. A. Murray, 2013). Most researchers would agree that research data on intelligence can only inform policy decisions, but the goals of the policy need to be determined through democratic means; we return to this issue in Section 6.6. Unfortunately, psychometric research on intelligence has often been portrayed as damaging to a progressive social agenda because there are substantial average test score differences among some racial and ethnic groups. These relative average group differences often motivate a general disregard for empirical research on intelligence although neuroscience approaches are advancing the field, as the following chapters discuss. Before we get to those, let's continue with more data about IQ scores and what they mean.

1.10.4 Longitudinal Studies of IQ and Talent

The predictive power of a single test score in childhood is also demonstrated dramatically in three classic longitudinal studies. Each one starts with children and tests their mental abilities and subsequent life successes at various intervals over decades. One study started in California the 1920s, one started in Scotland the 1930s, and one started in Baltimore in the 1970s.

Study 1. Professor Lewis Terman at Stanford University initiated a long-term study of high-IQ individuals in the 1920s. This is the same Louis Terman who brought Binet's IQ test to the United States and revised it into the Stanford–Binet intelligence test. Terman designed a straightforward study. It started by testing many school children with the Stanford–Binet test. Children with very high IQ scores were selected and then studied extensively for decades. Terman's study had two goals: to find the traits that characterized high-IQ children, and to see what kind of adults they would become. The common stereotype of intelligent adults was not so different then as it is now. Francis Galton, for example, wrote in his 1884 book, *Hereditary Genius* (Galton & Prinzmetal, 1884): "There is a prevalent belief that men of genius are unhealthy, puny beings – all brain and no muscle – weak-sighted, and generally of poor constitutions" (Galton, 2006: 321).

Here's how Terman's project started (Terman, 1925): In 1920–1921, 1,470 children with IQs of 135–196 were selected from over 250,000 children in California's public schools and they were retested and interviewed every seven years. Their average IQ was about 150, and 80 children had IQs over 170 (these were in the top 0.1 percent). This entire group became known unofficially as the Termites. They completed extensive medical tests, physical measurements, achievement tests, character and interest tests, and trait ratings, and both parents and teachers supplied additional information. A control group with average IQ scores was also tested. The results of Terman's study were published over time in five volumes. The data were quite extensive.

Here's a summary of the key findings about the lives of the Termites. Overall, they completely refute the stereotypes both for children and adults. The negative, nerdy attributes were basically unfounded. They were not odd or puny. On average, they were actually physically quite robust and more physically and emotionally mature than their age-mates. On average, the Termites were happier and better adjusted than the controls over the course of the study. Although they had their share of life problems, follow-up studies showed considerable achievement with respect to publishing books, scientific papers, short stories and poems, musical compositions, television and movie scripts, and patents (Terman, 1954). However, further follow-up indicated that high IQ alone did not necessarily predict life success. Motivation was also important, and Terman believed that while genes played an important role in high IQ, he also believed that exceptional ability required exceptional education to maximize a student's potential. This may not sound so radical, but even today there is a debate about whether any educational resources at all should be allocated to the most gifted students to develop their high ability.

Terman's project also demonstrated the predictive validity of the IQ score. That is, one IQ score in childhood can identify individuals who will excel in later life. Like all studies, however, there were some major flaws: (1) Terman intervened in the lives of these "subjects" and helped them with letters of reference for college and for employment; (2) strong sex bias in education and employment resulted in female Termites mostly becoming housewives, so valid male–female comparisons were not possible. Similarly, there are no data about minorities. Do these problems invalidate the main findings? Not likely (Warne, 2019). Overall, the level of success and the achievement of these very high-IQ individuals stand on their own. But fortunately, we have more data from a newer study that modified Terman's approach.

Study 2. The second longitudinal study is the Study of Mathematically and Scientifically Precocious Youth at Johns Hopkins University. This was an ambitious, longitudinal project initiated by Professor Julian Stanley in 1971 (Stanley, 1974). Dr. Stanley repeated Terman's approach, but instead of IQ scores he used extremely high SAT-Math scores obtained by junior high school students aged 11–13 in special testing sessions called "talent searches." So instead of general intelligence, Stanley focused on a very specific mental ability. This project also had two major goals. First, identify precocious students early, and second, foster their special talent.

I started graduate school at Hopkins in 1971 and I worked on this study in its early years. I must say that this experience was an early influence on my interest in intelligence, and Dr. Stanley was one of the most important and interesting mentors I had at Hopkins.

This project had its origins in the late 1960s. Dr. Stanley started working with a precocious student, and after he gave the student a battery of psychometric tests, Dr. Stanley helped the student to get into Hopkins at the early age of 13. Dr. Stanley subsequently referred to this young man as the first "Radical Accelerant," identified as Joseph B. In his first year at Hopkins, at age 13, Joseph took honors calculus, sophomore physics, and computer science, and his GPA was 3.69 out of 4.0. He lived at home during this time but he also made friends with other college students and adjusted well to his accelerated studies. In four years, he received a BA *and* a MSc degree in computer science. He began a PhD program in computer science at Cornell before he was 18 years old, and Joseph went on to a productive career.

From the beginning, a main goal for Dr. Stanley was to not only identify and follow such precocious students but also to select the best candidates for education acceleration, including early college admission. So was born the idea of using the SAT-Math test for screening junior high school students to find precocious individuals with talent for math and science. The Spencer Foundation provided multiple-year funding to Dr. Stanley beginning in 1971, and the first talent search was in 1972. For that search, junior high school students in the Baltimore area had to be nominated by their math teachers to participate. Actual SAT-Math tests were given in the standard way. In that first search, 396 seventh- and eight-grade students took the SAT-Math. Here are two fascinating results of that first talent search. Twenty-two of the 396 scored at least 660, which was higher than the average Hopkins freshman at the time. And all of these 22 were boys; none of the 173 girls scored over 600.

The male–female ratio has improved considerably over the years, but at the time, this huge disparity was surprising. What about the 22 boys who scored higher than a Hopkins freshman? What were they like? The early data analyses confirmed Terman's results with respect to stereotype. These mathematically precocious students were more physically and emotionally mature than age-peers. One of my first research projects was to give this precocious group some standardized tests of personality. On average, they scored more like college students than their age-peers (Weiss, Haier, & Keating, 1974).

Professor Stanley believed that enriched classes were not as productive as actual college classes, so he helped many of these very talented students go to college early. Over the years, many of the most precocious students did get early admission, usually living at home. And there was no evidence that they suffered any emotional harm from an accelerated program. Like the Termites, many went on to have successful and very productive careers (Bernstein, Lubinski, & Benbow, 2019; Lubinski, Benbow, & Kell, 2014; Makel et al., 2016; McCabe, Lubinski, & Benbow, 2020).

The original talent searches have evolved dramatically and now include many programs for enrichment in addition to early college admission, including summer camps that emphasize math and science experiences. You can find out more details about these programs using Google. Actually, one of the students associated with the talent searches co-founded Google: Sergey Brin. Mark Zuckerberg of Facebook was also identified in a talent search, as was Lady Gaga. Seriously. Look it up.

There are now detailed follow-up studies of thousands of the students who participated in several of the original searches. Follow-up results show that many of these mathematically precocious children, as determined by a single test score when they were in their early teens, became exceptionally successful in terms of occupational and life success (Bernstein et al., 2019; Lubinski et al., 2006; Lubinski et al., 2014; Lubinski, Schmidt, & Benbow, 1996; McCabe et al., 2020; Robertson et al., 2010; Wai, Lubinski, & Benbow, 2005). Figure 1.6 shows professional achievement based on a 25-year follow-up study of the top 1 percent of the original searches that included 2,385 students (Robertson et al., 2010). All these students in the top 1 percent are divided into quartiles – Q1, Q2, Q3, and Q4 – based on their SAT-Math score at age 13. On the x-axis, we have SAT-Math score at age 13. On the y-axis, we have the proportion of the quartile with an outcome such as getting a PhD, a JD, or an MD. Another outcome is having any peer-reviewed publications. Another would be getting a PhD and tenure in a STEM

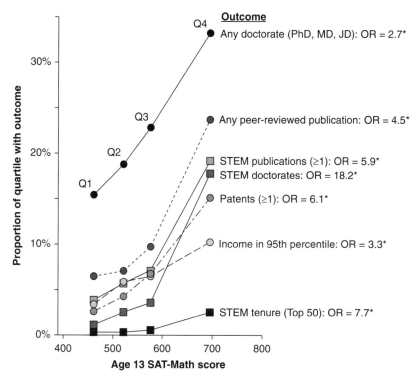

Figure 1.6 SAT-Math scores at age 13 predict adult outcomes of academic success. (Reprinted with permission (Robertson et al., 2010))

field, which includes science, technology, engineering, or math. Patents are another outcome and so is high income defined in the 95th percentile.

What we see in this chart is that for students with age 13 SAT-Math scores in the 400–500 range, which is in the top 1 percent for 13-year-olds but in the lowest quartile 1 for this sample, about 15 percent got a doctorate in any field, and this percentage increases with higher scores. In the top SAT-Math quartile 4, the percentage of advanced degrees is about 35 percent. This is all shown in the line with black dots at the top of the chart. You see this same trend for all the other outcomes.

The OR after each outcome stands for "odds ratio" and compares the top quartile proportion to the bottom quartile for each outcome. For example, the greatest disparity is 18.2 for getting a doctorate in a STEM field. This means the upper quartile within the top 1 percent were 18 times more likely to get a STEM doctorate than the bottom quartile within the top 1 percent. So even in this rarified group of the top 1 percent, the individuals with the highest scores did the best based on these outcomes.

Remember, a single test taken at age 13 identified these individuals. Again, you can see the predictive validity of this standardized test score is reasonably strong. Clearly, individuals in the top 1 percent of scores obtained in childhood have notable future achievements, but even within this rarified group, the higher the scores, the more likely there will be these kinds of achievements. The longitudinal study of the original talent search participants is continuing, with additional follow-ups conducted by researchers Professor Camilla Benbow and Professor David Lubinski at Vanderbilt University.

Study 3. The third longitudinal study is the Scottish Mental Survey. This was a truly massive project conducted by the Scottish government. All children born in Scotland in 1921 and in 1936 completed intelligence testing at age 11 years and were retested again in old age. This study differed from the other two in that it included virtually *all* children in the country on a test of general intelligence rather than identifying samples of very high scorers (von Stumm & Deary, 2013). The total number of children in the study was about 160,000.

At the time this study began in the 1930s, there was considerable debate around the world about national intelligence and eugenics. This had profoundly evil consequences in Nazi Germany. It's one of the reasons intelligence testing became a negative topic in academia following World War II. But another reason for using intelligence tests in some countries was the desire to open opportunities for better schooling to all social classes by using test scores as an objective evaluation to give all students an opportunity to attend the best schools irrespective of background or wealth. This actually happened in the United Kingdom after the war, and this motivation was important in the development and use of the SAT in the United States (Wooldridge, 2021).

But the Scottish survey was over after the second round of testing in 1936. It only became a longitudinal follow-up study somewhat by accident when the original records were rediscovered in an old storage room. A team of researchers, directed by Professor Ian Deary at the University of Edinburgh, used this database and follow-up evaluations to study the impact of intelligence on aging. Several years ago, Dr. Deary got a new grant from the Scottish government, restored the physical handwritten records as much as possible, and then computerized them all. He also identified 550 original participants who were still living and willing to be retested. So there is now follow-up data. Let's look at two interesting results from the longitudinal analyses:

1. IQ scores were fairly stable over time as demonstrated by showing scores at age 11 correlated to scores at age 80 (r = 0.72) (Deary et al., 2004). The intelligence test used at the beginning of the survey and for follow-up is called the Moray House Test. It gives an IQ score essentially equivalent to the Stanford–Binet or the WAIS. Recall that fluid intelligence decreases with age. Crystallized intelligence is more stable, and the IQ score from the test used in this study combined both fluid and crystallized intelligence. Although not part of this study, it should be noted that different components of IQ might rise and fall at different times across the lifespan (Hartshorne & Germaine 2015).

2. Individuals with higher intelligence scores at age 11 lived longer than their classmates with lower scores, as shown in Figure 1.7 (Batty, Deary, & Gottfredson, 2007; C. Murray et al., 2012; Whalley & Deary, 2001).

The top graph in Figure 1.7 shows the data for women, and the bottom graph shows men. Both show the same trends. On the x-axis, we see the ages of participants by decade from age 10 to age 80, and on the y-axis, we see the percentage of the group originally tested who are still alive at each age. The data are shown separately for the lowest and the highest quartile based on IQ.

So, for example, in Figure 1.7 let's look at the top graph of women, and let's focus on the data points at the far-right side of the graph (about 80 years old). You can see that more women are alive in the highest IQ quartile, about 70 percent compared to the bottom quartile, where about 45 percent are still alive. This is quite a large difference. And this is true starting around age 20. It's the same for men, but starting later at around age 40 and the trend is not quite as strong. Since the United Kingdom has universal healthcare, differential rates of insurance coverage do not influence these data. But why should IQ be related to longevity? Here is one possible explanation. Before age 11, several factors, both genetic and environmental, may influence IQ, and then higher IQ leads to healthier environments and behaviors and to a possibly better understanding of physician instructions, and these in turn influence age at death. However, there is compelling evidence that a better explanation is that mortality and IQ have genetic influences in common. An estimated 84–95 percent of the variance in the mortality–IQ correlation may be due to genes (Arden et al., 2015).

To recap the evidence from these three classic studies, Terman's project helped popularize the importance of IQ scores and demolished the

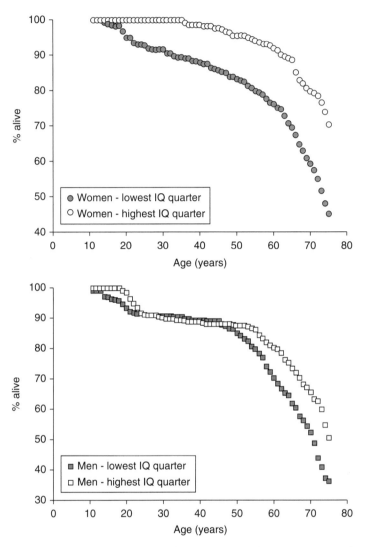

Figure 1.7 Childhood IQ scores predict adult mortality. Note that many more people in the highest IQ group are alive recently compared to the lowest IQ group. (Reprinted with permission (Whalley & & Deary, 2001))

popular but negative stereotype of childhood genius. Gifted student education essentially started with this study. Stanley's project went further and incorporated ways to foster academic achievement in the most gifted and talented students. Deary's analyses of the National Survey data in

Scotland provided new insights about the stability of IQ scores and the importance of general intelligence for a number of social and health outcomes.

These studies provide compelling data that one psychometric test score at an early age predicts many aspects of later life including professional success, income, healthy aging, and even mortality. The bottom line is that it's better to be smart, even if defined by test scores that have meaning only relative to other people.

1.11 Why Do Myths about Intelligence Definitions and Measurement Persist?

Given all this strong empirical evidence that intelligence test scores are meaningful, why does the myth persist that these scores have little if any validity? Here is an informative example. From time to time, a college or university admissions representative will assert that in their institution they find no relationship between GPA and SAT scores. Such observations are virtually always based on a lack of understanding of a basic statistical principle regarding the correlation between two variables. To calculate a correlation between any two variables, there must be a wide range of scores for each variable. At a place like MIT, for example, most students fall in a narrow range of high SAT scores. This is a classic problem of restriction of range. There is little variance among the students, so in this case, the relationship between GPA and SAT scores will not be very strong. Sampling from just the high end or just the low end or just the middle of a distribution restricts range and results in spuriously low or zero correlations. Restriction of range actually accounts for many claims about what intelligence test scores "fail" to predict.[1]

Here's another classic example of an erroneous finding due to restriction of range. In the 1930s, Louis Thurstone challenged Spearman's finding of a g-factor (Thurstone, 1938) and proposed an alternative model of "7 Primary Abilities" that he claimed were independent of each other. That is, they were not correlated to each other and there

---

[1] Nonetheless, this is one argument used by critics of testing, who have succeeded in eliminating the SAT and other standardized testing in the admissions process for many colleges and universities. Other arguments include alleged test bias and unequal education opportunities that lead to lower scores. The evidence, however, supports the idea that SATs can provide a fair evaluation that is useful for the admissions process (Wai, Brown, & Chabris, 2019; and see Textbox 1.2).

was no common g-factor. There's spatial ability, as measured by tests that require mental rotation of pictures and objects. There's perceptual speed, as measured by tests of finding small differences in pictures as fast as possible. There's number facility, as measured by tests of computation. There's verbal comprehension, as measured by tests of vocabulary. There's word fluency, as measured by tests that require generating as many words as possible for a given category within a time limit. There's memory, as tested by recall for digits and objects. And finally, there is inductive reasoning, as measured by tests of analogies and logic.

However, Thurstone's model was not supported by subsequent research. It turns out that the original research was flawed because the samples he used did not include individuals across the full range of possible scores. That is, the range was restricted, so there was no variance to predict any test from any other. When additional research corrected this problem, tests of the Thurstone "primary" abilities, in fact, were correlated to each other and there was a g-factor. Thurstone retracted his original conclusion (Thurstone & Thurstone, 1941). So why include this example from the 1930s in a modern book? As we will see in later chapters, a surprising number of studies still report erroneous findings because of restricted range.

Differences in factor structure among many models based on factor analysis have given some critics the idea that g is merely a statistical artifact of factor analysis methodology. We now have hundreds of factor analysis studies of intelligence on hundreds of mental tests completed by tens of thousands of people and using many varieties of factor analysis methods. The bottom line is that there is always a g-factor. Here's a key point: g-factors derived from different test batteries correlate nearly perfectly with each other as long as each battery has a sufficient number of tests that sample a broad range of mental abilities and the tests are given to people sampled from the wide range of abilities (Johnson et al., 2004; Johnson, te Nijenhuis, & Bouchard, 2008). A study of 180 college students reported that a g-factor derived from their performance on a battery of video games correlated highly (0.93) with a g-factor extracted from their performance on a battery of cognitive tests (A. M. Quiroga et al., 2015; M. A. Quiroga et al., 2019). Such studies provide strong evidence that g is not a statistical artifact, even though its meaning is limited as an interval scale. And, logically, if it were merely an artifact, g-scores would not correlate with other measures of the complexity of everyday life, as we noted, nor with genetic and brain parameters, as we detail in subsequent chapters.

Finally, perhaps the major motivation for diminishing the validity of intelligence tests, and other tests of mental abilities including the SAT, is the desire, shared by many, to explain away group differences in average scores as a mere artifact of the tests. In my view, this motivation is misplaced. The causes of average test score differences among groups are not yet clear, but the differences are a major concern in education and other areas. They deserve full attention with the most sophisticated research possible so causes and potential remediation can be developed based on empirical studies. Imaging studies of brain development and intelligence are beginning to address some issues, as detailed in Chapters 3 and 4, and the goal of enhancing intelligence, discussed in Chapters 5 and 6, is something to consider beyond science fiction (Haier, 2021).

Before we get into the brain itself, in Chapter 2 we summarize the overwhelming evidence that intelligence is strongly influenced by genetics and how genes may affect the brain. We also introduce the concept of the epigenetic influences of environmental/social/cultural factors on gene expression, all of which work through biological processes to affect the brain. Altogether, this evidence supports our primary assumption that intelligence is 100 percent biological.

Chapter 1 Summary

- Intelligence can be defined and assessed for scientific research.
- The *g*-factor is a key concept for estimating a person's intelligence compared to other people.
- It is surprising that intelligence is rarely considered explicitly in vigorous debates about why pre-college education appears to be failing many students. The best teachers cannot be expected to attain educational objectives beyond the capabilities of students.
- At least four kinds of study demonstrate the predictive validity of intelligence test scores and the importance of intelligence for academic and life success.
- Intelligence tests are the basis for many important empirical research findings, but going forward, the key problem for assessment is that there is no ratio scale for intelligence, so test scores are meaningful only relative to other people.
- Despite widespread but erroneous beliefs about definition and assessment, neuroscience studies seek to understand the brain processes that underlie intelligence and how they develop.

Review Questions

1. Is a precise definition of intelligence required for scientific research?
2. What is the difference between specific mental abilities that define savants and the *g*-factor?
3. Why is an intelligence test score not like a measure of length, liquid, or weight?
4. What is restricted range and why is it an important concept for intelligence research?
5. What are two myths about intelligence and why do they persist?
6. Why do you suppose this chapter begins with a quote from 1980?

Further Reading

Straight Talk about Mental Tests (Arthur Robert Jensen, 1981). This is a clear examination of all the issues surrounding mental testing, written without jargon by a real expert for students and the general public. It is still a classic, but you may find it only in libraries or from online sellers.

The g-Factor (A. R. Jensen, 1998). This is a more technical and thorough text on all aspects of the *g*-factor. It is considered the classic in the field but it may be hard to find.

IQ in the Meritocracy (Herrnstein, 1973). This controversial book put forth an early argument about how the genetic basis of IQ was stratifying society. The preface is a hair-raising account of the acrimonious climate of the times at Harvard University for unpopular ideas. This book is hard to find but try online sellers.

The Bell Curve (Herrnstein & Murray, 1994). This is possibly the most controversial book about intelligence ever written. It expands arguments first articulated in *IQ in the Meritocracy*. It is loaded with data and possible implications about what intelligence research could mean for public policy.

"The Neuroscience of Human Intelligence Differences" (Deary, Penke, & Johnson, 2010) This is a concise review article written by longtime intelligence researchers.

Coming Apart (C. A. Murray, 2013). This book advances Charles Murray's analysis of the implications of society segregating by levels of intelligence. It is prescient about today's political and social divisions.

The Nature of Human Intelligence (R. J. Sternberg, 2018). This edited book includes contributions from 19 of the most cited living intelligence researchers. Each chapter covers a different topic, including many not covered in this book, which is focused exclusively on neuroscience approaches.

The Aristocracy of Talent: How Meritocracy Made the Modern World (Wooldridge, 2021). This is a scholarly discussion of the social implications of intelligence and a broad defense of meritocracy.

The Cambridge Handbook of Intelligence and Cognitive Neuroscience (Barbey, Karama, & Haier, 2021). This is an edited book with chapters written by experts in cognitive neuroscience; many chapters are written for advanced students.

The Science of Human Intelligence (Haier & Colom, 2023). This is a textbook for students that covers a wide range of intelligence topics. It is the second edition of Hunt's classic textbook, *Human Intelligence* (Hunt, 2011).

Nature More than Nurture

The Impact of Genetics on Intelligence

Give me a dozen healthy infants, well-formed, and my own specified world to bring them up in and I'll guarantee to take any one at random and train him to become any type of specialist I might select--doctor, lawyer, artist, merchant-chief, and, yes, even beggar man and thief, regardless of his talents, penchants, tendencies, abilities, vocations, and race of his ancestors. I am going beyond my facts and I admit it, but so have the advocates of the contrary and they have been doing it for many thousands of years.

John B. Watson (1930)

[T]he Blank Slate is an empirical hypothesis about the functioning of the brain and must be evaluated in terms of whether or not it is true. The modern sciences of mind, brain, genes, and evolution are increasingly showing that it is not true.

Steven Pinker (2002)

The most far-reaching implications for science, and perhaps for society, will come from identifying genes responsible for the heritability of g … Despite the formidable challenges of trying to find genes of small effect, I predict that most of the heritability of g will be accounted for eventually by specific genes, even if hundreds of genes are needed to do it.

Robert Plomin (1999)

Finding genes brings us closer to an understanding of the neurophysiological basis of human cognition. Furthermore, when genes are no longer latent factors in our models but can actually be measured, it becomes feasible to identify those environmental factors that interact and correlate with genetic makeup. This will supplant the long nature/nurture debate with actual understanding.

Danielle Posthuma and Eco J. C. de Geus (2006)

It might be argued that it is no longer surprising to demonstrate genetic influence on a behavioral trait, and that it would be more interesting to find a trait that shows no genetic influence.

Robert Plomin and Ian Deary (2015)

[I]t's well established and uncontroversial among geneticists that together, differences in genetics underwrite significant variation in intelligence between people … intelligence is a classic polygenic human trait – just like many other cognitive and physical features, from mental disorders to height. … the genetic variation between individuals and its influence

on traits is more complex and subtle than scientists realized even at the start of this
century. What most people know about intelligence must be updated.

Intelligence Test, *Nature* (Editorial, 2017)

Learning Objectives

- Is the nature–nurture debate about intelligence essentially settled?
- What is the most compelling evidence that genes influence intelligence?
- What is the effect of age on environmental influences on intelligence?
- What are the key research strategies used in quantitative and molecular genetics?
- Why has it been so difficult to identify specific genes related to intelligence?

Introduction

Our brain evolved along with the rest of our body. It would be unlikely if genetics influenced all manner of human physiological differences but had no impact on the brain or the brain mechanisms that underlie intelligence. Nonetheless, genetic explanations of human attributes (even partial explanations) often arouse suspicion and unease. In part this comes from an assumption that anything genetic is essentially immutable, deterministic, and limiting. As we will see, this is not always a correct assumption, and the exact opposite may be true as we now have several powerful techniques for manipulating genes (see Sections 5.6 and 6.3). Some genes are deterministic – you have the gene, you get a specific characteristic – but for complex traits and behaviors such as intelligence, the genetic influences are best described as probabilistic rather than deterministic. That this, genes may increase the chances of having a characteristic, but whether you get it depends on multiple factors. For example, you may be at genetic risk for heart disease but you can lower your risk with diet and exercise.

In the most extreme scenario favoring genetics, if intelligence differences among people were due 100 percent to genetic factors inherited as a *random* mix of genes from a mother and a father, it would still be the case that some genes and their expression might be modified by environmental factors. The combination of genetic expression and nongenetic variables is one definition of epigenetics, discussed later in this section. In the 100 percent genetic scenario, it also would be the case that a person may well be liberated from a bad, suboptimal, or constraining environment by winning the genetic lottery for high intelligence (i.e., the random mix of genes from both parents including the salient ones for intelligence). The all (or mostly) gene scenario would also lead to a practical suggestion for maximizing your child's intelligence: find the smartest mate you can (simple but perhaps not so easy). In the 100 percent genetic scenario, a person who loses the genetic lottery and has low intelligence would be constrained in some important aspects of life success even if they had the best supportive or enriched environment money could buy.

In the other extreme scenario, if differences in intelligence involved no genetic mechanisms, each person's intelligence would be determined by the influences of their environment, especially during childhood when brain development is maximal and the child's ability to choose favorable environments is minimal. The all (or mostly) environment theory easily leads to a behaviorist or "Blank Slate" view that anyone could develop high intelligence, or any other psychological attribute, if only the right environmental ingredients were available (Watson, 1930). This view is quite popular despite the general demise of classical behaviorism. Moreover, the "Blank Slate" view of human potential has limited empirical support for most aspects of behavior (Pinker, 2002), and virtually no support at all for intelligence, as we will see later in this chapter.

The popular middle position holds that both genes (nature) and environment (nurture) explain differences in intelligence. The older, simple version of this position was that genetic and environmental factors both contributed about equally. We now recognize that gene expression during development and across the life span may be sensitive to non-DNA factors. This is the essence of epigenetics, which is the study of how environments, behaviors, and other biological factors (like aging) influence the ways genes function, possibly through a process of methylation or histone modification.[1] It's challenging to apportion intelligence variance to just genes when complex combinations of factors are part of the mix.

Epigenetics research in humans is a relatively new field, but there are some interesting findings. A longitudinal study of Romanian orphans, for example, has identified that the risk of cognitive and psychiatric problems is partially attributable to extreme social deprivation in early life experience. DNA analyses indicated that specific genetic alterations were

---

[1] Mechanism details are beyond the scope of this book, but see the Centers for Disease Control for more information: www.cdc.gov/genomics/disease/epigenetics.htm.

related to the degree of deprivation (Drury et al., 2012). Animal research suggests that some changes in gene expression related to environmental factors may actually be heritable (Champagne & Curley, 2009), but establishing causal relationships is difficult (Aristizabal et al., 2020). So far there is no strong connection to human intelligence, although there is emerging research on the epigenetics of memory, including a possible role for dopamine for both memory and intelligence (Heyward & Sweatt, 2015; Karalija et al., 2021; Lee et al., 2021; Kaminski et al., 2018). Environmental variables such as exposure to language in early childhood (Kuhl, 2000, 2004) influence brain neurobiology and development but whether these factors relate to intelligence has not been demonstrated. Specific epigenetic factors that may contribute to intelligence differences are not yet known, but the concept reinforces the assumption that any salient environmental variables work through biological mechanisms, genetic or not. For now, the weight of the evidence emphasizes the influence of genes on intelligence, with or without known epigenetic influences.

2.1 The Evolving View of Genetics

It may surprise you to know that the definition of a gene is not as precise as you may think (Silverman, 2004; Dawkins, 2016). Prior to the technology-driven Human Genome Project, genetic researchers expected to find about 100,000 genes because genes code proteins, which are the basic building blocks of life. Humans have at least 100,000 proteins. Each gene was thought to code one protein. But the Human Genome Project initially reported only about 25,000 protein-coding genes and that number has been revised downward to perhaps fewer than 20,000 (Ezkurdia et al., 2014). The number is not exact because the demarcation of where a gene segment begins or ends on a chromosome can be difficult to assess. Whatever the actual number, the fact of fewer than the number of proteins means that each gene can express itself in many different ways, and the mechanisms for controlling gene expression are largely unknown. Gene expression is just a way of saying that genes turn on and genes turn off over the lifespan. This results in a constantly changing mix of proteins influencing all aspects of biology in complex and dynamic combinations.

What exactly are the switches or triggers that turn genes on and off and how might the switches interact with other genes and with external environmental factors? How do the myriad gene protein products combine and interact with each other in multistep, cascading sequences that surely include a degree of randomness? The complexity inherent in such questions has caused many behavioral geneticists to conclude that the answers are beyond the reach of science, especially for complex traits such as intelligence. Perhaps they are motivated by an unwavering belief in "Blank Slate" assumptions or by social justice impulses that fear genetic explanations for differences among people or groups. I take the long-term view based on scientific advancement for the most complex questions in other fields, including molecular biology, and choose to be more optimistic, as discussed in Chapter 6.

Historically, most researchers have assumed that intelligence, no matter how it is defined, develops in childhood and is strongly influenced by environmental factors, especially home life and social culture. In this view, whatever role genes might play is minimized, and some even argue for a zero contribution from genes. Although this view about the importance of early environment seems reasonable, if not obvious, and even flattering to proud parents, the evidence for strong environmental effects on intelligence, especially in early childhood, is surprisingly weak, as we will see. Epigenetics provides a concept for the continued consideration of theories about the importance of environmental factors for intelligence, but epigenetic research on intelligence is just beginning (Haggarty et al., 2010; Kaminski et al., 2018) and its promise is still to be determined by the weight of evidence. By contrast, the data that support major genetic influences on intelligence are compelling.

Generally, we don't like to think about any constraints on potential life achievement, so the idea that intelligence has a major genetic aspect is not readily embraced as a good thing. This is especially so in some academic social science circles where there is a vested interest in the study of cultures. In fact, there is a decades-old concerted effort to undercut, deny, and impugn any and all genetic studies of intelligence (Gottfredson, 2005). A similar effort in the 1960s and 1970s regarding the "myth" of schizophrenia and other psychiatric disorders has all but disappeared. Much of the anti-genetic feeling originally arose as a moral response to eugenic movements in the nineteenth and early twentieth centuries, to the aftermath of Nazi atrocities during the 1930s and 1940s, and, most recently for our context here, to one specific paper published in 1969 by an educational psychologist at the University of California, Berkeley named Arthur Jensen. We discuss this infamous paper shortly.

Here is a crucial point to keep in mind as we introduce genetic studies: throughout this book, whenever we talk about the effects of any variable or factor on intelligence, we are actually referring to the effects on intelligence differences (variance) among people.

As the term implies, behavioral genetics generally refers to the study of behavioral traits (including cognitive abilities) and are of two basic kinds: quantitative and molecular. The former, with roots in Mendel's experiments with peas, deals with establishing whether a genetic component (a genotype) may exist or not for a behavior or characteristic (the phenotype) and if so, how much variance can be accounted for by genes. Quantitative genetics includes modeling a mode of gene transmission (e.g., dominant or recessive). Twin and adoption studies are primary methods of quantitative genetics and we will review key studies and some surprising findings that support a strong role for genes and a minimal role for environmental variables in explaining intelligence differences among individuals. Molecular genetics is a newer field and uses various DNA technologies and methods to identify genes that are related to variation in specific traits and, in the case of intelligence, how those genes work to influence brain development and brain function. This molecular-level ambition is as complex as any goal in any scientific field. So far, molecular genetic findings related to intelligence are quite tentative, with only limited replication of specific genes identified as possibly related to intelligence. Nonetheless, there is some progress and, in my view, the findings reviewed later in this chapter are somewhere between intriguing and amazing.

The early enthusiasm for molecular genetic techniques began about 20 years ago with the optimistic promise of imminent discovery of a few genes that accounted for considerable variance in intelligence. This has not happened, and more than a few critics of the genetic view emote a bit of glee in the failure to identify specific intelligence genes so far. Early indications, however, suggested that the hunt for intelligence genes actually was a hunt for "generalist" genes, each of which influenced multiple cognitive abilities that underlie intelligence. Kovas and Plomin (2006: 198) summarized this view simply: "genetic input into brain structure and function is general not specific." Two key concepts are: one gene can affect many dissimilar traits (pleiotropy) and many genes can affect one trait (polygenicity).

Although the concept of generalist genes is controversial, a broad consensus has emerged that intelligence is heritable and polygenetic. One study, for example, based on 3,511 unrelated adults, concluded that there are many intelligence genes that altogether may account for 40–50 percent of variance in general intelligence (Davies et al., 2011), although no one gene yet accounts for more than a tiny portion of variance. Additional research supports pleiotropy for diverse cognitive abilities (Trzaskowski et al., 2013). Researchers investigating schizophrenia, autism, obesity, and many other gene-rich targets (even height) find similar polygenetic and pleiotropic results. At this stage, the heritability of human intelligence is well established (Plomin, 2018; Plomin and von Stumm, 2018), and there is even supporting data in chimpanzees (Hopkins et al., 2014). There are interpretation issues regarding some aspects of the genetic data that are still unresolved (Nisbett et al., 2012; Shonkoff et al., 2000) and may remain so until specific genes related to intelligence are identified and confirmed. Generally, efforts to minimize the importance of genetic influences on intelligence in favor of environmental influences (Nisbett, 2009) do not stand up to scrutiny (Lee, 2010). Fortunately, there are even newer findings about intelligence that may signal real progress toward discovering specific genes and their effects. Before reviewing recent noteworthy studies in both quantitative and molecular genetics, let's start with some history to put the genetic story of intelligence into context by reviewing a surprising failure and an alleged fraud.

2.2 Early Failures to Boost IQ

The failure hit the fan in 1969 without warning. In the early 1960s, President Lyndon Johnson committed the United States to a war on poverty. One aspect of this admirable federal effort was aimed at a major concern that had been observed for decades. Poor children, especially from some minority groups, on average tended to score lower on cognitive tests, including IQ tests. At the time, the consensus among most educators, psychologists, and policy makers was that any cognitive gaps revealed by tests, especially for intelligence, were due mostly or entirely to educational disadvantages and therefore could be eliminated if poor children got the same early educational opportunities that middle- and upper-class families routinely provided. Such opportunities were virtually unavailable to the poor, especially prior to the 1954 Supreme Court decision striking down race-based separate but equal approaches to education. The solution for eliminating any cognitive gaps seemed obvious and the idea of compensatory education resulted in the federally funded Head Start Program. Prior to Head Start, several different compensatory education demonstration projects had been implemented on a limited basis. Some of these projects were reporting encouraging and even dramatically positive results at reducing cognitive gaps and increasing IQ scores. These efforts were the basis for the optimistic view that Head Start would be a great success at eliminating average IQ gaps.

The *Harvard Educational Review* asked Arthur Jensen, a noted educational psychologist, to review the claims of these early compensatory efforts (Head Start had not yet been implemented long enough to be included in this review). Jensen's article (Jensen, 1969) was entitled, "How Much Can We Boost IQ and Scholastic Achievement?" Here is the opening sentence: "Compensatory education has been tried and it apparently has failed." Jensen continued with over 100 pages of detailed analysis of intelligence research that revealed little if any lasting effect of the early compensatory efforts on either IQ scores or school achievement. That alone was bad enough in the political context of widespread enthusiasm for the nascent Head Start Program, but the article got worse when Jensen discussed genetics. He first reviewed studies of environmental effects on intelligence. He concluded that the empirical evidence for any major environmental effects on intelligence in general, and especially for the g-factor, was actually quite weak. He then argued that one reason for this would be that variance in intelligence, especially the g-factor, was mostly genetic, at least for individuals. He summarized genetic studies that appeared to validate this view. In 1969, concluding that there was a genetic component to intelligence was a bit of a stretch given the paucity of both environmental and genetic studies with large samples and solid research designs. But the article, which was already offensive to the majority view that intelligence derived mostly from environment, went even further with a controversial suggestion (and controversial is an understatement). Since IQ scores appeared to be impervious to compensatory efforts, and since genes apparently played an important role for individuals, Jensen asserted the hypothesis that the average intelligence differences found for some racial groups compared to whites (he focused on black–white differences) might also have a genetic component. And, with the publication of that hypothesis, research on intelligence all but ended for more than a generation.

The negative response to Jensen's review article was ferocious. The most vicious responses were directed to the inflammatory inference that blacks were intellectually inferior because of their genetic makeup and to the general idea that genes played a major role in intelligence and the environment did not. Jensen's concluding paragraphs about the importance of adjusting teaching methods to match the learning capabilities of individual students to maximize school achievement for all children received virtually no attention (see Section 6.6). In any case, critics have spent decades attacking Jensen personally and his arguments. As mentioned briefly in Chapter 1, another book published in 1973, *IQ in the Meritocracy* (Herrnstein, 1973), created a similar firestorm regarding the role of genetics in intelligence. Given the racial inferences and the hot emotional atmosphere, few researchers or their students opted to focus their careers on any questions at all about intelligence. Getting federal research support for researching intelligence became virtually impossible. Almost overnight, intelligence research became radioactive.

Head Start pushed ahead and similar compensatory research efforts included increasingly intensive interventions. In the 1970s and 1980s, Jensen's critics attacked the validity of IQ tests and scores, the existence of the *g*-factor, quantitative genetics in principle, and even the integrity and motivation of individual researchers. One simple argument was that any average group differences in intelligence test scores were most likely due to test bias and therefore had no meaning. The bias hypothesis, as noted in Chapter 1, has been studied extensively and has little empirical support (Neisser et al., 1996; Warne, 2020). As far as test scores being without real meaning, there is extensive evidence (as noted in Chapter 1) that scores predict many aspects of life (Deary et al., 2010; Gottfredson, 1997b). Moreover, in Chapters 3 and 4, neuroimaging shows that intelligence test scores are correlated to a variety of structural and functional measures of the brain, findings that would be impossible if the test scores were meaningless. Some critics challenged whether the *g*-factor was merely a statistical artifact, a view not supported by many sophisticated psychometric studies (Jensen, 1998a; Johnson et al., 2008). Other critics went beyond the debate about data and attacked Jensen and some behavior genetic researchers with *ad hominem* charges of explicit racism. Jensen was once asked directly if he was a racist. His answer was, "I've thought about this a lot and I have come to the conclusion that it's irrelevant" (Arden, 2003: 549). I knew Jensen for many years and I understand his point to be that his interpretation of the data, even if it was motivated by unconscious racism, was testable and falsifiable by objective scientific methods. He was confident that future research could potentially refute any of his hypotheses. He was, as far as most observers could perceive, unflappable in the face of personal attacks because he was completely driven by data. In my view, he would not have been disappointed at all if new data showed him to be wrong. For those of you interested in learning more about Jensen and his contributions to intelligence research, please see a special issue of *Intelligence* in 1998 that has several papers including these (Detterman, 1998; Jensen, 1998b).

The point in summarizing this incendiary period in the history of intelligence research is to help explain the origin of the negative valence that intelligence research still carries to some extent today. The modern neuroscience studies that are the focus of this book have helped the field move beyond these old destructive controversies. While the basis of average group differences on psychometric tests of intelligence and other cognitive abilities is still unsettled, the major role of genetics for explaining intelligence differences among individuals is firmly established, as detailed in Section 2.3.

It is also the case that the weight of evidence from modern studies of intensive compensatory education, now rebranded as early childhood education, still fail to find lasting effects on IQ scores, and even short-lived increases are not clearly related to the g-factor (te Nijenhuis et al., 2014), although one small-sample study reports an IQ increase in a specific comprehensive program that appears to be g-related (Pages et al., 2021). Newer, more intensive interventions, suggest that some important aspects of academic achievement, such as graduation rates, apparently improve in some studies (Barnett and Hustedt, 2005; Campbell et al., 2001; Ramey and Ramey, 2004) but not in others (Durkin et al., 2022), and there may be some long-term economic benefits (Bailey et al., 2021). Nonetheless, in my view, there still is no weight of evidence that Head Start is particularly effective in benefiting high-risk children's pre-academic skills (Miller et al., 2016) or other important school and adult outcomes (Pages et al., 2020), and the same holds true for similar interventions.

There are some quantitatively sophisticated estimated projections that IQ scores for disadvantaged children could potentially increase dramatically given the right program components at early ages (Duncan and Sojourner, 2013), although no such gains have been realized let alone tested for durability, with the possible exception of two recent small-sample studies (Pages et al., 2021; Romeo et al., 2021). It is my view that there are many good reasons to support early childhood education that do not depend on whether IQ scores change and are not due to genetics or other reasons. Including IQ in the discussion about justifying early education probably doesn't help make the case. More about the neuroscience potential for increasing intelligence is detailed in Chapter 5.

With respect to both a genetic basis for intelligence and the failure of early education to boost IQ, it is fair to say that Jensen's hypotheses have not yet been refuted by another 50 years of new data (see Johnson, 2012). The interested reader is referred to the references at the end of this chapter for sources that delve into the Jensen controversies in greater detail (Snyderman and Rothman, 1988). Steven Pinker's *The Blank Slate* is a comprehensive book for understanding the broader historical and philosophical context of intelligence research criticism. I also recommend that any student interested in pursuing a career in intelligence research using neuroscience or other approaches read Jensen's 1969 article. It is often cited, more often misrepresented, and in my view a classic work of psychology that still suggests important ideas and hypotheses to test with modern methods.

2.3 "Fraud" Fails to Stop Genetic Progress

Before moving on to modern advances in both quantitative and molecular genetics, we need to take one more historical side trip. Explaining this story also introduces basic strategies of quantitative genetics research. Following the 1969 article, another line of attack claimed that some of the genetic data Jensen cited to support his argument was fraudulent. These data came from identical twins reared apart as reported by Sir Cyril Burt, an eminent British psychologist in the mid–twentieth century.

The story of "fraud" begins with the undistinguished number 0.771. Here's the background. Since monozygotic (MZ) twins (i.e., identical twins) have 100 percent of their genes in common, any trait that was found in both twins was thought to have a genetic component. The more similar the twin pairs on the trait, the stronger the effect of genes. Of course, identical twins also share both the prenatal and the postnatal environment, so the fact that identical twins may have quite similar intelligence test scores does not rule out the possibility that the similarity is due to similar environments. Conceptually, this problem is easily addressed by comparing the similarity of a trait between pairs of identical twins who have 100 percent of their DNA in common to pairs of fraternal twins (i.e., dizygotic (DZ) twins). Fraternal twins share most of their early environment but only 50 percent (on average) of their DNA, so any similarities should not be as strong in the fraternal twins as they are in the identical twins.

Indeed, this is the undisputed case in many studies of intelligence, which collectively report that average correlations for identical twins are about 0.80 and 0.60 for fraternal twins (Loehlin and Nichols, 1976).

Adoption studies are even more powerful and compelling because they separate genetic and environmental influences more clearly than comparisons of identical and fraternal twins reared together. The Denmark Adoption Studies, for example, shifted the debate about the etiology of schizophrenia decidedly toward a genetic component because adopted children who had a biological parent with schizophrenia grew up with a higher risk of having schizophrenia than other adopted children with biological parents who were not schizophrenic. David Rosenthal was one of the principal investigators for the Denmark studies and I worked in his laboratory at the National Institute of Mental Health (NIMH) in my first job after graduate school. He once told me that although these studies did not elucidate much about schizophrenia other than that a genetic component of some kind was involved, the beauty of the adoption design was its conceptual simplicity. Basically, only two numbers counted. Anybody can see a higher rate of schizophrenia in the one group compared to the other, so in this case, it was hard to deny some role for genetics (although some anti-genetic critics certainly tried).

There are relatively few well-done adoption studies of intelligence. These studies are quite complex to do because so many variables are difficult to control (e.g., age at adoption, age at intelligence testing, indexing similarities of environments in a quantitative way, the rate of participant dropouts from the study, no random assignment to environments). Nonetheless, results consistently report higher intelligence test score correlations between adopted children and their biological parents compared to correlations with adoptive parents. In fact, the correlations with adoptive parents are very low and even near zero (Petrill and Deater-Deckard, 2004), especially as children grow older, which is another observation that is hard to explain for critics who argue against genetic effects on intelligence. Interestingly, one adoption twin study reports higher IQs in adopted children compared to their nonadopted siblings, suggesting that enriched educational opportunity in the adoptive home leads to an increase of about 3–4 IQ points in early adulthood (Kendler et al., 2015). The study is noteworthy for the large sample of sibling pairs and a replication in a large sample of half-siblings. These samples were identified in Sweden, a country that registers such information systematically. Some caution is warranted because the IQ measure consisted of only four subtests used by the Swedish military. As noted in Chapter 1, all IQ scores are estimates of an underlying construct and small differences between groups are difficult to attribute to any causation. This study suggests a small effect for the environment of the adoptive home and this finding does not contradict or impugn the other heritability data in any way. The heritability studies always demonstrate that some environmental effects must be at work. For example, a large study of 6,311 adoptees in the UK Biobank project found a statistical interaction between family environment and gene influences (Cheesman et al., 2020). DNA-based polygenic scores (explained in Textbox 2.2) showed stronger prediction of education attainment (i.e., years of education, an indirect measure of intelligence) in nonadopted individuals than in the adoptees, suggesting the home environment mediates the genetic influences. Whether IQ scores or *g* show the same mediation as education attainment is not yet determined. For a detailed discussion of adoption studies of intelligence, see (Haier and Colom, 2023).

An even more powerful design combines adoption and twins. Think about studying identical twins adopted away from their biological parents in early life with each one raised separately in a different family and exposed to different everyday environments, one twin often not even knowing of the existence of the other. Are identical twins reared apart still very similar to each other on things such as intelligence test scores?

This brings us back to 0.771. In mid–twentieth century Britain, Sir Cyril Burt did the first major studies of intelligence in identical twins adopted away from biological parents and reared in separate adoptive families. Over a number of years, Burt gave intelligence tests to pairs of identical twins who had been reared apart; an extremely rare group that is quite difficult to find and enter into a research study. He first mentioned a correlation of 0.77 in 15 pairs of identical twins reared apart (Burt, 1943), suggesting a strong genetic component to intelligence. Subsequently, he added six twin pairs and reported a correlation of 0.771 (Burt, 1955). His third report included 53 pairs with a correlation in the identical twins reared apart of 0.771 (Burt, 1966).

The three reports had different sample sizes ranging from 15 to 53, but each new, larger sample showed the same correlation of 0.771 (0.77 in the first report). Burt's results were a key element of Jensen's 1969 argument. Critics of Jensen's genetic view revisited Burt's publications looking for possible flaws and 0.771 caught their attention. They argued that the same correlation value to three decimal places based on different sample sizes was statistically improbable. They concluded that Burt surely committed scientific fraud, and this example is still cited today to undermine the idea that genes are important for intelligence. In the wake of the fraud charges, Jensen, who knew Burt (who died in 1971), examined Burt's original data files and found a number of serious concerns that he reported in detail (Jensen, 1974). Jensen was willing to exclude Burt's data from his argument but still maintained that other data supported a role for genetic influences on intelligence. Most independent investigations of Burt's data doubt the claim of intentional fraud (Mackintosh, 1995). We may never know for sure, but the main point is that the Burt data no longer matter.

Subsequent twin studies done by different investigators around the world with large samples arrive at an average value for the correlation of intelligence scores among identical twins raised apart of 0.75 (Plomin and Petrill, 1997). Burt's value was 0.77. For comparison, based on 19 studies ranging in sample sizes between 26 and 1,300 identical twin pairs, the average value for identical twins raised together is about 0.86 (see Loehlin & Nichols, 1976: 39, table 4.10). These values compare to the fraternal twin data (ibid.) that show average correlations for intelligence of

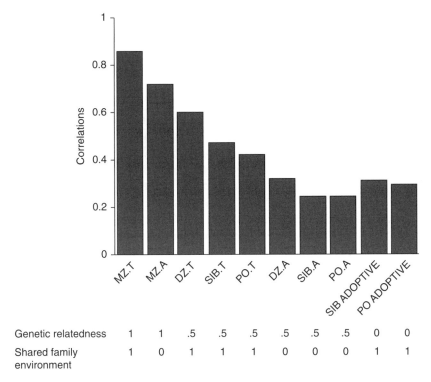

Genetic relatedness   1   1   .5   .5   .5   .5   .5   .5   0   0

Shared family   1   0   1   1   1   0   0   0   1   1
environment

Figure 2.1 Intelligence variance accounted for by genetics based on family, twin, and adoption data. T = reared together, A = reared apart, MZ = identical twins, DZ = fraternal twins, sib = sibling, PO = parent-offspring. (Reprinted with permission (Plomin and Petrill, 1997))

about 0.60 based on pair sample sizes of 26 to 864. The overall story from twin and adoption studies has been apparent for some time (Bouchard & McGue, 1981; Loehlin, 1989; Pedersen et al., 1992) is nicely summarized in Figure 2.1 (Plomin & Petrill, 1997). In subsequent research, however, a rather dramatic new insight has emerged that further informs the pattern in Figure 2.1. We now know that age at the time of intelligence testing makes quite a difference for heritability estimates. This is discussed in Section 2.4.

Thus, the 0.771 "fraud" ends with the recognition of overwhelming data from independent researchers that are fully consistent with Burt's analyses, flawed as they may have been. Any single study, or any one researcher, can be flawed, but the basic conclusion that genes play an important role in intelligence is consistently supported by data from numerous studies of twins, adoptees, and adopted twins. This is an excellent example of looking at the weight of evidence (recall my three laws from the Preface: no story about the brain is simple; no one study is definitive; it takes many years to sort out conflicting and inconsistent findings and establish a weight of evidence). Many issues about the role of genetics in intelligence remain unresolved. For example, cross-sectional and historical data show that average IQ scores have consistently increased about three points every decade around the globe. This is known as the Flynn effect (Flynn, 2013; Trahan et al., 2014). Some critiques of the genetic role in intelligence argue that such an increase cannot be attributed to the slow pace of genetic evolution; they are correct. The increase may be a partial *g* effect (Pietschnig & Voracek, 2015) and a main cause could be better general nutrition for infants (Lynn, 2009) or other general societal improvements. But the existence of the Flynn effect does not disapprove or negate a major role for genetic influences on intelligence. It just documents that a portion of intelligence differences among people are not due to genes alone, which is an old and uncontested finding. The weight of evidence summarized in this chapter leaves no reasonable doubt that genes have major influences on intelligence. Only ideologues are still in denial. Data from more recent twin studies, described later in this chapter, expand the basic genetic findings to a new level of focus for neuroscience with the addition of DNA assessments. But let us not ignore that genetic studies actually highlight a role for nongenetic factors with some surprising empirical observations.

2.4 Quantitative Genetic Findings Also Support a Role for Environmental Factors

While the atmosphere surrounding intelligence research was still toxic and Burt's data were under attack, a group of researchers at the University of Minnesota led by Professor Thomas Bouchard embarked on a new project to identify a large sample of the rare group: identical twins reared apart. Ultimately, 21 years of searching (1979–2000) yielded 139 twin pairs from around the world who participated in the project (including at least one set of triplets). Some twins had no contact with each other until they were reunited in Minnesota where all the twins completed an elaborate battery of tests for about 50 hours over a week. This battery included tests of intelligence, personality, attitudes, values, and many physical characteristics.

Genetic components were found for the variance of several personality traits such as extroversion and, surprisingly, even for some attitudes and values. But these identical twins reared apart were most similar on intelligence scores with a correlation of 0.70 (Bouchard, 1998, 2009). When correlations are computed in identical twins reared apart, the correlation is also one way to estimate hereditability, so a correlation of 0.70 indicates that 70 percent of the variance in intelligence is due to genetic factors and 30 percent was not. Although this result from a large, careful study did not end all skepticism about a role for genetics, it started to temper many critics who were suspicious of Burt's results and more inclined toward yet-to-be-identified environmental factors. Like the impact of the Denmark Adoption Studies of schizophrenia on psychiatry, the Minnesota study began to shift the tide toward a renewed objective interest in genetic contributions to intelligence.

All of the twin and adoption studies of intelligence that demonstrate an important role for genes also are consistent in showing that genes do not account for 100 percent of the variance. So, an important consequence of the genetic studies is demonstrating that nongenetic factors must be involved in some way. Prior to the current interest in epigenetics and gene–environment combinations, there were attempts to apportion the contributions of genetic and nongenetic environmental factors. The most common view was about 50–50. However, there is considerable variability among studies regarding this proportion and there's an interesting factor that accounts for much of this variability: the age when twins are tested (Haworth et al., 2010; McGue et al., 1993).

Based on cross-sectional data, in young twins who are 4–6 years old, the heritability of intelligence estimate is about 40 percent, and the heritability rises to a high of about 85 percent when the twins are older adults. In other words, the genetic influences on intelligence variance actually *increase* with age while environmental influences decrease. Note that cross-sectional means that different twin pairs participated in different studies at different times. Suppose we followed the same twins and retested them periodically as they got older. Would we see the same trend in such longitudinal data? The answer is yes. In a large Dutch twin study (Posthuma et al., 2003), the same identical twins were given mental test batteries repeatedly over time to assess general intelligence. The heritability estimate of general intelligence was 26 percent at age 5, 39 percent at age 7, 54 percent at age 10, 64 percent at age 12, and starting at age 18 the estimate grew to over 80 percent. The increases could be due to several factors, including more genes "turning on" with increasing age or gene–environment combinations. Heritability is a complex concept since it can include genetic and nongenetic factors for why a trait may run in families. Twin and adoption studies help separate these influences but a detailed discussion of heritability estimation and genetic modeling is beyond the intention of this book.[2]

Here, we are focused on an overview of genetic studies that provide a rationale for neuroscience approaches. Nonetheless, I want to present data that illustrate important findings about nongenetic factors that come from quantitative genetic studies. Until now, I have discussed environmental factors as a single category. One common quantitative genetic model divides the environment into two categories: shared and nonshared factors. A shared environment is what it sounds like. Twins and siblings grow up in the same family, live in the same neighborhood, and attend the same schools. They have many shared general experiences that may influence intelligence. There are also many experiences unique to each person, such as different friends, classes, and teachers. These unique influences are the nonshared environment.

In these models, genetic influences and shared and nonshared environmental factors added together account for 100 percent of the variance in any characteristic such as intelligence differences among people. The amount of variance attributed to each component can be distinguished and estimated statistically by comparing the similarities of intelligence scores for identical twins, fraternal twins, and siblings, with samples from each group reared together and reared apart. Differences in intelligence test score correlations in these groups are used to estimate how much variance each of the three components contribute (Plomin and Petrill, 1997). Although this basic three-component model does not incorporate gene–environment combinations or an error component that could include the influences of random events (Mitchell, 2018), it has provided important observations.

Let's consider additional data from the Dutch twin study described at the end of Section 2.3. The chart in Figure 2.2 shows the influence of genetics and both shared and nonshared environment on intelligence scores for people of different ages. The black part of the bar shows the genetic influence we noted at the end of Section 2.3. The white part of the bar shows shared environment, and the gray part shows nonshared environment influence. Shared environment influences, the white bars, peak at age 5 and then decrease to virtually zero by age 16. The nonshared environment, the gray bars, has a generally greater influence in the early years, but some nonshared influence continues

---

[2] See Hunt (2011: chapter 8) for a detailed presentation or www.nature.com/scitable/topicpage/estimating-trait-heritability-46889/.

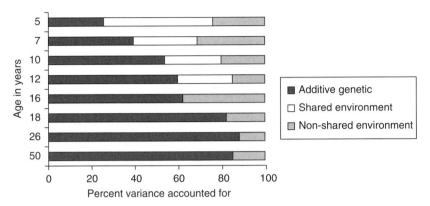

Figure 2.2 Genetic, shared, and nonshared environmental influences on intelligence variance for different age groups. (Reprinted with permission (from Hunt 2011, based on Posthuma et al., 2003))

through to at least age 50. Note the sources of nonshared influences likely change over time.

To recap this key piece of the genetic story, the heritability of general intelligence increases with age to about 80 percent by the end of the teenage years and the effects of shared environment on intelligence decrease to near zero a bit earlier. These findings are extraordinary and among the most powerful and important in all of psychology. They are difficult to explain if you are convinced that genes are unimportant for intelligence. They also give pause to the idea that enriching childhood family experiences, as pleasant as they may be for many good reasons, have a lasting effect on the development of intelligence. However, these data also show that both shared and nonshared environments have effects at different developmental stages and in different amounts, especially before age 18. They are just not as strong as was once believed, but they demonstrate clearly that genes alone are not the whole story. All genetic researchers know that genes always express their function within an environmental context that may influence expression in many ways. Specific sources of these nongenetic effects on intelligence are not yet determined, just like specific genes are not yet identified, although general factors such as schooling have some influence (Ceci, 1991; Ceci & Williams, 1997; Ritchie & Tucker-Drob, 2018; Tommasi et al., 2015). As noted, part of the complexity is that environmental factors such as SES often are confounded with genetic factors for intelligence since intelligence plays a role in income and other factors that define SES. This is discussed further in Textbox 2.1.

Textbox 2.1: Social class and intelligence

A widely cited study suggested that the heritability of intelligence is stronger in families with high SES and weaker in families with low SES (Turkheimer et al., 2003) but not all studies agree (Asbury et al., 2005; van der Sluis et al., 2008). Generally, SES is confounded with intelligence. On average people with high intelligence get higher-paying jobs and have more money to provide resources for their children. They attain a higher SES directly or indirectly due in part to intelligence along with other factors (including luck). To the extent that intelligence is passed along by genes, SES effects independent of genetic influence on intelligence are difficult to assess. This underscores one difficulty in assessing the combined influences of genes and environments. A recent meta-analysis of SES and heritability studies of intelligence suggests a more complex statistical interaction between SES and intelligence (Bates et al., 2013) and there is some evidence that SES, education, and general intelligence have genes in common (Marioni et al., 2014).

In this context, a fascinating study of social class in Poland during its socialist years addressed this issue in an unusual way. This is an older study but quite illustrative (Firkowska et al., 1978). Here is the summary quoted directly from the published report:

The city of Warsaw was razed at the end of World War II and rebuilt under a socialist government whose policy was to allocate dwellings, schools, and health facilities without regard to social class. Of the 14,238 children born in 1963 and living in Warsaw, 96 percent were given the Raven's Progressive Matrices Test and an arithmetic and a vocabulary test in March to June of 1974. Information was collected on the families of the children, and on characteristics of schools and city districts. Parental occupation and education were used to form a family factor, and the district data were collapsed into two factors, one relating to social marginality, and the other to distance from city center. Analysis showed that the initial assumption of even distribution of family, school, and district attributes was reasonable. Mental performance was unrelated either to school or district factors. It was related to parental occupation and education in a strong and regular gradient. It is concluded that an egalitarian social policy executed over a generation failed to override the association of social and family factors with cognitive development that is characteristic of more traditional industrial societies.

In the context of this chapter, the confounding of genetic and SES factors leads to a possible alternative conclusion: Any influence of social policy on mental performance failed to override the influence of genetic factors. The same confounding is apparent in new studies that suggest that SES accounts for the brain differences underlying cognitive/achievement gaps, which we detail in Chapter 6.

Nonetheless, by the beginning of the twenty-first century, the ascendance of the genetic view for intelligence was mirrored in three "laws" of quantitative genetics (Turkheimer, 2000): "First Law: All human behavioral traits are heritable. Second Law: The effect of being raised in the same family is smaller than the effect of the genes. Third Law: A substantial portion of the variation in complex human behavioral traits is not accounted for by the effects of genes or families." Turkheimer discusses the implications of these laws in the context of new challenges for explaining how genetic and environmental influences may work. Recently, Plomin and Deary (2015) offered their own version of three laws: All traits show significant genetic influence; no traits are 100 percent heritable; heritability is caused by many genes of small effect. This chapter is not the place to discuss these "laws" in detail. I cite them here to emphasize the sea change in thinking about the role genes play in complex traits, virtually all of which have high heritability estimates according to a comprehensive meta-analysis that included almost every twin and adoption study ever conducted (Polderman et al., 2015). Since genes work biologically, high genetic heritability is a primary reason that neuroscience research on intelligence is expanding so quickly. In fact, this expansion is also fueled by other replicated behavioral genetic finding relevant to intelligence research (Plomin et al., 2016).

Are the factors that influence intelligence consistent across the entire range? Here is an interesting example of progress in testing hypotheses not just about whether genes are involved in intelligence but about how genes are involved. An important question is whether the genetic basis of high intelligence and any influence of environmental factors on the salient genes are the same as for the average and lower parts of the normal distribution of intelligence test scores. High intelligence might result from different genetic and environmental factors than those that influence average and lower intelligence. This view is called the discontinuity hypothesis. One discontinuity hypothesis is central to the view that expertise associated with high intelligence is more a reflection of practice and motivation effects stemming from experiences rather than from inherited ability. Another discontinuity hypothesis is that different genes are involved in high intelligence than the genes involved in average intelligence. By contrast, the continuity hypothesis tests the view that the same genetic and environmental factors are at work throughout the intelligence distribution. The effects of each factor are additive so high intelligence reflects having more of the relevant genes and experiences.

In lieu of having specific intelligence genes to compare between a high-IQ group and an average group, twin studies can test these competing hypotheses by comparing the groups for the proportions of genetic, shared, and nonshared variance. Simply put, the discontinuity hypothesis predicts that the three components of variance would differ between high and average intelligence groups. A strong test of this was based on 9,000 twin pairs and 360,000 siblings sampled from 3 million 18-year-old males conscripted into military service in Sweden (Shakeshaft et al., 2015). All had completed a battery of cognitive tests from which a g-factor was extracted and the top 5 percent comprised the high-intelligence group (IQ estimated at greater than 125).

Several analyses were reported that were consistent in showing strong support for the continuity hypothesis and virtually no support for either the environmental or genetic discontinuity hypothesis. The authors concluded, "[s]tated more provocatively, high intelligence as we defined it appears to be nothing more than the quantitative extreme of the same genetic factors responsible for normal variation" (Shakeshaft et al., 2015: 130). They also cautioned that they did not have sufficient statistical power to determine whether the groups might differ if the high-intelligence group was defined more extremely, say at the upper 0.025 percent suggested for defining genius by Galton (Galton, 1869) rather than the upper 5 percent.

Another twin study tested the important issue of whether genes account mostly for the g-factor or for specific cognitive domains that comprise g (Panizzon et al., 2014). These investigators used a large sample of middle-aged veterans (average age 55) from the Vietnam era in a longitudinal study of aging and found 346 pairs of identical twins and 265 fraternal twin pairs. Everyone had completed a battery of 10 cognitive tests representing four basic and well-established cognitive domains (verbal ability, working memory, visual-spatial reasoning, processing speed). Several alternative factor analysis models of the relationship among tests and domains that did not place g at the peak of a hierarchy (see Chapter 1) were compared on estimates of variance accounted for by genes, shared environment, and unique environment. The model that best fit the data indicated that g in the hierarchical model was more heritable (86 percent) and accounted for most of the genetic effects in the specific domains than any other model. Although the researchers acknowledge some limitations in the study design, by directly testing alternative models, the results extend and strongly support the earlier research on g as the key common heritable factor underlying different mental abilities.

Additional progress is illustrated in elaborate new twin studies that not only have very large sample sizes but also combine DNA assessments and neuroimaging. These studies blend quantitative and molecular genetics and we review them in Chapter 4 after Chapter 3 has introduced neuroimaging. Before that, however, we continue here with some early studies of molecular genetics and the hunt for specific genes.

2.5 Molecular Genetics and the Hunt for "Intelligence Genes"

Technological advances in measurement drive scientific progress. Until DNA technology developed to a point where the double helix could be chopped into precise fragments using cost-effective methods and the billions of pieces (base pairs among the smallest units) could be characterized statistically, the hunt for human intelligence genes could not advance in earnest. For decades, breeding experiments in mice had produced tantalizing evidence that learning how to navigate a maze to find cheese had a genetic basis. Some mice learned faster than others, and when the "smart" mice were bred with other smart mice, the offspring learned the maze faster. In 1999, genetic engineering was used for the first time to create smart mice that could learn a maze more quickly (Tang et al., 1999). The researchers named the strain of these mice "Doogie," after a TV character that was a precocious teenager in medical school. This achievement (the mice not the TV show) was based on considerable previous animal work that showed a certain synaptic receptor, NMDA (N-methyl D-aspartate), was involved in learning and memory. A single gene (NR2B) was found to regulate a part of this receptor's function. The researchers spliced this gene into the DNA of ordinary mice embryos. The resulting Doogie strain of mice learned a series of tasks faster than controls.

All neurotransmitters and receptors work in complex balances. In the world of synapses, too much or too little of any component can have deleterious or fatal consequences so applying animal research findings to humans requires considerable patience and caution. Whether genetic manipulation of the NMDA receptor in humans might produce similar learning and memory enhancements, without serious side effects, is not yet known and it is not easy to find out. This example illustrates that finding a gene related to something like learning, memory, or intelligence is just a first step, albeit a challenging one even in animals. Determining how and why the gene functions, within a cascade of neurobiological steps and nonindependent combinations, is even more difficult. Manipulating genetic effects to produce a desired outcome is not for the faint of heart or for researchers with impulsive personalities or for short-term investors. Nonetheless, Doogie mice are a tantalizing example of the powerful potential for changing the determinism assumed when something has a strong genetic basis.

Separately from animal learning and memory research, the search for human intelligence genes began with a simple research strategy. DNA samples were collected from groups defined by IQ scores. Each participant's DNA was fragmented into small pieces where genes could be identified. Textbox 2.2 describes key terms and methods used in DNA studies. These fragments from high and low (or average) IQ groups were compared and differences were noted as candidates for genes related to intelligence. This is a needle-in-the-haystack strategy because the number of fragments and individual genes or base pairs was in the many millions, the cost per individual was quite high, and the IQ groups also differed on many difficult-to-control characteristics in addition to IQ. Nonetheless, researchers were optimistic that specific intelligence genes would be discovered, especially as new DNA assessment technologies were developed. In fact, many candidate genes were identified using various quantification techniques of increasing sophistication.

Despite the daunting challenges of this search, a number of research groups around the world used variations of this strategy in trying to identify specific genes that influence intelligence. One interesting approach in a Japanese study used multiple DNA assessment techniques in a sample of 33 identical twin pairs who were discordant for IQ scores (Yu et al., 2012). That is, the twins within a pair had at least a 15-point difference in scores (one standard deviation). Using discordant identical twins (reared together), even a small sample minimized irrelevant genetic and environmental factors and maximized the chances of finding salient differences in gene expression related to intelligence, even if differences might be the result of epigenetic influences. The use of multiple methods of DNA analysis allowed for independent replications within the same sample. The outcome identified several possible differences in gene expression that suggest brain mechanisms that might be related to intelligence. The findings illustrate the complexity of gene expression and regulation that are far beyond our discussion here, but they demonstrate that the importance of finding genes is only the first step toward understanding exactly what the genes do and how they influence or regulate neurobiology and brain function on the molecular level.

Even at the early stage of the search for intelligence genes, the emerging data were consistent in two important ways. First, none of the candidate genes accounted for much variance in intelligence test scores. This was a disappointment to those who had hypothesized that a few genes would account for considerable if not most variance given the high heritability estimates for intelligence from twin studies. Other researchers recognized that this reflected the complexity of the cognitive processes involved in intelligence and was consistent with Plomin's prediction about numerous generalist genes, each accounting for a tiny portion of the variance in intelligence. This view had early empirical support (Trzaskowski, Shakeshaft, & Plomin, 2013). The second consistency, which was far more distressing, was that none of the candidate genes identified in this early phase could be replicated in independent samples. Independent replication is a cornerstone and absolute requirement of scientific progress. Independent ideally means both a different investigator and a different sample. In this early phase there was considerable competition to find "the" genes and journals did not require independent replication for publication, even by the same investigators in a separate sample.

There is no reason to list all the early candidate genes here along with the subsequent failures to replicate. This disappointing state of affairs did not change for about two decades despite advances in the precision and cost-effectiveness of DNA technology and genomic information statistical analysis. One key problem was that most sample sizes were small and lacked the statistical power to replicate any tiny effects a gene might have on intelligence. By 2012, Chabris et al. (2012) summarized the search for intelligence genes in a comprehensive research paper titled: "Most Reported Genetic Associations with General Intelligence Are Probably False Positives" (i.e., they are wrong). This group attempted to replicate 12 candidate genes for intelligence from published studies. They had access to three independent samples totaling over 6,000 people who had completed DNA analyses and intelligence testing. Their analysis was decidedly negative. None of the 12 candidate genes were associated with intelligence in a robust statistical manner. This failure to replicate intelligence genes, in a large sample with sufficient statistical power to find small effects if effects existed, was offset by successful replication in the same sample for control candidate genes related to Alzheimer's disease and to body mass. The authors did not express discouragement at their failure to replicate genes for intelligence and concluded that even larger samples might be needed to find and replicate multiple genes if each one accounted for even tinier amounts of intelligence variance. They encouraged intelligence gene hunters to take part in multicenter consortia that could generate sample sizes of thousands of people.

Textbox 2.2: Basic genetic concepts
The technologies and methods of DNA analyses used in molecular genetics are diverse, complex, and evolving at a rapid pace (Mardis, 2008). The basic thrust of progress is that the costs decrease and the precision and scope of analysis increases. There are now many studies of intelligence using DNA analyses. Here are some key terms used in the representative studies summarized in this chapter (see Glossary for more). A gene is a unit of inheritance. There are an estimated 20,000 genes that make proteins, distributed on the two sets of 23 chromosomes (one from each parent) in the human genome. Genomics is the term used to describe the mapping of genomes using many different methods. Every person has a unique genome although most of the gene sequence is the same for everyone. Chromosomes are made of two strands of DNA molecules: the so-called double helix. During reproduction, offspring randomly inherit these strands from each parent. The strands all are constructed from a combination of only four base molecules, (A)denine, (G) uanine, (C)ytosine, (T)hymine. These four arrange themselves in pairs across the two strands of DNA in the double helix like rungs on a ladder. Each member of a pair is inherited from one parent. On each rung, A and T, and G and C, form pairs.

There are an estimated three billion of these "base pairs" (also called nucleotides) in the human genome. The order of these base pairs on a strand of DNA is the genetic code. All humans have nearly identical genetic codes and all the differences among individuals result from a relatively small portion of genetic variations. Genes create amino acids that form thousands of different proteins and proteins are the building blocks of life that determine how an organism develops and functions at the cellular level. The sequence from amino acid creation to protein formation is called gene expression. RNA is similar to DNA but RNA essentially translates the DNA code into amino acids and proteins. Genes can be active or inactive. They turn on and off during development and over the life span. The expression of a gene is regulated in part by methylation, one of several neurobiological mechanisms that can be influenced by nongenetic factors, including diet, illness, and stress (Jaenisch & Bird, 2003). Methylation is a process involved in many aspects of normal and abnormal cell development. Of particular interest is the fact that it can change the molecular structure of A and C in base pairs. These changes alter the expression of some human genes (Wagner et al., 2014), and importantly, the modified genes can potentially be inherited. Epigenetics studies how gene expression is modified by nongenetic factors.

DNA sequencing identifies the exact physical order of all the base pairs. Genes are contiguous segments of base pairs, although it is not always clear where one gene ends and another begins. Variations in base pairs at specific sites are called alleles (sometimes referred to as different forms of genes).

For example, a hypothetical gene for wrinkles on the chin may be expressed as WW and no wrinkles as ww. Each parent contributes either a W or a w, so the allele can be WW, Ww, wW, or ww. The inherited pair will determine whether the offspring has wrinkles or not.

The position of a gene on a chromosome is a locus. Quantitative trait locus (QTL) refers to a region of DNA related to a trait such as intelligence. There are often repeat copies of a gene at a locus and the number of copies can sometimes be related to normal or abnormal protein functions. DNA analysis generally breaks strands into fragments using any of several techniques. One technical breakthrough allowed the identification of small variations in the DNA sequence at any point on a strand where a base pair is changed or mutated. These errors are called single nucleotide polymorphisms (SNPs). For example, if a sequence at a locus is typically GTCGAATTGGAATTGG, sometimes the first T can be a C in some individuals. This variation in the general sequence is an SNP. Most SNPs are nonfunctional but some SNPs are related to diseases and possibly to traits such as intelligence. One estimate is that there are about 10 million SNPs in a person's DNA. SNPs can be compared between two groups, say defined by high or low IQ scores, in an effort to find segments of DNA, perhaps individual genes, that differentiate the groups. Early studies sampled thousands of SNPs. Now a person's genome can be sequenced and millions of SNPs are assessed in Genome Wide Association (GWAS) studies.

A large number of SNPs have been associated with intelligence measures in big samples and each apparently accounts for a tiny influence. Any individual person will have some of the relevant SNPs and how many will vary among people. The number of intelligence-related SNPs can be aggregated and generate a "polygenic score" (PGS), and this score can be correlated to the person's actual intelligence score. There are different ways to calculate PGSs and they have been used to predict height and many other traits and medical conditions (in the form of probabilities or percent of variance explained). As we will see in Section 2.7, PGSs predict an increasing amount of intelligence variance in GWAS analyses.

Such studies generate enormous data sets and the field of genomic informatics has developed statistical methods for sorting through all the possible combinations with the ultimate goal of identifying specific genes for diseases, medical conditions, and myriad inherited traits. There are now bioinformatic efforts to use cloud computing to accumulate, organize, and analyze Big Data sets of genetic information acquired from DNA analyses. Proteomics is the study of proteins and how they work. It is now possible to test gene expression for thousands of proteins and their varieties simultaneously, often using a small DNA sample on microarrays with different reactive agents. Altogether, the ever-evolving techniques and methods of molecular genetics provide detailed assessments deep into the neurobiology and neurochemistry of neurons, synapses, and brain development of function and structure. Despite the complexities and overwhelming amounts of DNA data, in my view, the challenges for understanding intelligence at this level are made finite by the DNA technologies and advances in genomic analyses. In fact, given the psychometric problems noted in Chapter 1, progress in understanding intelligence on the behavioral level might prove as much as a challenge as on the molecular level.

2.6 Eight Illustrative Studies of Molecular Genetic Approaches

The studies in this section are examined in chronological order to demonstrate how this field is progressing. The benefits of a consortium approach are nicely illustrated in a study co-authored by 59 investigators pooling data sets from around the world (Rietveld et al., 2014). This study was actually a combined effort from two consortia: the Social Science Genetic Association Consortium and the Childhood Intelligence Consortium (CHIC). These researchers used a conceptually simple and clever two-stage process that began with a sample of 106,736 individuals and ended with an indirect replication in an independent sample of 24,189 individuals. In the first sample, millions of SNPs were assessed (see Textbox 2.2) for each person's DNA and 69 were related to education attainment level (years of education). Education level is highly correlated to intelligence. In the second sample, these 69 SNPs were tested for any associations with a g-score derived from cognitive test scores. Although not every person completed the same tests, g-scores from different batteries are highly correlated (even over 0.95) if the test batteries and the sample are sufficiently diverse (Johnson et al., 2008). Several advanced statistical analyses revealed four promising genes of interest related to very small amounts of variance in cognitive performance. Interestingly, these genes (KNCMA1, NRXN1, POU2F3, SCRT; I am not responsible for how genes are named) are known to influence a glutamate neurotransmitter pathway related to brain plasticity and learning and memory. The pathway involves the NMDA receptor, glutamate binding, and synaptic changes. Despite the small amount of intelligence variance associated with these genes, this study demonstrates the statistical reality that large samples are necessary to find small effects. Such findings also provide hints about molecular mechanisms that may be associated with intelligence that could be the basis for hypotheses about the salient neurobiology.

About the same time, the second study (Hill et al., 2014) used genome-wide analyses in 3,511 individuals to investigate the small effects of 1,461 individual genes on intelligence by finding associations with cognitive ability in aggregated networks of functionally related genes. They started with a specific hypothesis that focused on genes related to post-synaptic functioning. After replication in independent samples, proteins related to the NMDA receptor were associated specifically with fluid intelligence. Other aspects of post-synaptic functioning were not related to variation in any other cognitive abilities. NMDA was also implicated indirectly in the Rietveld et al. (2014) study, but these post-synaptic findings tied the genetic variation of specific proteins to individual differences in fluid intelligence. The key protein was guanylate kinase, which is fundamentally important for converting neuronal action potentials into biological signals that underlie information processing throughout the brain. This study provides additional hints about the neurobiology of intelligence.

Whereas the Rietveld et al. (2014) study used a very large a-theoretical shotgun approach to find needles in a haystack, and the Hill et al. (2014) study focused on a network of functionally related genes, another group of researchers used a different strategy that focused on a specific gene and its effects on intelligence after traumatic brain injury (TBI) (Barbey et al., 2014). There is a neurochemical called brain-derived neurotropic factor (BDNF) that promotes and regulates well-functioning synapses. BDNF is related to cognitive functioning in healthy people, especially to aspects of memory and to impaired cognition in Alzheimer's disease and other brain disorders. Val66Met, a gene associated with BDNF, also is implicated in neural repair mechanisms that stimulate neuro-regeneration in the prefrontal cortex after recovery from TBI. Is BDNF related to intelligence? After TBI to the frontal lobes, some patients show persistent deficits in g-loaded tasks whereas in other patients there is a preservation of g-loaded task performance. The genetic basis for BDNF is the Val66Met polymorphism (see Textbox 2.2) that has two main variations: Val/Met and Val/Val. The question for this study was whether either of these variants was related to the preservation of intelligence after TBI.

Unfortunately, there are a large number of TBI cases. Many are treated in Veterans Administration hospitals. Participants in this study came from a group of 171 male veterans who suffered penetrating head injuries during the Vietnam War. For 151 of these individuals the sites of brain lesions were located by computerized tomography scans and confirmed in the frontal lobes. Each participant completed 14 subscales of the WAIS III and each person also had completed the battery of tests in the Armed Forces Qualification Test (AFQT) when they entered the military, prior to the TBI. Both these tests allowed the computation of a *g*-factor score along with other subfactors. Two groups were defined based on genotyping: Val/Met (n = 59) and Val/Val (n = 97) and the intelligence factor scores were compared between these two groups with sophisticated psychometric analyses.

The results were rather striking. Scores derived from the AFQT did not differ between the groups. In other words, prior to the TBI the veteran's genotype (Val/Met or Val/Val) had no impact on general cognitive ability. However, there was a substantial difference following TBI. The Val/Val group showed average diminished factor scores for *g* and other primary factors including verbal comprehension, perceptual organization, working memory, and processing speed. The average score differences were relatively large at about half a standard deviation. The authors concluded that having the Val/Val genotype was associated with cognitive susceptibility to TBI whereas the Val/Met genotype may help preserve cognitive functioning following TBI. These results may have implications for cognitive rehabilitation strategies that might be more effective than others, although there is a paucity of such research at present. The results also tie variations in the BDNF gene to intelligence and demonstrate some progress toward identifying specific gene variants related to intelligence. Such data can help generate hypotheses about the step-by-step cascade of neurochemical events at the molecular level that lead from the genetics of BDNF expression to explaining a small amount of variance among individuals in intelligence. It is likely that there are many steps in the cascade and multiple complex combinations with other genetic or biological factors, some of which might occur in an epigenetic context. And BDNF is only one of many factors probably involved.

The fourth study is from CHIC. They reported a genome-wide association study of intelligence in children aged 6–18 years old with a combined discovery cohort total of 12,441 and a replication cohort total of 5,548 (Benyamin et al., 2014). No single SNP was associated with intelligence but the aggregate of common SNPs accounted for 22–46 percent of variation in intelligence in the three largest cohorts. The FNBP1L gene was associated with intelligence, accounting for small amounts of variance in three separate replication cohorts (1.2 percent, 3.5 percent, and 0.5 percent, respectively). Despite the large sample, the authors concluded that even larger samples might be necessary to detect individual SNPs with genome-wide significance.

The fifth example of interest is from China. In my view, it is also a landmark exercise. It reports a broad systems biology approach (Zhao et al., 2014) that is designed to elucidate complex regulation and interactions and generate hypotheses about the mechanisms that drive them. Whereas, Chabris and colleagues had started with 12 candidate genes from previous studies and failed to replicate any of them, these researchers selected 158 genes that had been associated with IQ scores. They mapped the locations of these genes on chromosomes and found some clustering in seven regions of chromosome 7 and the X-chromosome. Many of these genes were known to be involved in various neural mechanisms and pathways. Using a type of network analysis, "IQ-related pathways" were constructed. These pathways primarily involved dopamine and norepinephrine; neurotransmitters involved in many brain functions. The details of this analysis are far beyond our intentions in this chapter but this report illustrates how molecular genetics can generate testable hypotheses about specific neural mechanisms related to intelligence, and how those mechanisms might be tweaked by drugs or other means. This kind of analysis fuels my optimism that the genetic basis for intelligence is not a retreat to determinism and immutability. Rather the opposite: the genetic basis, once understood, can lead to the remarkable ability to treat or prevent brain disorders that result in low IQ and to the holy grail of increasing intelligence across the whole range, as discussed in Chapter 5.

The sixth study demonstrating progress takes another approach (Davis et al., 2015). These researchers also focused on one molecular factor, a protein called DUF1220 that is associated with brain size and brain evolution. DUF1220 has two main subtypes, CON1 and CON2. Many gene sequences have multiple copies in a person's DNA and the number of copies can be related to diseases and other traits. In this study, not only was the number of copies of CON2 associated with IQ scores, the association was linear. That is, the more copies of CON2, the higher the IQ scores. Brain size, assessed by magnetic resonance imaging (MRI), also was correlated to IQ scores, especially for the surface area of the temporal cortex bilaterally, and right frontal surface area was related to increased dosage of CON1 and CON2. These findings came from a sample of 600 young North American people and were replicated in a smaller sample of 75 individuals living in New Zealand. Although both samples are quite small compared to the previous studies just summarized, the linear nature of the CON2 copy–IQ finding is intriguing, especially because it was strongest in males that are 6–11 years old. There are a number of reasons to be cautious about this finding, as acknowledged by the authors, and it is too early to accept it at face value. Nonetheless, it represents another example of how the search for intelligence genes is pushed forward with a priori hypotheses about specific genetic factors.

The seventh study comes from another multi site consortium (Cohorts for Heart and Aging Research in Genomic Epidemiology) of 31 cohorts (N = 53,949). They reported a meta-analysis based on a genome-wide association study of middle-aged and older adults who had completed a battery of four cognitive tests (Davies et al., 2015). This is the largest such study of general cognitive ability to date. Across all the samples, 13 SNPs were associated with general cognitive ability, together accounting for 29 percent and 28 percent of the variance in two of the largest samples, respectively. Three genomic regions were associated with these SNPs with special focus on the HMGN1 region. Four genes previously associated with Alzheimer's disease also were associated with general cognitive ability (TOMM40, APOE, ABCG1, MEF2C). Consistent with the polygenetic model of inheritance, these genes individually accounted for small proportions of variance. These researchers also conclude that even larger samples will be required to identify more genome-wide associations. No one yet knows how many genes may contribute to variations in intelligence, but the very existence of these multicenter collaborations is a giant step toward finding out. In fact, just after this book went into production, an elaborate collaborative study identified two networks of genes (1,148 genes in one network and 150 genes in the other) that were related to general cognition (Johnson et al., 2016). Many of these genes were related to specific synaptic functions that could potentially be manipulated to influence intelligence. I do not have space to elaborate more details of this landmark study but it illustrates another step forward.

The eighth study is an illustrative meta-analysis of almost 80,000 children and adults in different cohorts, which reported that about 4.8 percent of the variance in intelligence test scores could be predicted by association with 47 genes, including 336 SNPs from over 12 million that were assessed (Sniekers et al., 2017). Some of the top genes were implicated in synapse formation (SHANK3), axon guidance and putamen volume (DCC), and neuron differentiation (ZFHX3). Using a variety of methods, a total of 47 genes were identified as possibility related to intelligence or educational attainment. This was an early demonstration that intelligence could be predicted from DNA using PGSs, as described in Textbox 2.2, even though the percentage (4.8 percent) was small. This study was also noteworthy for the editorial that accompanied its publication in *Nature*, which clearly articulated the importance of genetic research for understanding intelligence (Editorial, 2017). This editorial helped bring intelligence research back into mainstream psychology. The actual prediction of intelligence using PGSs stimulated considerable interest in this approach, as described in Section 2.7. For advanced students, there is an excellent summary of genetic and molecular studies of intelligence by Deary et al. (2022).

2.7 Predicting Intelligence from DNA Using Polygenic Scores

PGSs are most powerful when based on GWASs with very large samples. For example, Plomin and von Stumm have written an accessible discussion of how they relate to intelligence and educational attainment data (Plomin and von Stumm, 2018). They estimated that a GWAS sample size of one million people would likely be necessary to build a polygenic score that predicts 10 percent of intelligence variance in an independent sample. How long would it take researchers to reach such an astonishing number? Not long.

A PGS study related to intelligence hit this landmark GWAS goal of over one million participants just a few months later (Lee et al., 2018). The dependent measure was educational attainment (i.e., the number of years of education, an easily obtained variable available for most participants in research consortia that is correlated with intelligence test scores). These researchers identified 1,271 SNPs related to educational attainment that implicated as many as 1,838 genes, many involved in brain development and neuron communication. Using subsamples, polygenic scores predicted 11–13 percent of variance in educational attainment and 7–10 percent of the variance in cognitive test performance. These values are essentially what Plomin and von Stumm had predicted.

Similarly, another study of 7,026 children and adolescents from the UK Biobank project reported that PGSs predicted up to 11 percent of intelligence and 16 percent of educational attainment (Allegrini et al., 2019). Predictive power increased with age (as does heritability, discussed in Section 2.4) and there were no sex differences. PGSs yielded the strongest predictions compared to multi-trait genomic methods.

But wait. There's more. A recent study dramatically topped the 1.1 million sample with an analysis of 3 million individuals! It reported 12–16 percent of explained variance in educational attainment. This was only a small increase from the values obtained in the 1.1-million-person study. As the authors concluded, "the sample size of the GWAS of educational attainment reported in this paper is the largest published to date. For some purposes, such as attaining greater predictive power for the polygenic score, there are clearly diminishing returns" (Okbay et al., 2022: 444). It should be no surprise by now that the genetic story is complex and unsettled.

It must be noted that predicting 16 percent of the variance using DNA is an achievement that validates an important role for genetics. But, based on the behavioral genetics research on twins and adoptees discussed earlier, we could expect that 50 percent of intelligence variance should be predicted with PGSs. Whether there is "missing heritability" is an open question since estimates based on PGSs may be underestimates (Willoughby et al., 2021). Perhaps educational attainment has too many other factors that confound any g effects and predictions may be better for some cognitive abilities than others (Genc et al., 2021). There are also other possible considerations about PGSs. One is that predictions might vary across populations of dissimilar genetic ancestry and even geographic regions can influence results (Abdellaoui et al., 2022; Mitchell et al., 2022). Another one is whether PGSs can give meaningful insights into the molecular dynamics of how gene expression influences intelligence, but there are other methods to use for mechanism questions. Despite these considerations, in my view the full potential of PGSs for intelligence research is still progressing and represents an exciting new approach (Visscher, 2022; von Stumm and Plomin, 2021; see also Haier and Colom, in 2023: chapter 6).

Studies like these are the reason this chapter has not dwelt on older (and a few current) criticisms about whether genes are important for intelligence. Although the full role of genes is not yet known, the evidence for major genetic involvement in intelligence is overwhelming. No one ever believed that understanding intelligence on the molecular level would be simple, but the studies and their complex analyses summarized here indicate to me that the challenge is not impossible.

On a final note, genetic studies are logistically complex and expensive, especially when large samples are involved. DNA-sequencing machines alone, for example, cost about $1–2 million each. Reportedly, as early as 2012 a single research institute in China, the Behavioral Genetics Institute, had 128 of them, along with supercomputers. Finding intelligence genes is a high priority. This one institute has over 4,000 scientists and technicians working there and a poster on the wall reportedly says: "Genes build the future." Consider the race to find intelligence genes and how they work. At the end of the twentieth century, Plomin (1999: c27) stated: "The most far-reaching implications for science, and perhaps for society, will come from identifying genes responsible for the heritability of g." On the one hand, China has substantial investment in this hunt, and on the other hand, a majority of members currently in the United States Congress apparently do not believe in evolution. Seriously.

All the studies outlined in this chapter illustrate how quantitative genetic research strategies and sophisticated DNA analyses are being used to establish the genetic basis of intelligence and search for specific gene effects and how they work. There are now several worldwide consortia working on this effort that add a third methodological element: quantitative neuroimaging to measure brain structure and function. This combination of three research elements targets the identification of genes that influence brain characteristics related to intelligence. In my view, these studies represent a new phase in the search for how genes influence intelligence. There are exciting findings and we review them in Chapter 4. First, however, to understand the full impact of the newest DNA studies using twin research designs that target the brain, we now introduce the third element – neuroimaging – in Chapter 3.

Chapter 2 Summary

- Sir Cyril Burt and Professor Arthur Jensen were early advocates of the importance of a genetic role in intelligence but their views were attacked and widely rejected.
- Modern quantitative genetic studies overwhelmingly support a major role for genes in explaining the variance of intelligence test scores among individuals.
- The same studies indicate that environmental factors play a role in early childhood, especially nonshared factors, but this role diminishes almost entirely by the early teen years.
- The weight of evidence from modern studies of intensive compensatory education, now rebranded as early childhood education, still fails to find lasting effects on IQ scores.
- Progress in the search for specific genes involved in intelligence has been slow and disappointing, leading to the conclusion that many genes, each having a small effect, must be involved.
- Advanced DNA technologies applied to molecular genetic studies are beginning to identify genes that influence intelligence and how they might work on a neurobiological level.

Review Questions

1. Why are genetic explanations of behavior so controversial?
2. What were the immediate and long-term impacts of Jensen's 1969 article?

3. What is the most compelling evidence from quantitative genetic studies that genes are involved in intelligence differences among people?
4. How does the influence (or relative contribution) of genetic and environmental factors on variation in intelligence change across development?
5. Why may there be an advantage to intelligence involving many genes of small influence?
6. What is an example of a recent finding regarding intelligence from a specific molecular genetics study?

Further Reading

"How Much Can We Boost IQ and Scholastic Achievement?" (Jensen, 1969). This is possibly the most infamous paper in psychology and is the basis for much modern intelligence research.

The IQ Controversy, the Media and Public Policy (Snyderman and Rothman, 1988). Based on survey data, this is a controversial book that argues that liberal bias systematically distorted the reporting of Jensen's work and other genetic research on intelligence.

Cyril Burt: Fraud or Framed? (Mackintosh, 1995). This is a collection of essays on all sides of the Burt controversy.

Intelligence, Race, and Genetics: Conversations with Arthur R. Jensen (Jensen and Miele, 2002). This book offers an update of Jensen's views by Jensen himself in his own words.

"Genetics and Intelligence Differences: Five Special Findings" (Plomin and Deary, 2015). A good review of key genetic findings related to intelligence and what they mean.

Blueprint: How DNA Makes Us Who We Are (Plomin, 2018). A highly readable overview of genetic research written by a legendary behavioral geneticist who has studied intelligence for decades.

Innate: How the Wiring of Our Brains Shapes Who We Are (Mitchell, 2018). An accessible account of modern genetic research and what the data mean.

CHAPTER THREE

Peeking Inside the Living Brain

Neuroimaging Is a Game Changer for Intelligence Research

The brain is a black box – we cannot see in it and must ignore it.

Attributed to B. F. Skinner in the 1950s (but not corroborated)

[I]f Freud were alive today, he'd trade his couch for an MRI machine.

Richard Haier (2013)

Because there are not two individuals on Earth with the same genome, there will not be two individuals with the same brain … This fact cannot be ignored from now on and should be properly weighted for achieving reliable answers to the question of why some people are smarter than others.

Kenia Martinez and Roberto Colom (2021)

Learning Objectives

- How has neuroimaging technology advanced the study of human intelligence beyond psychometric methods?
- How do the basic technologies of positron emission tomography (PET) and MRI differ?
- What was a surprising finding from the early PET studies of intelligence?
- Does imaging research indicate that there is an "intelligence center" in the brain?
- What brain areas are included in the Parieto-Frontal Integration Theory (PFIT) model of intelligence?

Introduction

This chapter and Chapter 4 review brain-imaging studies of intelligence. This chapter is written to give a somewhat personal historical perspective on the early studies from 1988 to 2006, a period I describe as phase one in the application of modern brain-imaging technology to intelligence research. This phase began with the first PET study of intelligence published in 1988 and ends with a review of the relevant literature published in 2007. The 37 studies during this period reported several unexpected results and set the direction for current imaging research related to intelligence. This chapter is written in roughly chronological order of publication to demonstrate how the early research unfolded, including my research. This perspective helps students understand how researchers advance from one set of findings to new questions. There are also basic descriptions of how the main imaging technologies work. In Chapter 4, we will see the subsequent and more sophisticated phase two of worldwide brain-imaging research on intelligence. Brain-imaging technology has significantly helped advance intelligence research from mainly psychometric methods (described in Chapter 1) to neuroscience approaches that can quantify brain characteristics. Brain imaging is a key development in this field and that is why we devote two chapters to it.

The early quantitative genetic research described in Chapter 2 provided the rationale for a biological component to intelligence and laid the foundation for neuroscience research just as powerful new neuroimaging methods were becoming available. Prior to the introduction of neuroimaging in the early 1980s, brain researchers were limited to indirect measurements of brain chemistry by-products found in blood, urine, and spinal fluid. Electroencephalogram (EEG) and evoked potential (EP) research allowed millisecond-by-millisecond measurements of brain activity but technical issues such as distortion of electrical signals by the scalp and poor spatial resolution limited the scope and interpretation of data. Today, EEG-based techniques are more sophisticated and include ways to map cortical activity (Euler et al., 2017; Euler and Schubert, 2021; Euler et al., 2015) (see Chapter 4). Inferences from studies of patients with brain damage and from autopsy studies similarly had only limited success in identifying brain–intelligence relationships. For example, some studies of patients with brain damage concluded that the frontal lobes were the seat of intelligence (Duncan et al., 1995), a conclusion we now know is overly simplistic based on newer lesion studies (see Chapter 4).

3.1 The First PET Studies

In the early 1980s PET was a game changer. Twenty years before the wide availability of MRI (to be discussed later in this chapter), PET technology allowed researchers to see inside the brains of living people and make relatively high-resolution measurements about what brain areas are more or less active during mental activity. This differs dramatically from X-ray technology that had been available much earlier, including

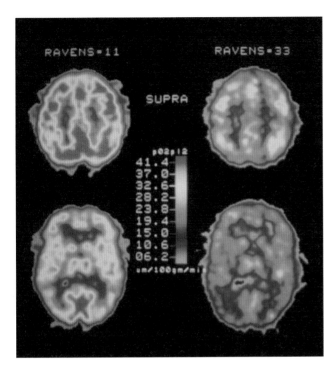

Figure 3.1 Brain activity assessed with PET during the RAPM test. Red and yellow show greatest activity in units of GMR. The person with the highest test score (images on right) shows lower brain activity during the test, consistent with brain efficiency related to intelligence. (Courtesy Richard Haier)

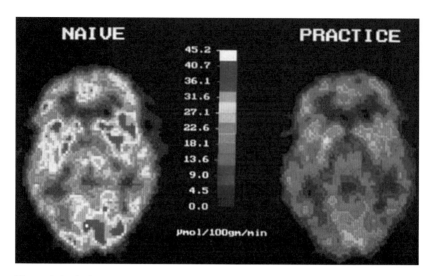

Figure 3.2 Playing Tetris naïve vs. practiced PET images. Red and yellow show greatest activity in units of GMR. Brain activity decreases with practice, consistent with the brain becoming more efficient. (Courtesy Richard Haier)

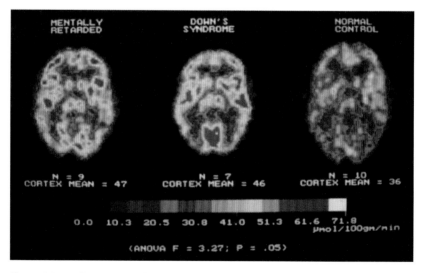

Figure 3.3 PET images in two low-IQ individuals showing higher brain activity than a control with average IQ. Red and yellow show greatest activity in units of GMR. (Courtesy Richard Haier)

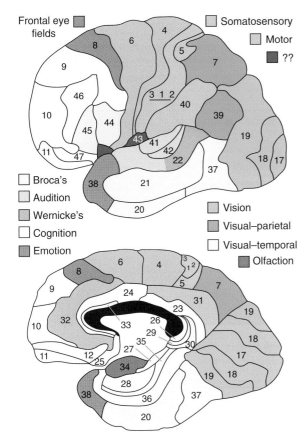

Figure 3.6 BAs are a standard way to label different brain regions based on early autopsy studies of neuron organization. (Public domain)

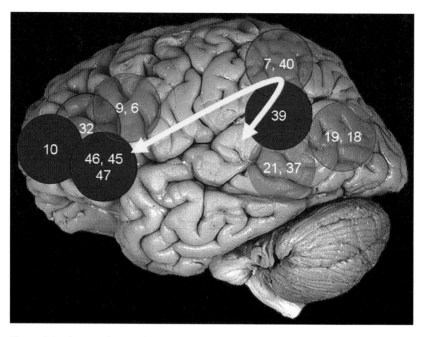

Figure 3.7 The PFIT showing brain areas associated with intelligence. (Courtesy Rex Jung)

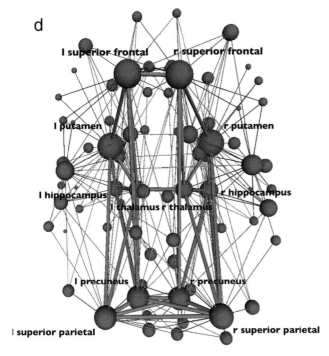

Figure 4.1 Brain connections determined by graph analysis. The red nodes show brain areas with many connections (larger nodes indicate more connections). Blue lines called edges indicate the strength of connections among areas (thicker lines indicate stronger connections; dark blue lines show rich club connections to other brain areas). (Adapted with permission from van den Heuvel and Sporns (2011)

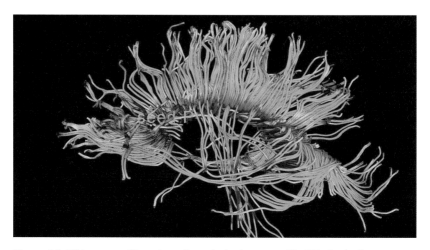

Figure 4.2 White matter fibers throughout the brain assessed by DTI. Seed refers to a spot selected for determining connections from that spot to other areas. (Courtesy Rex Jung)

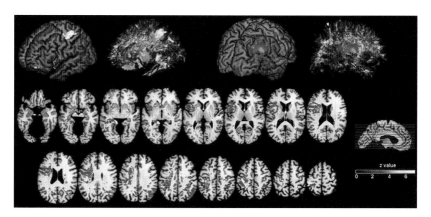

Figure 4.3 3D renderings show cortical and subcortical regions with a statistically significant relationship (red/yellow) between lesion location and the g-factor (top row). Bottom rows: Axial (horizontal) slices are shown for a more detailed inspection. (Reprinted with permission (Glascher et al., 2010: 4707, figure 2))

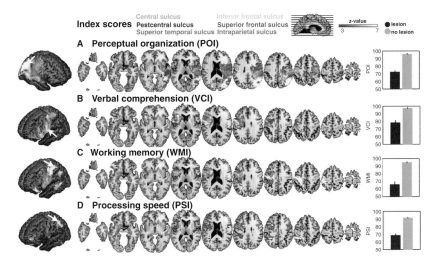

Figure 4.4 Effect of lesion location on four indices of mental ability. Row (A) is perceptual organization, (B) verbal comprehension, (C) working memory, and (D) processing speed. The red/yellow colors show where the location of lesions significantly interferes with scores on the indices. The graphs on the right show the mean difference on each index score between patients with and without lesions at the area of maximum effect (white arrow on the 3D projection). (Reprinted with permission (Glascher et al., 2009: 684, figure 2))

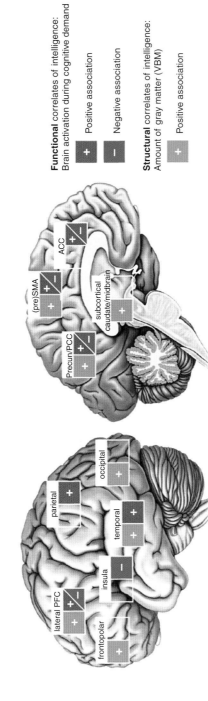

Figure 4.5 Brain areas related to intelligence from a 2015 review are shown on lateral (left) and medial (right) surfaces of the brain. ACC = anterior cingulate cortex; PCC = posterior cingulate cortex; PFC = prefrontal cortex; (pre) SMA = (pre-)supplementary motor area; VBM = voxel-based morphometry. (Reprinted with permission (Basten et al., 2015))

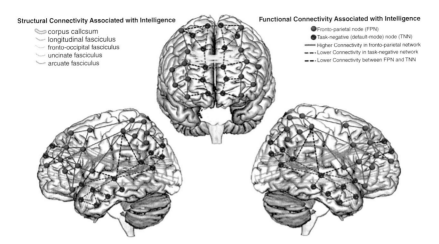

Figure 4.6 The brain bases of intelligence – from a network neuroscience perspective. (Reprinted with permission (Hilger and Sporns, 2021))

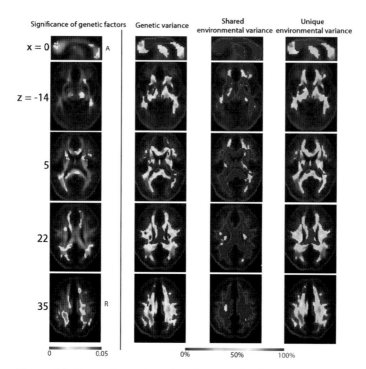

Figure 4.8 Maps of genetic and environmental influences on white matter integrity (measured by FA). Each row shows a different axial brain view (horizontal slice). Red/yellow shows strongest results. Left column shows significance of genetic influences. Other columns show the strength of the FA measure for genetic, shared, and nonshared environment, respectively. (Adapted with permission (Chiang et al., 2009: figure 4))

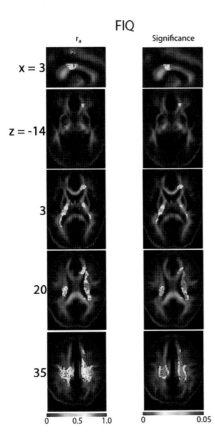

FIQ

r_a Significance

x = 3

z = -14

3

20

35

0   0.5   1.0   0   0.05

Figure 4.9 Overlap of common genetic factors on FA and FSIQ (left column) based on cross-trait analysis of areas shown in Figure 4.8. Right column shows statistical significance. Each row shows a different axial (horizontal slice) brain slice. (Adapted with permission (Chiang et al., 2009: figure 7))

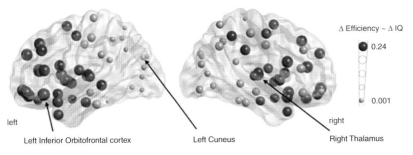

Δ Efficiency ~ Δ IQ

0.24

0.001

left   right

Left Inferior Orbitofrontal cortex   Left Cuneus   Right Thalamus

Figure 4.10 Correlations between three-year change in IQ score and change in local brain efficiency measured with FA. The largest purple spheres show the strongest IQ/efficiency change correlations. (Adapted with permission (Koenis et al., 2015))

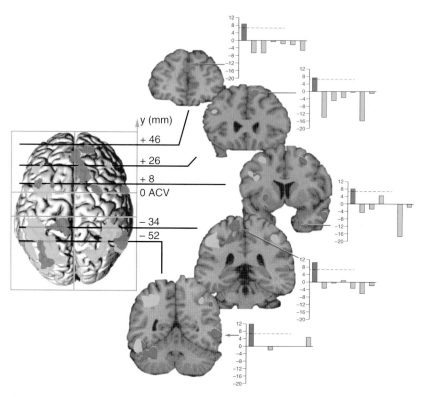

Figure 6.1 PET scans of an expert memory champion performing complex mental calculations compared to six nonexpert controls. Brain areas uniquely activated by the expert are shown in green; areas activated both by the expert and nonexperts are shown in red. Bar graphs show activations in each area for each person (red bar is the expert). (Reprinted with permission (Pesenti et al., 2001))

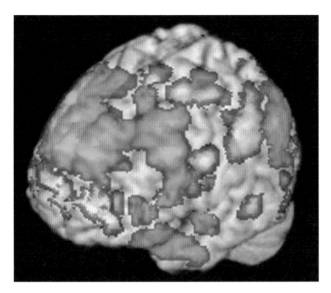

Figure 6.2 Brain activations (red/yellow) and deactivations (blue/green) in jazz pianists during improvisation. (Adapted from Limb and Braun (2008), open access)

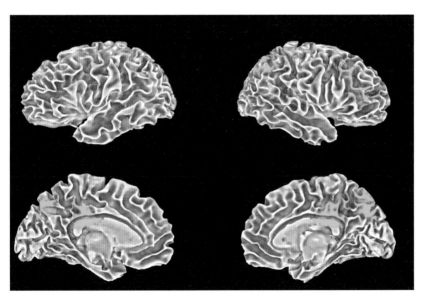

Figure 6.3 fMRI comparison of rappers during improvised and conventional conditions. Yellow represents significant increases in fMRI blood flow during improvisation; blue represents significant decreases. Top row shows cortical surface; bottom row shows medial (inside) surface. (Adapted from Liu et al. (2012) open access)

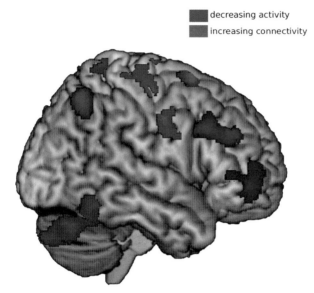

Figure 6.4 fMRI in pianists with varying degrees of improvisation experience. More training is related to less brain activity (blue) during creative expression and to increased functional connectivity among other areas (red). (Reprinted with permission from Pinho et al. (2014: figure 3), free access)

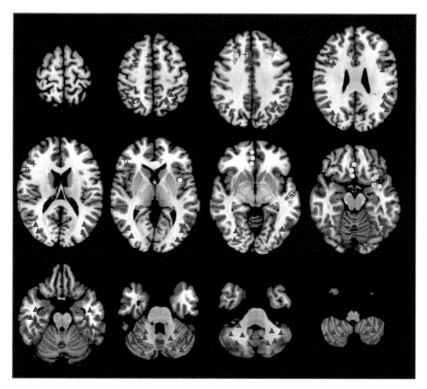

Figure 6.5 Different creativity findings from seven MRI studies. Each colored symbol shows activated brain areas related to creativity from a different study. There is little overlap of areas across studies. Arden et al. (Reprinted with permission from (2010: figure 1))

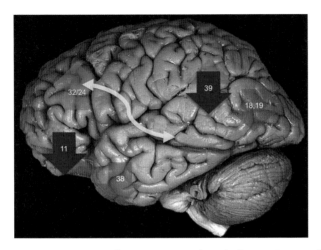

Figure 6.6 F-DIM of creativity. Numbers indicate BAs associated with increased (up arrows) or decreased (down arrows) brain activity based on a review of studies. Blue is left lateralized; green is medial; purple is bilateral; the yellow arrow is anterior thalamic radiation white matter tract. (Reprinted with permission (Jung & Haier, 2013))

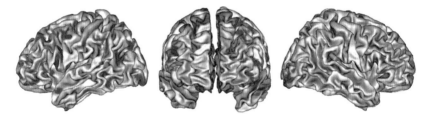

Figure 6.7 Summary findings from 34 functional imaging studies of creativity. Common brain areas of activation are shown revealing distributed networks related to creativity. From Gonen-Yaacovi et al. (2013: figure 1), open access)

computerized tomography (CAT) scans. Whereas X-rays pass through the head and show brain tissue structure, they are silent as to brain activity. A CAT scan of a person looks the same whether the person is awake, asleep, doing mental arithmetic, or dead. Since the brain is soft tissue, X-rays pass through easily and brain pictures are not very detailed. By contrast, PET can quantify brain activity as glucose metabolism, blood flow, or in some cases neurotransmitter activity. This is accomplished in a conceptually simple way. Radioactive tracers are injected into a person while they perform a cognitive task and the brain areas that are most active during the task take up the most tracer. The radiation exposure is within limits set for medical uses. The subsequent PET scan detects the radioactivity and mathematical models allow an image to be constructed showing the spatial locations where the varying amounts of radioactivity have accumulated.

For example, a positron-emitting isotope such as fluorine[18] can be attached to a special glucose called flurodeoxyglucose (FDG). Since glucose, a sugar, is the energy supply of the brain, the harder any area of the brain is working, the more radioactive glucose is taken up and metabolically fixed in that part of the brain and the more positrons are accumulated. The positrons collide with electrons, which are naturally plentiful everywhere, and each collision releases energy in the form of two gamma rays, always at 180 degrees from each other. The 180-degree angle is a fact of physics and millions of gamma rays are released from the FDG tracer. When the head is placed inside the PET scanner, which contains one or more rings of gamma ray detectors, the spots in the brain where the gamma rays originated can be reconstructed mathematically based on detection of a gamma ray and at the same moment in time a coincident detection of another gamma ray 180 degrees away. Somewhere on the straight line connecting these two simultaneous events, a positron decayed. With millions of these coincident detections, the spatial location of the accumulated positrons can be determined and the areas releasing the most gamma rays can be quantified. These are the areas most active during the FDG uptake and the activation patterns in areas will be different depending on the mental activity during the uptake. It takes about 32 minutes for the brain to take up the FDG tracer. This means that brain activity is summed over the 32 minutes so the time resolution of FDG PET scans is very long. You cannot see how brain activity changes from second to second. Radioactive oxygen instead of glucose, however, can be used in PET to image blood flow with a time resolution of minutes. Other imaging techniques based on MRI have time resolutions of about 1–2 seconds and newer methods such as the magneto-encephalogram

(MEG) show changes millisecond by millisecond. Compared to PET, MRI and MEG techniques also are far less intrusive (no injections or exposure to radioactivity), as we detail in due course as they have been applied to intelligence research.

An advantage of PET is that the rate of glucose metabolism can be calculated from measurement of radioactivity decay in the blood periodically after the injection of a tracer. The PET image shows a quantitative map of the glucose metabolic rate (GMR) while the cognitive task was performed. The physics of fluorine[18] gives the radioactive glucose a half-life of about 110 minutes, so the logistics of a PET study are formidable. The steps include manufacturing the fluorine[18] in a cyclotron, attaching it to glucose in a nearby hot lab, injecting it into a person while they perform a cognitive task for about 32 minutes, and then scanning for about 45–60 minutes to acquire millions of coincident gamma ray detections. The glucose is metabolically fixed so scanning happens after the task is complete and the image shows glucose uptake during the task. The expense is similarly formidable, usually about $2,500 per scan. There are other isotopes that can be used to create tracers that show blood flow and some neurotransmitter activity. The PET images are constructed as slices that cover the entire brain. Color coding shows rates of glucose activity. In the same person, PET images will differ depending on whether the person is awake or asleep or doing any cognitive task such as solving problems on the RAPM test of abstract reasoning described in Chapter 1.

I first learned about PET when I worked in the Intramural Research Program at NIMH in the early 1980s and recognized the potential for intelligence research. Before NIMH took delivery of one of the very first PET scanners, however, I left for Brown University where I did rudimentary EEG/EP mapping of brain activity (proudly with an APPLE II Plus) and related it to Raven's scores (Haier et al., 1983). When the opportunity came to join my former NIMH colleague, Monte Buchsbaum, when he relocated to the University of California, Irvine (UCI) and acquired a new PET scanner, I moved to California. In the early 1980s, most of the first PET research was on schizophrenia and psychiatric disorders. PET scans for psychological studies were rare. The first research project I was able to undertake in 1987 was based on only eight scans that were provided without charge as a reward for a successful fundraising effort (the politics and cost of scan access also were formidable challenges; publicly available databases with thousands of brain scans were in the distant, unimaginable future). I used those eight scans to ask a simple question: Where in the brain is intelligence?

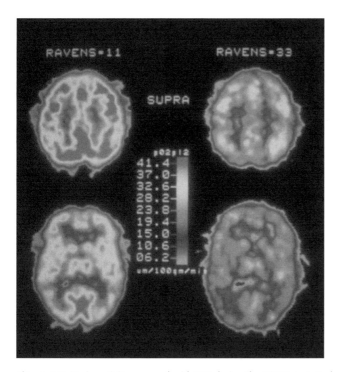

Figure 3.1 Brain activity assessed with PET during the RAPM test. Red and yellow show greatest activity in units of GMR. The person with the highest test score (images on right) shows lower brain activity during the test, consistent with brain efficiency related to intelligence. (Courtesy Richard Haier)
Note: A black and white version of this figure will appear in some formats. For the color version, please refer to the plate section.

In 1988 we published the first PET study of intelligence (Haier et al., 1988). We had the eight male volunteers take the RAPM test with 36 items. These included some very hard items to create sufficient variance in a college sample in order to avoid the problem of restricted range. Remember, RAPM is a nonverbal test of abstract reasoning that is one of the best single estimates of the *g*-factor. After each participant completed a practice set of 12 items and began working on the 36 test items, we injected the radioactive glucose used to label the parts of the brain working the hardest while the person was solving the problems. After 32 minutes of working on the items, we moved the person into the PET scanner to see where in the brain there was increased activity compared to other control individuals doing a simple test of attention that required no problem solving.

When we did the typical analysis and compared the GMR between the group doing the RAPM and the group doing the attention task, several areas across the brain cortex were statistically different. We went a step further that was not typical but it was logical from the perspective of individual differences. There was a range of RAPM scores so we correlated the scores to the glucose rate in each brain area that was different from the attention control group. There were significant correlations, but to our surprise all of the correlations were negative. In other words, the individuals with the highest test scores showed the lowest activity in the brain areas that differed between the groups. This inverse relationship is shown in Figure 3.1.

The two images on the right are from one person doing the RAPM test and the other two images on the left are from another person doing the RAPM test. These are horizontal (axial) slices through the top and center of the brain. All images are shown with the same color scale of glucose metabolism so you can compare them easily. Red and yellow show the highest activity; blue and black show the lowest. The person on the left shows much more activity in both slices than the person on the right (top of image is front of the brain). But the person on the left with the very active brain actually had the lowest RAPM test score of only 11; the person on the right had the highest score of 33. No one saw this coming. It seemed backward. More brain activity went with worse performance. What could this mean?

3.2 Brain Efficiency

At the time, this counterintuitive result suggested to us that it's not how hard your brain works that makes you smart; it's how efficiently it works. Based on this result, we proposed the brain efficiency hypothesis of intelligence: higher intelligence requires *less* brainwork. About the same time, another group reported inverse correlations in multiple areas of the cortex between GMR and scores on a test of verbal fluency, another test with a high *g*-loading (Parks et al., 1988). They scanned 16 subjects while performing a verbal fluency test. During the test, GMR increased compared to another 35 controls scanned in a resting state. The correlations between GMR and scores on verbal fluency were negative in frontal, temporal, and parietal areas. Similarly, a third group of researchers (Boivin et al., 1992) scanned 33 adults also performing a verbal fluency test. They found both positive and negative correlations between scores and GMR across the cortex. Negative correlations were found in frontal areas (left and right) and positive correlations were in temporal areas, especially in left hemisphere. Their participants included a wide age range (21–71 years old) and combined males and females but removing age and IQ statistically had little apparent effect on the results (although no sex-specific analyses were reported). It should be noted that by today's standards of image analysis, all these studies used rudimentary methods for defining cortical regions. Nonetheless, the negative correlations found during cognitive activation were unexpected and, for many cognitive psychologists, hard to believe.

Since this surprising finding, many researchers have been trying to understand how exactly brain efficiency might relate to intelligence. We return to the efficiency concept in Chapter 4 where I detail recent studies that show the concept is still viable. Back in 1988 we started thinking about how learning, a key component of intelligence, might make the brain more efficient. When you learn something like driving a car, for example, doesn't your brain get more efficient so you now can drive in traffic and have a conversation at the same time, something not possible that very first day you were concentrating on driving back and forth in a big empty parking lot?

We decided to do a PET study of learning so we turned to Tetris, a computer game just out at the time, and now one of the most popular games of all time. We scanned another 8 volunteers before and after 50 days of practice on the original Tetris version (Haier et al., 1992a). The volunteers, all college males, used my office computer to practice because almost no one had computers at home in the early 1990s. Since access to PET was so limited, there was not much data about brain changes after learning a complex task. The natural expectation was that after learning to perform a complex task, brain activity would increase to reflect the harder mental work necessary to perform at a higher level. Based on our RAPM finding and the interpretation of efficiency, we hypothesized the opposite: after learning to perform better, brain activity would decrease.

In case you don't know Tetris, here's how the original version works. Different shapes made from the arrangement of four equal squares (there are five different shapes) appear one at a time at the top of the screen and slowly fall to the bottom. You can move them right or left or rotate them or drop them immediately by pressing buttons on the keyboard. The object is to place each shape so they form perfect rows with no gaps at the bottom of the screen. When you complete a row, it disappears and all the shapes above drop down, changing the configuration as the shapes continue to drop. The main object is to complete as many rows as possible before the shapes not in complete rows stack up

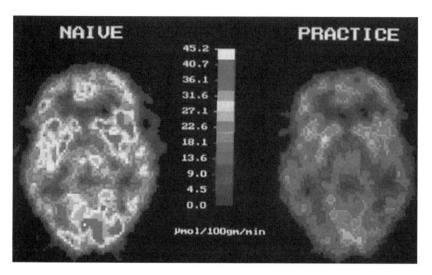

Figure 3.2 Playing Tetris naïve vs. practiced PET images. Red and yellow show greatest activity in units of GMR. Brain activity decreases with practice, consistent with the brain becoming more efficient. (Courtesy Richard Haier)
Note: A black and white version of this figure will appear in some formats. For the color version, please refer to the plate section.

to the top of the play space, which ends the game. The better you do (the more rows you complete), the faster the shapes drop, so with practice the game is faster and harder. Although the rules are quite simple to learn, playing and improvement are based on complex cognition including visual-spatial ability, planning ahead, attention, motor coordination, and fast reaction time.

On day 1, the first time any of the students ever played Tetris except for 10 minutes of practice to be sure they understood the game, they completed 10 rows per game on average while the radioactive glucose was labeling their brains during the first PET scan. This increased to nearly 100 rows per game during their second scan after the 50-day practice period. At the end of the practice period, some of the games were moving so fast, you could scarcely believe a human being could make and execute decisions so quickly.

Figure 3.2 shows what we found. The image on the left shows the scan of a person's first Tetris session. Notice all the high activity in red. The scan on the right is the same person after the 50 days of practice. There is less brain activity after practice even though the game was faster and harder. Our interpretation was that the brain learned what areas *not* to use and became more efficient with practice. We also noticed a trend in this study for the people with the highest intelligence test scores to show the greatest decreases in brain activity after practice (Haier et al., 1992b). In other words, the smartest people became the most brain efficient after practice. Other subsequent studies have shown inconsistent results on this observation, so the jury is still out on what the weight of evidence will show. Many other subsequent studies, however, have replicated decreased brain activity after learning, consistent with the brain efficiency hypothesis. Other studies have not shown this effect so the conditions and variables relevant to learning/brain activity are still open questions. From the perspective of individual differences, the important variables may be within the person rather than within the task.

All this time, I was applying for federal grant money to fund a brain-imaging program to study possible influences on intelligence. As explained in Chapter 2, intelligence research from a biological viewpoint was viewed with some suspicion, and my applications were going nowhere. So I decided to shift emphasis a bit and I was able to get a grant to study Down's syndrome, a genetic disorder typically associated with low IQ. These individuals would be of inherent interest and so would the requisite normal control group. Federal agencies are more inclined to fund research on disease and syndrome categories (and stupidity is not yet a category recognized by the National Institutes of Health so there is no national institute to study it), especially if the grant application barely mentions IQ. By the way, this is still largely the case in the United States although an exception has emerged for projects that propose to increase IQ in disadvantaged children by means of cognitive training. We discuss this more in Chapter 5.

We had been wondering if low-IQ individuals might have inefficient brains, possibly due to a failure of neural pruning, the normal developmental reduction in excess or extraneous synapses starting about age 5. We were interested in scanning people with Down's syndrome who had IQs between 50 and 75, and of course, control groups of people without Down's syndrome who also had IQ scores in the same low range for no apparent genetic or brain damage reason. We also had other controls with IQs in the average range (Haier et al., 1995).

At the time, most researchers predicted that PET scans of low-IQ individuals, especially those with known brain abnormalities such as those found in Down's syndrome, would show lower activity because some kind of brain damage was assumed to be responsible for low IQ. A failure of neural pruning, however, was consistent with earlier research in Down's syndrome showing a higher density of synapses (Chugani et al.,

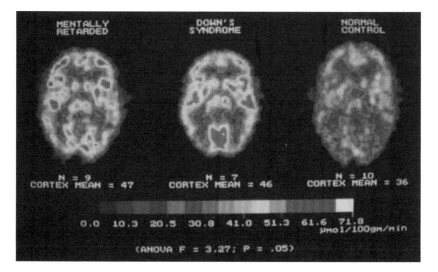

Figure 3.3 PET images in two low-IQ individuals showing higher brain activity than a control with average IQ. Red and yellow show greatest activity in units of GMR. (Courtesy Richard Haier)
Note: A black and white version of this figure will appear in some formats. For the color version, please refer to the plate section.

1987; Huttenlocher, 1975). Based on the efficiency hypothesis and a possible lack of neural pruning, we were open to the possibility that we might see higher activity in the low-IQ groups.

Figure 3.3 shows that this is what we found. The two PET images on the left show more activity (red and yellow) throughout the brain in both low-IQ groups compared to normal controls on the right. We saw this as more evidence for the efficiency hypothesis, although we recognized alternative interpretations including compensation for possible brain damage (Haier et al., 1995).

3.3 Not All Brains Work in the Same Way

By this time, I had negotiated for one free PET scan for every scan I paid for with grant money, so we turned to a different way to investigate the efficiency hypothesis. Recall that in Chapter 1 I talked about the Hopkins study of mathematically precocious students that Professor Julian Stanley started in the 1970s. The early talent searchers found many more young boys than girls with high SAT-Math scores. In 1995

I decided to use PET to see if men and women both showed equal brain efficiency in the same brain areas while they solved mathematical reasoning problems. Mathematical reasoning is a more specific mental ability than the g-factor so this would expand the bounds of the efficiency hypothesis. I worked on this project with Professor Camilla Benbow, another former Hopkins graduate student who had worked with Professor Stanley.

We recruited 44 male and female college students from my university (UCI) based on their SAT-Math scores at admission (Haier and Benbow, 1995). We selected four groups: men with high SAT-Math scores over 700; women with equally high scores over 700; men with average SAT-Math scores in the 410–540 range; and women with average scores in the same range. There were 11 students in each group (44 participants was all we could afford but this still was one of the largest PET studies at the time). Each person completed a PET scan while they solved actual SAT-Math reasoning problems. We expected to see lower brain activity in both the high SAT-Math men and the high SAT-Math women compared to the average groups, consistent with brain efficiency. We also thought that the men and women matched for high math reasoning might show efficiency in different brain areas because there are sex differences in brain size and structure, although at the time the evidence for these differences was not as compelling as it is today (Halpern et al., 2007; Luders et al., 2004; Martinez et al., 2017; Ritchie et al., 2018; van der Linden et al., 2017).

Here's what we found. In the 22 men, statistical analysis showed that high math ability went with *greater* activity in the temporal lobes (the lower side parts of the brain that include important memory areas such as the hippocampus) during the problem solving. This was just the opposite of efficiency. In the 22 women, we found no systematic statistical relationship between mathematical reasoning ability and brain activity. How the brains in the high SAT-Math women were working to solve the problems could not be determined, even though they were solving the same problems as the men equally well. And the men showed the opposite of what we expected. That is how research often goes.

Actually, this finding was one of the first clear indications from imaging data that men and women may process information and problem solve with different brain networks. Remember, in this study the men and women were equally matched on SAT-Math scores, and they solved the same problems during the scan equally well. Their brains, however, showed apparently different patterns of activity. To us, this meant that not all brains work the same. This may seem obvious and even trite to you, but most cognitive researchers are interested in discovering how brains work in general, assuming that all brains basically work the same way. A focus on individual differences and the idea that not all brains work the same way was not so popular then, although now there is increasing interest among cognitive researchers (Barbey et al., 2021). Also, remember that mathematical reasoning ability is a more specific factor; it's not *g*. Brain efficiency may be related to *g*, but for specific abilities such as mathematical reasoning, better performance may require more brain activity. Along these lines, another PET study about the same time in eight middle-aged individuals reported increased activation during the performance of a perceptual maze task, a measure of visual-spatial reasoning, which is also a more specific factor of intelligence than *g* (Ghatan et al., 1995).

Confused? My purpose in describing these studies in the chronological order they occurred is to give you a feel for how researchers go about their work and sort through apparently discrepant findings. Remember my three laws. Repeat after me: No story about the brain is simple, no one study is definitive, and it takes many years to sort out conflicting and inconsistent findings and establish a weight of evidence. Chapter 4 brings some clarity to imaging results and, unsurprisingly, which raise new questions.

But before we continue to other early imaging studies of intelligence, I want to mention one more PET study we did. By the year 2000, it was still the case that very few other intelligence researchers were using PET or other imaging. We were still interested in brain efficiency, but we also started to wonder about whether efficiency would be related to intelligence even when the brain was not solving problems. In other words, could a smart brain be distinguished even when it was not working to be smart?

Our next PET study looked at eight new college students while they passively watched videos with no problem solving required (Haier et al., 2003). This was a project on emotional memory so some videos were more emotionally loaded than others, but as a separate analysis we looked at whether intelligence, assessed by the *g*-loaded RAPM test of abstract reasoning, was related to watching the videos irrespective of their emotional content. We correlated brain activity during this nonproblem-solving condition to RAPM scores. Significant correlations were apparent in several areas. None were in the frontal lobes. Most were in the posterior areas of the brain where basic information is perceived before it is processed by association areas more toward the front of the brain. This suggested that people with higher RAPM scores seemed to be viewing videos with different brain activity than lower RAPM people.

We think this means that smarter people are more engaged and actively processing the video information differently. In other words, the smarter brains were not so passive. This is more evidence that not all brains work the same way, perhaps even while watching television.

Several other PET studies related to intelligence were done in this early period. Collectively they reported activations in areas throughout the brain while performing different tests of deductive/inductive reasoning (Esposito et al., 1999; Goel et al., 1997; Goel et al., 1998; Gur et al., 1994; Wharton et al., 2000). The individual differences approach, that is, looking for correlations between test scores and degree of activation, was not systematically reported but all these studies found that multiple areas across the entire brain were activated during reasoning. The evidence was mounting that intelligence was not just a function of frontal lobes.

3.4 What the Early PET Studies Revealed and What They Did Not

The PET studies we've covered in this chapter so far represent the first attempts to use high-tech functional brain imaging to investigate intelligence. The overall point is that even these earliest studies helped shift intelligence research away from predominately psychometric approaches, and the controversies about them, to a more neuroscience perspective, because imaging provided a way to determine how psychometric test scores were related to measurable brain characteristics such as glucose metabolism.

Here's a summary of four key observations that emerged from these early functional imaging studies:

1. Intelligence test scores are related to brain glucose metabolism. This helps validate that the test scores were not meaningless numbers representing a statistical artifact. In fact, as neuroimaging studies of intelligence continue to increase, old criticisms about intelligence test scores having no meaning are less and less meaningful, if they were ever meaningful at all.

2. Early on, we had the unexpected and counterintuitive finding that higher intelligence test scores were associated with less brain activity. The resulting efficiency hypothesis encouraged many subsequent studies and it is still viable, although as we will see in Chapter 4, the story gets more complex as more studies are done; the same progression of progress found in all science.

3. Learning some tasks is associated with the brain becoming more efficient as indicated by lower brain activity after practice. This raises the
   question of whether intelligence can be enhanced by mental training.
   We discuss this possibility and our deep skepticism of recent efforts in
   detail in Chapter 5.
4. PET scan differences between men and women solving problems, and
   PET differences between high and average intelligence watchers of
   videos, indicate that not all brains work the same way. We discuss this
   concept in Chapter 4.

There's another important inference based on something we did not see.
These early data did not show any one area in the brain that could be called
the center of intelligence. In fact, the early PET imaging data reported that
many areas distributed throughout the brain were associated with intelligence test scores. In 2000, however, one group of researchers claimed their
PET study showed that the neural basis of the g-factor was derived from a
specific lateral frontal lobe system and downplayed the importance of other
regions (Duncan et al., 2000). They imaged blood flow over a two-minute
period as 13 subjects (with the wide age range of 21–34) performed a small
number of problems that varied in g-loadings. Blood flow increased during
the tasks but only the frontal activations were noted as common to the
high and middle g-loaded tasks. This publication, in *Science*, received considerable attention, but many researchers in the field were quick to point
out several major flaws in design and interpretation (Colom et al., 2006a;
Newman and Just, 2005). Design questions included the omission of any
description of the subjects in terms of sex and IQ. Also, they were apparently recruited at a distinguished university so a severe restriction of range
of g-scores is likely, limiting correlations. Imaging occurred during a few
problems while subjects worked at their own pace so the task reliability
was low and averaging over subjects could minimize any differences due
to differences in speed of responding to the problems. As for interpretation, none of the previous PET studies showing distributed areas related
to high g-scores were cited so there was no acknowledgment or discussion of inverse correlations or a distributed network of areas other than in
the frontal lobe. Furthermore, their own data for the high-g task showed
activation in multiple areas outside the frontal lobes. Subsequently, Dr.
John Duncan, the research leader, apparently abandoned the frontal lobe–
centered model of intelligence and came to the view that other areas also
were involved (Bishop et al., 2008; Duncan, 2010), consistent with virtually
all other studies available at that time (Jung and Haier, 2007). So, we will
not dwell on this short-lived detour other than to note that its publication in
Science brought important attention to imaging/intelligence research and

validated again that *g*-scores could be studied scientifically, a proposition that had been surprisingly controversial. So even this flawed paper had some positive effect. At the time, journals such as *Science* were reluctant to publish intelligence research, owing in large part to the controversies of the 1970s and 1980s concerning average group differences (see Chapter 2). The late Constance Holden, a science writer working at *Science*, lamented the prejudice against intelligence research she saw from the inside and did her utmost to cover intelligence research with a journalist's combination of integrity and skepticism. Following her untimely accidental death, the International Society of Intelligence Research sponsors a presentation by a journalist as the Constance Holden Memorial Lecture at its annual meeting in recognition of her efforts.

3.5 The First MRI Studies

By 2000, a new imaging technology was becoming available much more rapidly than had PET. PET is based on positrons colliding with electrons and requires injection of radioactive tracers. MRI does not require radioactive injections so no cyclotron or hot lab is required, making MRI scans considerably less expensive than PET (about $500–$800 versus $2,500). MRI is based on the effect of magnetic fields on spinning protons and hydrogen molecules. Because it produces high-resolution images of the entire body that have many important clinical uses without radiation exposure, MRI quickly became a must-have technology for most hospitals, especially ones associated with universities. This allowed many cognitive psychologists access and, in fact, within a dozen years or so, most psychology departments at major universities had their own MRI scanner, a multimillion-dollar expense once unthinkable for a psychology department (although acquisition of imaging equipment by psychology departments was predicted by at least one prescient researcher) (Haier, 1990). MRI studies of cognition have grown exponentially since the year 2000 and MRI analyses are now a mainstay of cognitive neuroscience research (Barbey et al., 2021).

Here's how MRI works. Protons naturally spin around an axis and the spinning creates a weak magnetic field. Each proton axis has a different, random north–south orientation. If protons enter a strong magnetic field, they snap into the same north–south alignment. When a radio wave is pulsed on and off rapidly into the magnetic field, the protons snap out of alignment and then back in. This pulsing can be done many times per second. As the protons snap in and out of magnetic alignment, the shifts give off weak energy, and this energy can be detected and

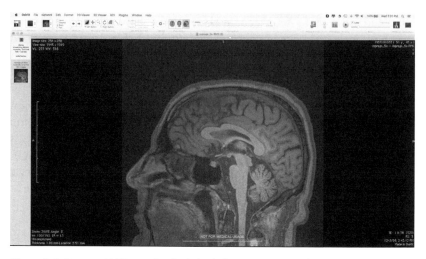

Figure 3.4 Structural MRI scan (sagittal view). (Courtesy Rex Jung)

mapped showing where the protons are if the magnetic field is applied along a gradient of different intensities. This sequence of events is called "magnetic resonance imaging" (the original name of this technology was "nuclear magnetic resonance" but was changed to avoid a "nuclear" connotation). Hydrogen protons are abundant in water and most of the body, especially soft tissue, is made of water, so MRI gives beautifully detailed images of the body and the brain.

MRI scanners are large donut-like devices that contain a very powerful magnet. When a person lies on the scanner bed and the head or whole body goes into the central tube-like area surrounded by the magnet, radio waves are rapidly pulsed into the magnetic field, and the protons in the body snap in and out of alignment. The person has no sensation of this snapping. The shifting energy patterns formed by all this snapping are detected and mathematically turned into a picture.

The illustration in Figure 3.4 shows an example of a basic MRI of brain structure in great detail. It shows a side view slice (sagittal slice). Even whole brain 3D images can be viewed. These, of course, are mathematical slices not actual slices. Like a picture printed in a newspaper, each brain image is made up of many individual dots called pixels. With MRI, the pixels actually have three dimensions so they are called voxels and they have volume rather than just area. There are millions of voxels in a brain image. Each voxel has a value determined by the imaging technique. In this case the value is the amount of energy detected by protons snapping and this can be interpreted as the amount of gray matter, for example.

The structural image in Figure 3.4 shows gray matter, where neurons work, and the white matter fibers that link brain areas and carry information around the brain. Gray matter and white matter tissue have different water content, so they can be distinguished in these images. Note that structural images do not contain any functional information, so you cannot look at a structural MRI and tell if the person is awake or asleep, solving math problems, or even alive or dead. You can see tumors, strokes, and many kinds of brain damage. MRI can also be used to show brain function. Very rapid sequential images can show regional blood flow as a function of hemoglobin determinations and blood flow is an indirect measure of neuron activity. The more a brain area is active, the more blood flows to it. It is functional MRI (fMRI) that has been used widely in cognitive neuroscience to show brain activity during specific cognitive tasks.

The basic MRI technique is quite versatile. By changing various parameters of the scanning sequence, for example, like the frequency of the radio wave pulses, different kinds of pictures can be made that emphasize different brain characteristics. As noted, the two main kinds of MRI are structural and functional. Structural MRI methods include the basic scan which shows gray and white matter in anatomical detail and other methods that maximize the imaging of white matter fibers and tracts such as spectroscopy (MRS) and diffusion tensor imaging (DTI). Such structural images are not affected by what the brain is doing during the scan. We'll now review the early intelligence studies that used structural and functional MRI. We start with the basic structural MRI since this was the first way MRI was applied to intelligence research.

3.6 Basic Structural MRI Findings

The first question about intelligence addressed by MRI had to do with whole brain size. Numerous previous studies had reported a positive correlation between brain size and intelligence test scores. The correlation was typically modest, but the main problem was that the measures of brain size were estimates based on indirect measures such as head circumference (or in the 1800s the number of metal pellets required to fill a skull). MRI provided a much more exact measurement of brain size/volume *in vivo*, so it was not surprising to find confirmation of a positive correlation with intelligence test scores when accurate MRI-based measurements of brain size were used (Willerman et al., 1991). This is a straightforward finding that has been replicated many times. A comprehensive meta-analysis of this literature (37 studies, 1,530 subjects) reported an average correlation between whole brain size/volume and intelligence test scores to be about 0.33 overall (McDaniel, 2005), including adults and children. The correlation was higher in females (about 0.40 compared to 0.34 in males). In female adults and children, the correlations were 0.41 and 0.37, respectively. In male adults and children, the correlations were even more different at 0.38 and 0.22 respectively. These data essentially resolve the earlier debate and show definitively that bigger brains are modestly associated with higher intelligence. Moreover, a genetically informative study (with replication) using GWAS data indicated that brain size had a causal effect on intelligence (Lee et al., 2019).

Of course, questions remain. Are the volumes of some specific brain areas more related to intelligence than other areas? What influences the development of brain size, and can the developmental mechanisms be accentuated? We discuss the latter question in Chapter 5 when looking at enhancing intelligence. The former question was addressed soon after structural MRIs were augmented with image analysis methods that segmented or "parcellated" cortical and subcortical areas into regions of interest (ROIs). ROIs were typically derived either by applying a simple algorithm based on an arbitrary proportion of voxels thought to define a region or by human observers tracing ROIs on each image to the best of their ability using various brain landmarks. These early segmentation methods varied among research groups and all were rudimentary by today's standards, but the results did indicate that the size/volume of some areas was more related to intelligence than it was in other areas. One group (Andreasen et al., 1993), for example, reported small positive correlations between FSIQ and volume of the temporal lobes, hippocampus, and cerebellum and (Flashman et al., 1997) further reported small correlations with performance (nonverbal) IQ in frontal, temporal, and parietal lobes. None of these correlations exceeded whole brain–IQ correlations but they hinted at the importance of regional analyses.

3.7 Improved MRI Analyses Yield Mixed Results

By the time the next structural MRI studies of intelligence were reported, image analyses had improved spatial localization by replacing ROI segmentation with methods that quantified gray and white matter voxel by voxel (Ashburner & Friston, 1997, 2000) with spatial resolution of millimeters rather than lobes or parts of lobes. Software for the application of voxel-based morphometry (VBM) became available in about 1999 (Statistical Parametric Mapping, SPM) and the field moved dramatically away from customized image analysis based on ROIs that differed in their boundaries among research groups to a more standardized approach. Typically, the results of voxel-based analyses were reported as spatial locations in the brain using a standard set of coordinates developed at the Montreal Neurological Institute (MNI coordinates) and the locations were described additionally using a standard nomenclature based on Brodmann areas (BAs) derived from early autopsy descriptions of different cellular organization among cortical regions (Brodmann, 1909). Textbox 3.1 describes the VBM method and includes an illustration of BAs. SPM is updated periodically with improvements and additional options for analyses.

Textbox 3.1: Voxel-based morphometry
One main method for analyzing MRI used on structural or functional images is called "voxel-based morphometry." There are three basic steps, as shown in Figure 3.5. First, we start with an image like the MRI on the left. Next, mathematical algorithms determine the boundaries of gray and white matter tissue. Finally, values are calculated that reflect the amount of gray or white matter tissue in each voxel in the whole brain. Since there are millions of voxels in the whole brain image, you get a very large data set. You can then correlate a test score, for example, to every one of these voxels and identify where the correlations are statistically significant. The location of any finding from any image analysis can be described with a system of standard spatial coordinates (height, width, depth) or by a standard nomenclature of brain areas differentiated by cellular structure, originally determined by Brodmann. BAs are shown in Figure 3.6. Often both BAs and spatial coordinates (usually based on the Talairach brain atlas or on the MNI's system) are included in research reports.

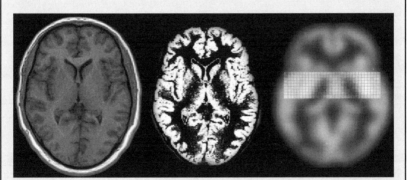

Figure 3.5 VBM technique starts with an image (left), automated algorithms then separate gray and white matter (middle), and then a value reflecting density is assigned to each voxel in the image. This value can be correlated to IQ, age, or other variables. (Courtesy Rex Jung)

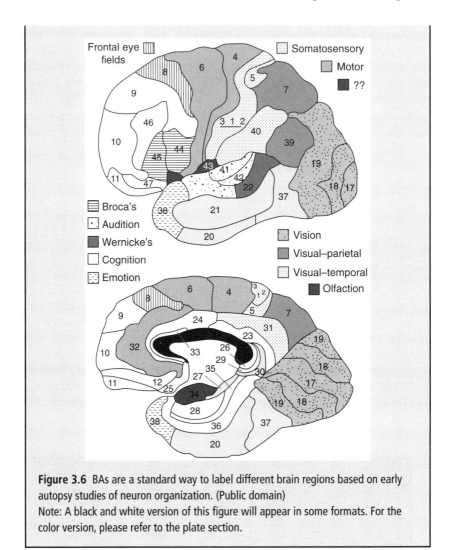

Figure 3.6 BAs are a standard way to label different brain regions based on early autopsy studies of neuron organization. (Public domain)
Note: A black and white version of this figure will appear in some formats. For the color version, please refer to the plate section.

Some of the first voxel-based MRI analyses of intelligence were reported in children. One research group obtained MRIs in 146 children (mean age 11.7, standard deviation 13.8) and reported a correlation of 0.30 between FSIQ and gray matter volume of the anterior cingulate gyrus (BA 42) (Wilke et al., 2003). Another group studied 40 children (mean age 14.9, standard deviation 2.6) and reported that gray matter volume was correlated to different parts of the cingulate (BAs 24, 31, 32) and to areas in the frontal lobes (BAs 9, 10, 11, 47) and in the parietal lobes (BAs 5, 7) (Frangou et al., 2004). More recent and advanced imaging studies of children are reported in Chapter 4, but these early studies in children were important for demonstrating the potential of imaging to elucidate relationships between brain development and IQ scores.

One additional early study exemplifies this. In the largest and most representative sample of children studied up to that time with MRI, a research group (Shaw et al., 2006) introduced another method of image analysis that determined the thickness of the cortex. This was not VBM. Rather, it used numerous cortical landmarks and Euclidian geometry to calculate thickness at many points around the cortex. They had a sample of 307 normal children (mean age 13, standard deviation 4.5) who completed IQ tests and MRI scans on multiple occasions over time. Cortical thickness (CT) was correlated with IQ, but there was a clear developmental sequence showing a dynamic relationship between regional CT and intelligence as the brain matures through childhood and adolescence. The strongest correlations between IQ and CT were found in late childhood (approximately 8–12 years). These correlations were positive and they were found in areas throughout the brain. However, there was a difference between high and average IQ individuals. The high-IQ subjects showed "an initial accelerated and prolonged phase of cortical increase, which yields to equally vigorous cortical thinning by early adolescence" (Shaw et al., 2006: 676). This interesting finding, however, requires replication and we discuss new studies in Chapter 4. The Shaw et al. (2006) study, published in *Nature*, a prestigious science journal, and funded by the National Institute of Child Health and Development, added further credence to investigations of brain–intelligence relationships.

About the same time, VBM was applied in studies of intelligence in adults for the first time. We obtained MRIs for 47 adults across a wide age range (18–84) and correlated gray and white matter to IQ scores, correcting for age and sex (Haier et al., 2004). The results showed gray matter correlations in several areas distributed across the brain in all four lobes and in both hemispheres. One correlation between IQ and white matter was prominent in the parietal lobe (near BA 39). When we reanalyzed the data separately for males and females (Haier et al., 2005), we were surprised to see different results. In men the largest brain areas where more gray matter went with higher IQ were in posterior regions, especially in a part of the parietal lobe related to visual-spatial processing. But in women, almost all the areas where gray matter correlated to IQ were in the frontal lobes, especially around a part of the brain related to language called Broca's Area.

As with our previous PET study of mathematical reasoning, males and females showed different patterns of correlations. An unsettled issue is whether different male/female patterns are statistically significant. Nonetheless, these findings reaffirmed our view that all imaging studies of intelligence should analyze data separately for males and females just as routinely as groups of different ages are analyzed separately. Finding age and sex differences underscore one of our basic assumptions: not all brains work the same way.

Even with the application of more standard VBM methods, however, many inconsistent results were reported. For example, one group (Lee et al., 2006) studied 30 older adults (mean age 61.1, standard deviation 5.18) and found only a correlation between performance IQ and volume in the posterior lobe of the right cerebellum. Another group (Gong et al., 2005) studied 55 adults (mean age 40, standard deviation 12) and reported that correlations between gray matter and FSIQ were limited to areas in the anterior cingulate and the medial frontal lobes. As we have pointed out, results were often based on analyses that did not separate males and females, and restricted ranges may have limited correlations, as discussed in Chapter 1. Another key issue is the assessment of intelligence using IQ tests. Although the standard IQ tests provide a good estimate of the g-factor, IQ scores combine g and other specific intelligence factors. Would imaging results be more consistent if a better estimate of g was used?

We addressed this in two studies (Colom et al., 2006a; Colom et al., 2006b) based on a reanalysis of our 2004 VBM data using the method of correlated vectors (Jensen, 1998b). In this case, this method correlates the rank of g-loadings for each subtest of the WAIS to the rank of the same test correlation to gray matter. We found that g accounted for many of the FSIQ correlations with gray matter in the anterior cingulate (BA 24), frontal (BAs 8, 10, 11, 46, 47), parietal (BAs 7, 40), temporal (BAs 13, 20, 21, 37, 41), and occipital (BAs 17, 18, 19) cortices (Colom et al. 2006b). Moreover, in a separate analysis, we found a nearly perfect linear relationship between the g-loading of each subtest of the WAIS and the amount of gray matter correlated to each subtest score (Colom et al. 2006a). Thus, we come to another important observation. IQ tests have the advantages of a standardized test battery but the scores combine the general factor along with other specific factors. So, the question of how intelligence correlates to brain structure and function depends on whether the question is about g or about more specific mental abilities. Inconsistent results among these early studies likely result from confusion on this issue as well as from issues about sampling and image analysis.

3.8 Imaging White Matter Tracts with Two Methods

A different kind of structural MRI is called DTI. Here, the MRI sequences are optimized to image the water content (i.e., hydrogen molecules) of white matter fibers and, when combined with special mathematical algorithms, the resulting images show white matter tracts in great detail. DTI measures can assess the density and organization of the tracts, which relate to how well they transmit signals. DTI is an excellent technique for identifying brain networks. Most DTI studies of intelligence are more recent and are detailed in Chapter 4. Here we note the first DTI study of intelligence by Schmithorst and colleagues (Schmithorst et al., 2005). They studied 47 children aged 5–18 years. After correcting for age and sex, the strongest correlations between IQ and the density/organization of white matter fibers were found in frontal and parietal/posterior areas. They noted that these findings were consistent with the Wilke et al. study that had used VBM methods. As with the other early MRI studies, this one added to the excitement of using new imaging techniques to quantify brain–intelligence relationships.

Whereas DTI can quantify white matter density and organization, MRS can make neurochemical determinations of white matter integrity, which is another measure of how well signals are transmitted through the fibers. For example, MRS can determine N-acetylaspartate (NAA), a marker of neuronal density and viability. The early MRS methods, however, were limited to single-voxel analysis so the entire brain could not be studied at once. There were three early studies of intelligence using MRS. Jung's research group studied 26 college students and placed the NAA measurement voxel in white matter underlying BAs 39 and 40 in the left parietal lobe (Jung et al., 1999). They found a correlation between NAA and FSIQ of 0.52. They replicated and extended these findings in a new sample of 27 college students where the same region showed the NAA–IQ correlation and control regions in bilateral frontal lobes did not (Jung et al 2005). They also showed that the NAA–IQ correlation was higher in the subsample of women. In the third MRS intelligence study, another group reported a sample of 62 adults in a wide age range (20–75 years). Scores on the high-g vocabulary subtest of the WAIS-R were correlated to NAA in voxels underlying left frontal BAs 10 and 46 ($r = 0.53$) and left BAs 24 and 32 ($r = 0.56$) in the anterior cingulate gyrus (Pfleiderer et al., 2004). All these early MRI studies of gray and white matter structure were exciting because they found correlations between various psychometric test scores of intelligence and quantifiable brain characteristics both in specific locations and in the connections among them. This increased optimism for the potential of discovering not only "where" in the brain intelligence is located but also "how" intelligence is related to brain function.

3.9 Functional MRI

fMRI uses scanning parameters that image aspects of hemoglobin in red blood cells because hemoglobin contains iron and iron molecules are quite sensitive to the magnetic fields used in MRI. A sequence of very rapid images is made: thousands per second. These are interpreted as showing blood flow in the brain. Those brain areas that are most active during a task (often compared to a no-task resting condition) have greater blood flow; less active areas have reduced blood flow. Whereas glucose PET scans show the accumulation of brain activity over 32 minutes, fMRI scans show that activity changes almost second by second. fMRI is now the most widely used imaging technology in cognitive psychology research.

The first intelligence study using fMRI was from a group at Stanford University (Prabhakaran et al., 1997). They used individual items from the RAPM test chosen for three different types of reasoning required to solve the problem. They found that blood flow increases in frontal and parietal brain areas while seven young adults (aged 23–30) solved each item. They did not look for correlations between amount of activation and task performance because only a few problems were used and each person answered all items correctly. By design, this eliminates individual differences in task performance. This approach is typical in many cognitive studies where any individual differences among subjects related to intelligence are essentially ignored. More recently, there are excellent discussions of the individual differences approach and its potential for advancing cognitive imaging studies (Anderson & Holmes, 2021; Biazoli et al., 2017; Kanai & Rees, 2011; Parasuraman & Jiang, 2012).

Even though fMRI was used in hundreds of cognitive studies by 2006, only 17 studies included any measure of intelligence or reasoning. Of these 17 fMRI studies, all but three had sample sizes of 16 or fewer and there were a variety of control tasks (or the lack of any control task in some studies) and a variety of intelligence/reasoning measures. None of the measures in these early studies were based on a battery of tests to estimate the g-factor. Some of the tests used in these studies during the imaging included working memory (Gray et al., 2003), chess (Atherton et al., 2000), analogies (Geake & Hansen, 2005; Luo et al., 2003), visual reasoning (Lee et al., 2006), deductive or inductive reasoning (Goel &

Dolan, 2004; Fangmeier et al., 2006), and verb generation (Schmithorst & Holland, 2006). This last study was unique for its impressive sample size of 323 children (mean age 11.8 years, standard deviation 3.7). Given all these various findings and methods in the early studies, could any consistent threads be identified?

3.10 The Parieto-Frontal Integration Theory (PFIT)

In December 2003, I hosted an invited symposium at the annual meeting of the International Society for Intelligence Research. It was the first time imaging researchers had come together to discuss intelligence studies. In addition to myself, the participants included Jeremy Gray, Vivek Prabhakaran, Rex Jung, Aljoscha Neubauer, and Paul Thompson. With the exception of Aljoscha Neubauer, it was the first time I had met these researchers in person. Rex Jung's presentation was a compelling review of several studies that emphasized the distributed nature of brain areas associated with intelligence. Based on his clinical background as a neuropsychologist and his MRS research on white matter and IQ, he also emphasized the importance of white matter connections among the salient brain areas. It was apparent that he and I had similar interests so we undertook a comprehensive review of the entire brain-imaging/ intelligence literature. It took us over two years to write the review which was published in 2007 along with commentaries from other researchers (Haier and Jung, 2007; Jung and Haier, 2007).

From our first PET study of only eight subjects in 1988 to the larger fMRI studies through 2006, there were 37 imaging studies of intelligence from different research groups around the world. Given the wide disparity of methods and measures, and the number of potential brain areas involved, a typical meta-analysis was not appropriate. Instead we followed a method used to review the emerging literature from cognitive neuroimaging studies (Cabeza & Nyberg, 2000). We reviewed structural MRI results, PET results, and fMRI results. We focused on findings common among studies irrespective of different imaging and assessment methods. Several brain areas were common in 50 percent or more of the 37 studies. This may seem a rather weak proportion but it is similar to the proportions found in the Cabeza and Nyberg (2000) review of well-controlled cognitive experiments.

The salient brain areas we identified were distributed throughout the brain but were mostly in parietal and frontal areas. We called our model the "Parieto-Frontal Integration Theory" (PFIT) of intelligence. Note that "Integration" emphasizes that communication among the salient

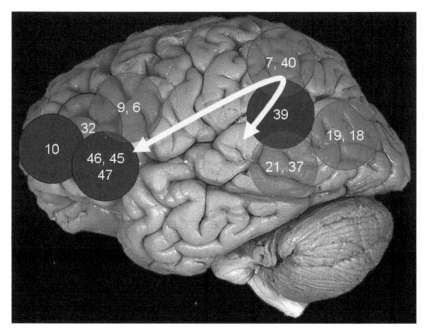

Figure 3.7 The PFIT showing brain areas associated with intelligence. (Courtesy Rex Jung) Note: A black and white version of this figure will appear in some formats. For the color version, please refer to the plate section.

areas was key to the model since we have always recognized that identifying specific brain areas was only the beginning of a useful brain model of intelligence. Understanding the temporal and sequential interactions among networks that link the areas would be key. The illustration in Figure 3.7 shows all the areas we included in the model.

The circles in Figure 3.7 show brain areas and the numbers refer to the standard BA nomenclature (Brodmann, 1909). We proposed that these areas define a general brain network and subnetworks that underlie intelligence. Most of the areas are in frontal and parietal lobes, some in the left hemisphere (blue circles), and some in both hemispheres (red circles). A major white matter tract of fibers (yellow arrow) connects the frontal and parietal lobes like a superhighway. It's called the arcuate fasciculus and we have proposed that it is an important tract for intelligence.

The brain areas in our model represent four stages of information flow and processing while engaged in problem solving and reasoning. In stage 1, information enters the back portions of the brain through sensory perception channels. In stage 2, the information then flows forward to association areas of the brain that integrate relevant memory; and in stage 3, all this continues forward to the frontal lobes that consider the integrated information, weigh options, and decide on any action; so in stage 4, motor or speech areas for action are engaged if required. This is unlikely to be a strictly sequential, one-way flow. Complex problems are likely to require multiple, parallel sequences back and forth among networks as the problem is worked out in real time.

The basic idea is that the intelligent brain integrates sensory information in posterior areas, and then the information is further integrated to higher-level processing as it flows to anterior areas. The PFIT also suggests that any one person need not have all these areas engaged to be intelligent. Several combinations may produce the same level of general intelligence, but with different strengths and weaknesses for other cognitive factors. For example, two people might have the same IQ, or g-level, but one excels in verbal reasoning and the other in mathematical reasoning. They may both have some PFIT areas in common, but it is likely they will differ in other areas.

Cognitive studies show that some PFIT areas of the brain are related to memory, attention, and language, suggesting that intelligence is built on integrating these fundamental cognitive processes. Our hypothesis is that individual differences in intelligence, whether the g-factor or other specific factors, are rooted both in the structural characteristics of the specific PFIT areas and in the way information flows around these areas. Some people will have more gray matter in important areas or more white matter fibers connecting areas and some people will have more efficient information flow around the PFIT areas. These brain features lead some individuals to score higher on intelligence and mental ability tests, and other individuals to be less efficient and less good at problem solving. How the salient brain features may develop is a separate issue for future longitudinal studies of children and adolescents. In Chapter 4, we will see newer imaging methods that show millisecond changes in information flow throughout the brain so that hypotheses about efficient information flow and intelligence can be tested.

While we were formulating the PFIT, we were unaware of a similar review published in a book chapter by two cognitive psychologists (Newman and Just, 2005). These authors also favored a distributed network for intelligence rather than a model concentrated only in the frontal lobe. They also noted the importance of white matter connections among brain areas. Efficient information flow and the importance of computational load were prominent features of their model. Independently, we arrived at similar views, although we came from different perspectives. Their work is listed at the end of this chapter under Further Reading. I highly recommend it.

And there is one more thing to mention. Many of the gray and white matter areas related to IQ first reported in these phase-one studies that contributed to the PFIT appeared to be under genetic control to some degree (Pol et al., 2002; Posthuma et al., 2002; Posthuma et al., 2003, Thompson et al., 2001; Toga & Thompson, 2005), and we discuss these and newer, even more compelling findings in Chapter 4 in relation to studies that combine advanced genetic analysis, including DNA, and neuroimaging in very large samples.

Since our review of 37 studies was published in 2007, there are now hundreds of additional imaging studies of intelligence from research groups all over the world, as more researchers appreciate the connections between general intelligence and fundamental cognitive processes. We refer to these post-2006 studies as phase two in the application of neuro-imaging to intelligence research (Haier, 2009a). This new wave of studies includes many that are far more sophisticated with respect to large, representative samples, multiple measures of intelligence to estimate the g-factor, and advanced image analysis techniques that include better anatomical measurement and localization methods. We detail important phase-two studies in Chapter 4. But first, let's peek into a famous brain.

3.11 Einstein's Brain

Before closing this chapter, let me draw your attention briefly to Einstein's brain. It was removed after his death and preserved in a jar by a physician who kept it at home and then in his car as he moved across the country. He was reluctant to share it, but eventually samples were made available to researchers. The main findings (Witelson et al., 1999a; Witelson et al., 1999b), not without technical issues that could influence interpretation of results (Galaburda, 1999; Hines, 1998), were that Einstein's brain showed more tissue and more neuron-support cells in a posterior part of the brain. This area was pretty much the same parietal area where the men showed correlations with IQ and the women didn't (Haier et al., 2005). A detailed analysis of photographs of Einstein's brain also suggested differences in frontal and parietal areas (Falk et al., 2013). Anything about how Einstein's brain may differ from other brains is inherently interesting, but perhaps the most remarkable thing about his brain is that it is not all that remarkable from a purely anatomical analysis. In fact, at autopsy it is often the case that a person who had an IQ under 70 may have no remarkable anatomical brain features to distinguish it from brains of people with high IQs. This is why functional neuroimaging and quantitative image analysis have provided many new insights.

During the first phase of applying new medical neuroimaging technologies, intelligence researchers had limited access to expensive equipment and the first studies were characterized by small samples, single measures of intelligence, and rudimentary image analysis methods that typically ignored individual differences. Nonetheless, slow but steady progress from 1988 to 2006 allowed a literature review based on 37 studies that concluded there were a finite number of identifiable areas distributed across the brain where structure and/or function were related to scores on intelligence and reasoning tests. Phase two of imaging/intelligence studies builds on these findings with advanced methods and the latest progress is the focus of Chapter 4.

Chapter 3 Summary

- This chapter laid out the early history of neuroimaging studies of intelligence, a period from 1988 to 2006 we refer to as phase one that indicated surprising findings.
- From the first studies, it was apparent to most researchers that intelligence was not centered solely in frontal lobes but instead involved networks distributed across the brain.
- A surprising early finding was an inverse correlation between intelligence test scores and brain activity determined by GMR, suggesting a hypothesis that efficient information flow was an element of higher intelligence.
- Imaging studies showed that not all brains worked the same way. Individual differences required examination rather than being ignored when group data were averaged.
- Despite the limitations of phase-one studies, some consistent results across studies suggested the PFIT of intelligence that emphasized both the structural and functional characteristics of specific brain areas and the connections among them.

Review Questions

1. Explain the difference between structural and functional neuroimaging.
2. What are the main differences between the PET and MRI technologies?

3. What is the basis for the brain efficiency hypothesis of intelligence?
4. What is the evidence about whether there is an "intelligence center" in the brain?
5. List key limitations of the early brain-imaging studies of intelligence.

Further Reading

Looking Down on Human Intelligence (Deary, 2000). This is a sophisticated and comprehensive account of intelligence research. Clearly written with wit and without jargon, it ranges from early thinkers and philosophers to the end of the twentieth century, including the early neuroimaging studies.

"The Parieto-Frontal Integration Theory (PFIT) of Intelligence: Converging Neuroimaging Evidence" (Jung and Haier, 2007). This is the original, somewhat technical review of 37 imaging/intelligence studies. It includes a broad range of commentaries from other researchers in the field (Haier and Jung, 2007).

"Human Intelligence and Brain Networks" (Colom et al., 2010). This is a more general description of the PFIT model.

"The Neural Bases of Intelligence: A Perspective Based on Functional Neuroimaging" (Newman and Just, 2005). This chapter is clearly written and presents a brain model of intelligence similar to, but developed independently of, the PFIT.

IQ and Human Intelligence (Mackintosh, 2011). This is a thorough textbook that covers all aspects of intelligence written by an experimental psychologist. It has a chapter that is a good summary of early imaging studies of intelligence (chapter 6).

The Cambridge Handbook of Intelligence and Cognitive Neuroscience (Barbey et al., 2021). This is an edited book with imaging chapters written by experts in cognitive neuroscience; many chapters are written for advanced students.

CHAPTER FOUR

50 Shades of Gray Matter

A Brain Image of Intelligence Is Worth a Thousand Words

There are more than enough brain-injured people in the modern world to permit resolution of every fundamental question concerning the human mind, could this material but be brought under adequate study.

Ward C. Halstead (1947)

The data are intriguing. The field is maturing. The pace is quickening. As intelligence research engages 21st century neuroscience, new hypotheses and new controversies are inevitable. What a terrific time to work in this field.

Richard Haier (2009a)

From the fact that some of our abilities – … to see, hear, feel, and move – can quite specifically be traced back to the contributions of distinct brain regions … one might derive the expectation that there must be another part of the brain responsible for higher cognitive functioning and intelligence. But, … there is no single 'seat' of intelligence in our brain. Instead, intelligence is associated with a distributed set of brain regions.

Ulrike Basten and Christian Fiebach (2021)

Learning Objectives

- How has neuroimaging revealed brain networks related to intelligence?
- What is the empirical support for the PFIT framework?
- Does the weight of evidence support a relationship between brain efficiency and intelligence?
- What are the key issues for predicting intelligence test scores from brain images?
- Does imaging research on intelligence differ from imaging research on reasoning?
- What brain structures share genes with intelligence test scores?
- How have neuroimaging studies advanced the search for gene influences and brain mechanisms related to intelligence?

Introduction

Any lingering doubts that intelligence is a rich topic for neuroscience should have melted away given the genetic and neuroimaging studies described in Chapters 2 and 3. If not, please suspend any remaining disbelief until the end of this chapter wherein even more compelling findings are described. They come from more advanced neuroimaging studies, including those obtained in conjunction with genetic methods. I include studies of adults and children, and patients with brain damage, and introduce more advanced methods of brain image acquisition and analysis, including brain connectivity analyses. Such studies are now part of the emerging field of network neuroscience and include studies of intelligence. These studies continue to intrigue and motivate researchers worldwide to push assessment technologies to even greater precision, increase the sample sizes to previously unimaginable levels, and offer new testable hypotheses about intelligence and the brain. One word of caution: most of the studies in Chapter 3 and some in this chapter have sample sizes that are too small for conclusive interpretations. Remember our second law: no one study is definitive. As this field continues to mature, sample sizes are increasing rather dramatically. Over time, the weight of evidence always favors studies with sufficient sample sizes that maximize the stability of findings and minimize unreliable ones. This is especially so for the early studies seeking to identify gene influences related to intelligence. I am including such studies for historical context and for illustrating how the weight of evidence evolves.

The PFIT framework described in Chapter 3 proposed that intelligence was related to 14 specific areas distributed throughout the brain (Haier & Jung, 2007; Jung & Haier, 2007). These areas formed a broad network of frontal-parietal communication along with subnetworks involving several other temporal and occipital areas. How information flowed through these networks was proposed as a basis for individual differences in mental abilities, and especially for the g-factor. The model also proposed that individuals with the same IQ might achieve their level of g from different combinations of PFIT areas. In other words, there may be multiple, even redundant, neuro-pathways to the g-factor just like there are multiple routes driving from New York City to Los Angeles. Efficient information flow, through whichever subnetworks are relevant for an individual, was hypothesized to relate to high g, and subnetworks of the PFIT were hypothesized to relate to a person's pattern of mental ability strengths and weaknesses.

At the time the PFIT was proposed in 2007, testing these hypotheses was difficult. Methods of neuroimaging and analysis were limited with respect to their ability to assess structural or functional brain network connections and how well information was processed in networks during problem solving. This state of affairs improved quickly and dramatically for intelligence research with the application of new mathematical/statistical ways to assess connectivity among brain areas, new image analysis techniques to assess the integrity of white matter transmission of information, and the use of MEG technology to assess regions of neuron activity dynamically every millisecond during the performance of a cognitive task. Adding to the increased pace of intelligence research, neuroimaging is now combined with genetic methods in several large-scale consortia. These advances are the focus of this chapter. There are numerous recent neuroimaging studies to choose from that illustrate the momentum of these advances in intelligence research. I cannot summarize them all, but let's start with some key studies of brain network connectivity and what they find. These studies are mostly presented in chronological order so that the story is told as it has unfolded. All the studies reported in this chapter implicate many brain areas. You will get a sufficient feel for the general findings without memorizing these areas. It may be helpful to refer to the brain area maps in Figures 3.6 and 3.7 as you read this chapter.

4.1 Brain Networks and Intelligence

Every brain image is constructed from many thousands of small voxels. As explained in Chapter 3, each voxel is assigned a value based on the type of imaging used. In the FDG PET studies described in Chapter 3, the value was the GMR. In structural MRI, the value can be the density of gray or white matter. In fMRI, the value is based on blood flow. To determine how one brain area may be related to all other brain areas, correlations can be computed between any individual voxel, or group of voxels, defining an ROI, and all other voxels (or ROIs) throughout the entire brain. The starting voxel is called the "seed." Multiple seeds can be placed wherever the researchers wish depending on the hypothesis to be tested. The pattern of correlations indicates how the seed areas are connected to other brain areas. The connections are statistical and may or may not reflect actual anatomical connections.

This kind of connectivity analysis was applied to fMRI data in 59 individuals who also had completed the WAIS IQ test (Song et al., 2008). Usually, fMRI is acquired while the participants perform a cognitive task. Since different studies use different cognitive tasks, comparing results is often problematic since each task has its own cognitive requirements that involve different brain areas. In this case, the functional connectivity was determined using fMRI data acquired during a rest condition. In other words, no cognitive task was performed while fMRI data were obtained. The idea was to test whether brain activity at rest might reveal functional connections related to IQ. A consistent pattern of resting-state brain activity has been characterized as a "default network." That is, the pattern of brain activity when a person is not engaged in a cognitive task tends to be a stable pattern of maintenance involving specific brain areas, rather than a completely random pattern of uncorrelated, chaotic activity.

In this study, the seed was placed in a part of the frontal lobes corresponding to where BAs 46 and 9 come together (see PFIT, Figure 3.7); one seed was in each hemisphere. In the first step of the analysis, the resting-state functional connectivity between the seeds and the rest of the brain was determined statistically by correlating the blood flow value in the seeds to blood flow values in all other voxels. Several statistically significant connections were identified. As expected, some connections between the frontal seeds and other brain areas were stronger than others (i.e., there were stronger correlations). In the second step, the strength of the connections was correlated to IQ scores. The strongest IQ correlations were for connections between areas noted in the PFIT model. Moreover, this study indicated that individual differences in resting-state default network activity were related to IQ.

Soon thereafter, several studies reported the use of a better statistical method for inferring brain networks and how they relate to intelligence. The method is called graph analysis (Reijneveld et al., 2007; Stam and Reijneveld, 2007), a more mathematically sophisticated approach that determines how every voxel (also called a node in graph analysis) is correlated to all other voxels and how strong the connections are (connections are called edges). Graph analyses can be computed on structural or functional imaging data. Some nodes are hubs with many connections to other nodes. Networks in the brain tend to be "small-world" connections in that most clustering of connectivity is around adjacent brain areas or "neighborhoods." There are also connections among more distant regions in the brain through hubs that are connected to each other; these are the so-called rich clubs (van den Heuvel et al., 2012). Small-world networks tend to allow more efficient transmission of information across shorter distances with less wiring (white matter fibers) and rich clubs foster faster communication across more distant brain areas. These networks develop at different times and rates from infancy through early adulthood. The factors that influence how networks develop are not yet understood but they likely are related to individual differences in cognitive abilities. Graph analysis is illustrated in Textbox 4.1.

Van den Heuval and colleagues applied graph analysis to fMRI data collected during a resting state in a small sample of 19 adults (van den Heuvel et al., 2009). They calculated a measure of global efficient communication among multiple brain areas based on the overall length of pathway connections. This measure was inversely correlated with IQ scores. In other words, higher IQ scores were related to shorter pathways indicative of greater efficiency of information transmission within the entire brain. Path length of frontal-parietal connections had the strongest inverse correlations to IQ. Similarly, another group (Song et al., 2009) reported a graph analysis targeted specifically at the default network and how it differed between high and average IQ subgroups (N = 59). They too found that differences in the overall global efficiency of connections in the default network were related to IQ. The high-IQ group showed greater efficiency. A different research group reported a graph analysis of global efficiency using fMRI obtained from 120 participants (Cole et al., 2012). After a whole brain analysis, they reported that efficient connections involving only the left dorsal lateral prefrontal cortex with other frontal-parietal connections were correlated to intelligence test scores. Other researchers used graph analysis on resting-state EEG data in 74 participants and reported that efficient connections centered in the parietal lobe were most correlated to intelligence test scores (Langer et al., 2012).

Santarnecchi and colleagues also used graph analysis based on resting-state fMRI obtained in 207 individuals across a wide age range and found that IQ scores were related to connections distributed around the brain including PFIT areas (Santarnecchi et al., 2014). Both strong local connectivity and weak distant connectivity were found but their study added a new and surprising observation. IQ was most related to the strength of the weaker, long-distance connections than to the stronger, shorter connections. These researchers also reported quite a clever experiment using graph analysis and "damage" created mathematically. First, they constructed a measure of brain resilience based on functional connectivity related to IQ scores. Then they tested the impact of "damage" to specific areas or to random ones (Santarnecchi, Rossi, & Rossi, 2015). They concluded that higher intelligence was related to brain resilience to targeted damage and that the key areas were consistent with the PFIT. Supporting this general conclusion, recall from Chapter 2 that the Val/Met gene related to BDNF might play a role in preservation of IQ after TBI.

Textbox 4.1: Graph analysis

Graph analysis is a mathematical tool that is used to model brain connectivity and infer networks. The idea is to establish how each voxel in a brain image is correlated to all other voxels throughout the brain. These connections, called edges, can be computed for structural or functional images. A voxel, or a cluster of voxels, that shows correlations to many other voxels is called a hub. Hubs that show correlations to many other hubs are called rich clubs. The strength of any connection is determined by the magnitude of the correlation between voxels or hubs. The efficiency of any connection can be estimated by determining its length. Most of the brain has local connectivity in that many nearby voxels are connected to each other via a neighborhood hub. This makes for efficient information transfer. Rich clubs connect more distant brain areas and this makes for faster communication. This is illustrated in Figure 4.1 from van den Heuvel and Sporns (2011). Psychometric test scores

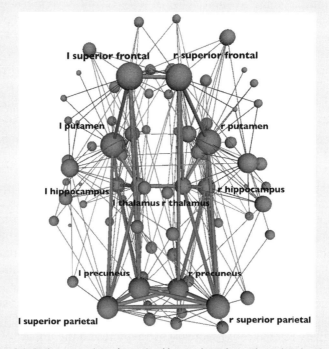

Figure 4.1 Brain connections determined by graph analysis. The red nodes show brain areas with many connections (larger nodes indicate more connections). Blue lines called edges indicate the strength of connections among areas (thicker lines indicate stronger connections; dark blue lines show rich club connections to other brain areas). (Adapted with permission from van den Heuvel and Sporns (2011) Note: A black and white version of this figure will appear in some formats. For the color version, please refer to the plate section.

> can be correlated to the strength of hubs and connections to indicate which
> brain networks are related to intelligence, as described in Section 4.1. Graph
> analysis has evolved with different ways to compute connectivity and con-
> nectivity analyses have defined the emerging field of network neuroscience
> (Hilger and Sporns, 2021).

Using the same sample of 207 and fMRI data, this same group has
reported a different type of connectivity analysis based on the functional
correlations between the same brain areas in the right and left hemispheres (Santarnecchi et al., 2015b). This is called homotopic connectivity and the results were counterintuitive. Higher IQ was correlated to
brain areas that show weaker interhemispheric homotopic connectivity,
suggesting that decreased interhemisphere communication is related
to higher intelligence. Several of the homotopic areas are included in
the PFIT but this study adds a new dimension of interhemisphere communication. Age and sex differences were also reported. For example,
higher-IQ females showed less homotopic connectivity in prefrontal
cortex and posterior midline regions. Younger participants (below age
25) with higher IQs also showed increased homotopic connectivity patterns. These age and sex analyses were done on smaller subsamples so
must be viewed with caution, but they illustrate the potential importance of using these variables as a matter of routine. Since the results are
based on resting-state data, one wonders whether functional homotopic
relationships with IQ might be even stronger if based on fMRI during a
cognitive task.

Wonder no longer. In a 2014 study of 79 participants, networks were
identified from both resting-state fMRI and fMRI during problem-
solving tasks from the RAPM test (Vakhtin et al., 2014). This was
the only imaging study of intelligence at the time that investigated
both resting-state and task-activation conditions in the same subjects.
Connectivity was determined with a statistical technique called
Independent Component Analysis prior to the homotopic analysis
reported subsequently (Santarnecchi et al., 2015b). Functional
connectivity during the problem solving overlapped with functional
connectivity during the resting state. The overlapping networks were
consistent with the PFIT.

The PFIT framework has also received strong support from other methods of voxel-wise analyses of brain connectivity (Shehzad et al., 2014)
and from both an evolutionary perspective (Vendetti & Bunge, 2014)
and a developmental perspective (Ferrer et al., 2009; Wendelken et al.,

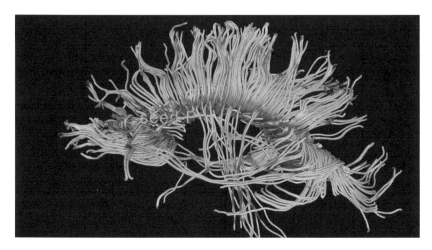

Figure 4.2 White matter fibers throughout the brain assessed by DTI. Seed refers to a spot selected for determining connections from that spot to other areas. (Courtesy Rex Jung) Note: A black and white version of this figure will appear in some formats. For the color version, please refer to the plate section.

2015); both perspectives emphasize the importance of parietal/frontal connectivity for reasoning ability. The PFIT hypothesis about subnetworks underlying different cognitive functions has received strong support from experiments using nonverbal reasoning tasks during standard fMRI (Hampshire et al., 2011). Overall, it is clear that results from numerous network analyses that use different methods converge substantially and support the existence of intelligence-related networks that are distributed across the brain. The findings are generally consistent with the PFIT framework, although that framework is subject to modification and elaboration, or even disproof as new data emerge.

Many analyses identify networks mathematically, irrespective of actual brain anatomy. White matter fibers are the tangible structural units of the brain that transmit information from one area to another. Figure 4.2 shows white matter fiber throughout the brain as determined by DTI. An early study reported that the thickness of the corpus callosum, the white matter fibers that connect the left and right hemispheres, was positively related to intelligence (Luders et al., 2007).

One group of researchers (Li et al., 2009) reported a graph analysis specifically of white matter connections to assess brain efficiency. In Chapter 3 we introduced DTI as a special variety of MRI that assessed the integrity of white matter. The Li group used DTI in 79 young adults.

Among other findings, global white matter efficiency was greater in the high-IQ subgroup. They concluded that "higher intelligence scores corresponded to a shorter characteristic path length and a higher global efficiency of the networks, indicating a more efficient parallel information transfer in the brain … Our findings suggest that the efficiency of brain structural organization may be an important biological basis for intelligence" (Li et al., 2009: 1).

Another research group assessed white matter with DTI in 420 older adults (Penke et al., 2012). They did not find any one white matter tract to be highly correlated to intelligence scores. However, they reported that 10 percent of the variance in intelligence test scores could be explained by a general factor of global white matter integrity computed from all tracts combined. This effect was due entirely to a factor of information-processing speed. Another group (Haasz et al., 2013) reported similar findings in middle-aged and older adults. Other researchers calculated white matter/intelligence correlations separately for males and females in a small sample of 40 young adults (Tang et al., 2010). The pattern of correlations with IQ differed between the sexes. Although their sample sizes were too small for generalization, from the perspective of individual differences and known sex differences in the brain (Luders et al., 2004; Luders et al., 2006), there is a strong argument for always computing separate analyses for males and females, especially when both groups are matched for intelligence.

These early studies had relatively small samples but, although results were often inconsistent, they collectively indicated links between intelligence and white matter brain connections. A detailed review can be found in (Genc & Fraenz, 2021). A large sample study of white matter and intelligence was reported in 2022 (Stammen et al., 2022). It was intended to address previous, inconsistent findings using advanced imaging and analysis methodologies in four independent samples totaling over 2000 healthy individuals. They only reported results that replicated in all four samples and the key finding was that specific white matter bundles were related to the g-factor, controlling for age and sex. They concluded with a detailed discussion of how their results fit with previous research, including the PFIT. Not surprisingly, some results were consistent and others pointed to the importance of subcortical areas that went beyond the PFIT. This study is a terrific example of how imaging research data are advancing theories and proposing new testable hypotheses. Hopefully, the use of multiple samples for replication *in the same report* will become more common (possibly required for publication) and contribute to fewer inconsistent findings.

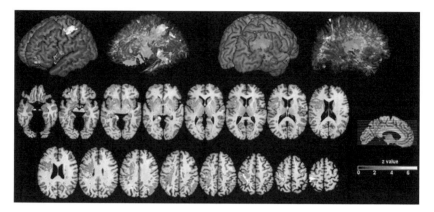

Figure 4.3 3D renderings show cortical and subcortical regions with a statistically significant relationship (red/yellow) between lesion location and the g-factor (top row). Bottom rows: Axial (horizontal) slices are shown for a more detailed inspection. (Reprinted with permission (Glascher et al., 2010: 4707, figure 2))
Note: A black and white version of this figure will appear in some formats. For the color version, please refer to the plate section.

Another kind of study that examines brain networks is based on patients with brain lesions and the pattern of cognitive deficits that result. Prior to the advent of neuroimaging, the study of brain lesion patients was a primary, if inexact, source of data for inferring brain–intelligence relationships. Neuroimaging advanced this approach by providing exact localization of lesions and mapping correlations between cognitive test score deficits and brain parameters. For example, Glascher and colleagues assessed primary factors of intelligence including g in a sample of 241 neurological patients with brain damage (Glascher et al., 2009; Glascher et al., 2010). The main finding was that damage in frontal and parietal areas was related to deficits in the g-factor and that other intelligence factors (verbal comprehension, perceptual organization, and working memory) showed deficits when damage occurred in different parts of the frontal-parietal network (See Figures 4.3 and 4.4). Aron Barbey and colleagues also reported lesion mapping studies with similar results (Barbey et al., 2012; Barbey et al., 2014; see also an excellent review of intelligence and brain networks in Barbey, 2021). Brain–intelligence relationships were also tested by other researchers with structural MRI, fMRI, and DTI in a small sample of people with Shwachman–Diamond syndrome, a rare genetic disorder characterized in part by various cognitive impairments (Perobelli et al., 2015). They found brain abnormalities consistent with the PFIT.

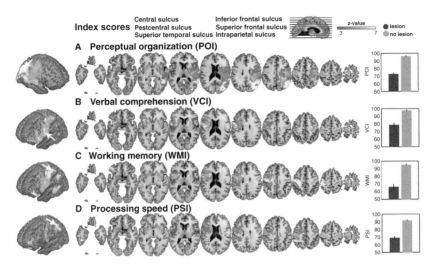

Figure 4.4 Effect of lesion location on four indices of mental ability. Row (A) is perceptual organization, (B) verbal comprehension, (C) working memory, and (D) processing speed. The red/yellow colors show where the location of lesions significantly interferes with scores on the indices. The graphs on the right show the mean difference on each index score between patients with and without lesions at the area of maximum effect (white arrow on the 3D projection). (Reprinted with permission (Glascher et al., 2009: 684, figure 2))
Note: A black and white version of this figure will appear in some formats. For the color version, please refer to the plate section.

Another comprehensive lesion study is particularly informative in providing evidence that g-scores (obtained by factor analysis of multiple cognitive tests) and working memory are uniquely related psychometrically and by common association with the same white matter tract – the arcuate fasciculus (Bowren et al., 2020) – highlighted in the PFIT. This study analyzed data from 402 people with chronic focal lesions and cross-validated the key findings in another sample of 101 acute stroke patients, where the g–working memory anatomical localization correlated 0.42 (p < 0.001) with actual g-scores. The authors concluded that "working memory is a key *mechanism* contributing to domain-general cognition" (Bowren et al., 2020: 8924, emphasis added). They further suggest "that we can build on our understanding of the mechanisms of domain-general cognition by reframing individual differences in g as being largely driven by individual differences in working memory" (Bowren et al., 2020: 8934). Working memory and g have been linked for some time (Kyllonen and Christal, 1990), but the reasons for this linkage are becoming more interesting with this kind of brain data.

As we can see, the PFIT framework has generated support from a number of studies and additional data like those just described are providing refinements to it. Earlier, one research group expanded neuroimaging beyond the cortex to subcortical areas (Burgaleta et al., 2013). They analyzed the shape of several subcortical structures based on MRI in 104 young adults who had completed a battery of cognitive tests. Fluid intelligence scores, highly correlated to the g-factor, were related to the morphology of the nucleus accumbens, caudate, and putamen, all in the right hemisphere only. These areas and the morphometry of the thalamus were also related to the factor of visual-spatial intelligence. Another study reported that the volumes in the basal ganglia were correlated to different intelligence factors and there were some sex differences (Rhein et al., 2014). Both these studies expanded the PFIT framework to subcortical areas and there is now supporting evidence from the newest white matter data discussed earlier (Bowren et al., 2020; Stammen et al., 2022).

In a comprehensive report, a German group of researchers led by Ulrike Basten completed a detailed meta-analysis of neuroimaging studies of intelligence through 2014 with the explicit purpose of testing the PFIT (Basten et al., 2015). In their final analysis, they only considered studies where individual differences in intelligence could be assessed directly; studies of average group comparisons were excluded. Jung and Haier had included both kinds of studies and their PFIT analysis was based on a qualitative assessment of areas common across studies. The German group compared structural and functional imaging results in an empirical voxel-by-voxel analysis (VBM, as described in Chapter 3) to identify common brain areas related to intelligence across 28 studies totaling over 1,000 participants. They concluded that the results generally supported the primary involvement of the parietal-frontal network. They also found evidence that suggested revising the PFIT to include areas of the posterior cingulate/precuneus, caudate, and midbrain. Their revised framework is shown in Figure 4.5. See also a recent excellent review in Basten and Fiebach (2021).

Whereas the PFIT was a good start and generated many tests of its hypotheses, more advanced models are necessary so that more specific predictions can be tested (Hilger et al., 2022). Frameworks such as the original and revised PFIT have conceptual problems related to a reliance on correlations that are fundamentally not interpretable regarding cause and effect between brain measures and cognitive measures (Kievit et al., 2011). One promising possibility for addressing this limitation that might advance the study of "neuro-g" may be the use of analyses based on multiple indicators and multiple causes (Kievit et al., 2012). These

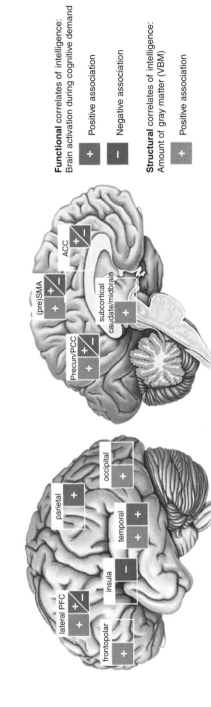

Figure 4.5 Brain areas related to intelligence from a 2015 review are shown on lateral (left) and medial (right) surfaces of the brain. ACC = anterior cingulate cortex; PCC = posterior cingulate cortex; PFC = prefrontal cortex; (pre) SMA = (pre-)supplementary motor area; VBM = voxel-based morphometry. (Reprinted with permission (Basten et al., 2015))

Note: A black and white version of this figure will appear in some formats. For the color version, please refer to the plate section.

advanced statistical approaches, which are too complex to detail here, can generate more specific hypotheses about brain variables and they are particularly important for the large data sets. They offer the potential for clarifying the weight of evidence regarding how brain physiology specifically relates to cognitive measures, especially for identifying the different brain variables relevant for individual differences in cognitive test performance (see Section 4.3).

There are also PFIT findings related to brain development. Prior to the PFIT and the advent of network analyses, early imaging studies had groups of children and young adults at different ages (cross-sectional design) and other studies imaged the same individuals over time as they aged (longitudinal design), a particularly informative research design; see review by (Kievit and Simpson-Kent, 2021). Cross-sectional studies inferred patterns of brain development and their relationship to intelligence scores from groups of different ages (Shaw et al., 2006). A series of studies of children and adolescents in a large national representative sample reported associations between CT and intelligence (Burgaleta et al., 2014; Estrada et al., 2019; Karama et al., 2009b; Karama et al., 2014; Román et al., 2018). This series now includes an MRI-based white/gray matter contrast as an indirect measure of myelination which provides additional information beyond CT (Drakulich et al., 2022). This measure was associated mostly with performance IQ scores in areas across the brain, including ones in the PFIT. Another study reported that efficient structural brain networks related to the PFIT are related to perceptual reasoning and to one high g-loaded measure in a sample of 99 children aged 6–11 years old (Kim et al., 2016). More about efficiently later in this chapter.

More recently, Ruben Gur and colleagues specifically tested the PFIT in a sample of 1,601 young people aged 8–22 who were tested with MRI measures and cognitive assessments (Gur et al., 2021). This is a fascinating report with a set of complex findings that generally support the PFIT but extends the model to incorporate "cortical, striatal, limbic, and cerebellar regions and networks … that support motivation and affect." They also concluded that "[a]ssociations of brain parameters became stronger with advancing age group from childhood to adolescence to young adulthood, effects occurring earlier in females. This Extended PFIT network is developmentally fine-tuned, optimizing abundance and integrity of neural tissue while maintaining a low resting energy state" (Gur et al., 2021: 1444). This is another good example of expanding theory from new empirical observations.

Let's digress a bit and talk about sex and the PFIT (not yet a movie or book title as far as I know). The Gur et al. (2021) study was not the first to suggest sex differences for some brain–intelligence relationships, although the early studies that did so had small samples and rudimentary imaging (Haier & Benbow, 1995; Haier et al., 2005; Jung et al., 2005). More sophisticated intelligence studies also indicate that sex matters for brain–intelligence relationships (Dreszer et al., 2020; Ryman et al., 2016). Here are two more studies using the connectivity analyses we discuss later in this chapter, but I mention them now because they report interesting sex differences.

In the first study, a group from China used fMRI connectivity patterns to predict IQ scores and replicated the predictive equations in two other samples (Jiang et al., 2020). A number of brain areas were predictive across the samples, including PFIT and other areas which also suggest expanding the PFIT. But this study also reported an intriguing sex difference: The correlation between the predicted IQ scores and the actual scores in the discovery sample for males was 0.63 and for females 0.77. These are quite strong correlations but, not surprisingly, they dropped appreciably in the two replication samples (to 0.25 and 0.23 for males; to 0.29 and 0.40 for females). Note the prediction was stronger for females. But interestingly, the predictions for males and females were based on connectivity patterns involving different brain areas and these were consistent among the three samples. This is a stronger empirical observation than similar ones from the early underpowered studies (Haier & Benbow, 1995; Haier et al., 2005; Jung et al., 2005).

In the second study, another group used functional connectivity measures based on resting-state fMRI to specifically test the PFIT and its relationship to a high-g measure of matrix reasoning (Fraenz et al., 2021). They studied 1,489 individuals including both a discovery and a replication sample. Brain connectivity patterns were associated with test scores in both samples for PFIT areas BA 8, 10, 22, 39, 46, 47 (left hemisphere) and 44 and 45 (right hemisphere). These findings were present in the total sample but when analyzed separately by sex, only the findings in females were significant. Thus, the connectivity of the original PFIT areas assessed during a resting state were associated with a good measure of fluid intelligence, but only for one sex. Sex remains a mystery.

At this point, you may be finding it difficult and confusing keeping in mind all the different brain areas and networks related to intelligence. I know the feeling. Here is something to help. It would be nice to have a table showing what each area does. There used to be such tables but as more data became available, it became clear that any one area or network is typically involved in more than one function. This had been observed for the g-factor in early neuropsychology studies of lesion patients (Basso et al., 1973) and more recently formalized as multiple demand theory (Duncan, 2010). So how could it be helpful if there is no simple correspondence between one brain area or network and one cognitive function? Doesn't that make a complex situation worse? It is also the case that the way brain areas are defined is not exact and boundaries can be quite different from one brain to the next. Remember our first law: Nothing about the brain is simple. In my view, there is no need for you to memorize all the things any brain area or network does. Just think about the fact that we are at the point where we can identify a set of brain areas and networks (including subnetworks) related to something as complex as intelligence. As you can see from reading this far, there are many intriguing findings that inspire even more questions. We are identifying the individual instruments and sections in the orchestra. Learning how they work together to create the symphony of intelligence is a challenge that requires even better technology and methods of analysis. We discuss more progress in the next sections and in Chapters 5 and 6.

Before we get to more of the recent findings based on newer methods, let's recap the early brain network findings. Two main hypotheses proposed from the first phase of neuroimaging studies from 1988 to 2007, discussed in Chapter 3, were that intelligence was related to brain efficiency and involved multiple areas distributed throughout the brain, especially in a parietal-frontal network. The second phase of neuroimaging studies summarized so far in this chapter has applied more sophisticated image acquisition and analysis in much larger samples to test these ideas. It is fair to summarize these early studies as building a weight of evidence that provides support for the parietal-frontal distribution hypothesis (albeit with some modifications) and some tentative support for the efficiency hypothesis based on measures of brain connectivity.

So, what do the latest studies show about the PFIT? Connectivity analyses of intelligence were just being introduced as the first edition of this book was going to production (Finn et al., 2015; Smith et al., 2015). The Finn paper was an intriguing first report from the Human Connectome Project related to intelligence. It was based on resting-state fMRIs from 461 participants. Functional connectivity computations among 200 brain areas incorporated 158 demographic and psychometric variables in a single analysis. A g-factor was not derived, but the main result showed that intelligence variables were among the strongest related to overall connectivity among brain areas such that greater connectivity was associated with higher test scores.

I found this study breathtaking not only for the findings but also because it fulfilled a dream I had for 40 years about using brain profiles to describe individuals and their mental abilities. Here's what they reported based on analyses of connectivity patterns among brain areas (like those shown in Figure 4.1). They started with fMRI data from 126 people collected during six sessions, including four task and two resting conditions. The typical analysis would have compared the average connectivity for the entire group among the task and rest conditions. These researchers, however, focused on individual differences. The simple question was whether connectivity patterns were stable within a person. To address this question, functional connectivity patterns among 268 brain nodes (making up 10 networks) were calculated for each person separately for each session. Not only was the connectivity pattern stable within a person when the two resting conditions were compared, it was also stable across the four different tasks. In addition, each person's pattern was unique enough that it could be used to identify the person. Because these remarkable results combined stability and uniqueness, the connectivity pattern was characterized as a brain fingerprint (I would have called them brainprints).

Of particular interest to us, individual brain fingerprints predicted individual differences in fluid intelligence. It gets even better. The strongest correlations with fluid intelligence were in parietal-frontal networks. And, best of all, cross-validation was included in the report. The authors note that "[t]hese results underscore the potential to discover fMRI-based connectivity 'neuromarkers' of present or future behavior that may eventually be used to personalize educational and clinical practices and improve outcomes" (Finn et al., 2015: 6). They conclude:

Together, these findings suggest that analysis of individual fMRI data is possible and indeed desirable. Given this foundation, human neuroimaging studies have an opportunity to move beyond population-level inferences, in which general networks are derived from the whole sample, to inferences about single subjects, examining how individuals' networks are functionally organized in unique ways and relating this functional organization to behavioral phenotypes in both health and disease. (Finn et al., 2015: 7)

I believe this is a landmark study. I wish I had done it. As we will now see, it ushered in a trove of new studies that further explicate the brain networks related to intelligence.

There are now many studies of brain connectivity and intelligence (e.g., Santarnecchi, Emmendorfer, & Pascual-Leone 2017a; Santarnecchi et al., 2017). We already discussed two other studies of connectivity related to sex differences (Fraenz et al., 2021; Gur et al., 2021). They help define the emerging category of network neuroscience (Barbey, 2018, 2021; Girn

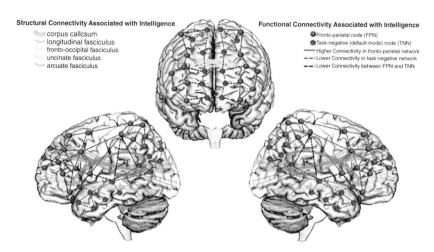

Figure 4.6 The brain bases of intelligence – from a network neuroscience perspective. (Reprinted with permission (Hilger and Sporns, 2021))
Note: A black and white version of this figure will appear in some formats. For the color version, please refer to the plate section.

et al., 2019). In my view, these studies define a third phase of imaging/intelligence research. As far as the PFIT, there is additional evidence that differences in the connectivity of functional networks, including the frontal-parietal, are related to WAIS IQ scores (Hilger et al., 2017b). This study was based on resting-state fMRI in 309 individuals. It grouped individual brain areas (nodes) into modules within networks for a more detailed analysis of localized components (modules are groups of connected nodes) within global networks. Building on these findings, Hilger and colleagues looked at the temporal dynamics of functional changes over time in brain modules (Hilger et al., 2020). They found less variability over time (more stability) in individuals with higher IQ scores, and this was most noteworthy in a key attention network (also see Shi et al., 2022). As we see, neuroimaging-based connectivity studies continue to identify more and more aspects of brain networks associated with intelligence. A compelling summary illustration shows connectivity networks and intelligence findings in Figure 4.6.

In the next sections, we discuss how connectivity analyses have advanced intelligence research aimed specifically at the brain efficiency hypothesis and at predicting intelligence from neuroimages. As we discuss in Chapter 5, such findings could become the basis for experimental studies that probe and manipulate networks with the not-so-small goal of enhancing intelligence.

4.2 Functional Brain Efficiency: Does Imaging Show that Less Is More?

Following the observation of inverse correlations between intelligence test scores and GMR in the cortex (Haier et al., 1988), we formulated the hypothesis that high intelligence was related to efficient brain activity. In that report, the concept of efficiency was general and included possible characteristics of brain networks, neurons (especially mitochondria), and/ or synaptic events. We also speculated that the decreased cortical activation we observed following task practice might result from the brain learning which areas not to use while task-relevant areas worked harder (Haier et al., 1992a). From this rather inexact beginning, it is not surprising that demonstrating a relationship between brain efficiency and intelligence has produced inconsistent results over the years. A subsequent review of the research literature concluded that brain efficiency was moderated primarily by type of task and by sex (Neubauer & Fink, 2009). Up to that time, most brain efficiency studies were based on EEG methods (see also (Dreszer et al., 2020). The graph analyses summarized in Section 4.1 provided indirect evidence that structural and functional brain network efficiency was related to intelligence, but the story became more complex as more variables were identified that apparently influenced efficiency (see Dreszer et al., 2020; Ryman et al., 2016; and a review by Euler & Schubert, 2021).

Two early small sample fMRI studies investigated brain efficiency by comparing cortical activations between high and average IQ participants (Graham et al., 2010; Perfetti et al., 2009). These studies are noteworthy for selecting participants for differences in intelligence; high- and low-IQ groups are compared. Most cognitive imaging studies avoid using intelligence as an independent variable because there is a general assumption that all human brains basically work the same way, so comparing groups with different IQs would not be meaningful. The validity of this assumption, however, is quite doubtful. When intelligence is considered in the research design of imaging studies, differences are apparent. Both these studies reported generally consistent results. One study concluded that:

[W]hen complexity increased, high-IQ subjects showed a signal enhancement in some frontal and parietal regions, whereas low-IQ subjects revealed a decreased activity in the same areas. Moreover, a direct comparison between the groups' activation patterns revealed a greater neural activity in the low-IQ sample when conducting moderate task, with a strong involvement of medial and lateral frontal regions thus suggesting that the recruitment of executive functioning might be different between the groups. (Perfetti et al., 2009: 497)

Similarly, the other study concluded that:

Whether greater intelligence is associated with more or less brain activity (the 'neural efficiency' debate) depends therefore on the specific component of the task being examined as well as the brain region recruited. One implication is that caution must be exercised when drawing conclusions from differences in activation between groups of individuals in whom IQ may differ. (Graham et al., 2010: 641)

Unfortunately, cognitive studies like these that use intelligence as an independent variable are still exceptions (one additional example is discussed in Section 4.4).

Two other early studies tested the efficiency hypothesis directly with fMRI data. The first one studied 40 teenagers (20 males and 20 females) and incorporated sex, task difficulty, and intelligence in their research design (Lipp et al., 2012). These 40 were selected from a pool of 900 so that the male and female samples were matched on intelligence scores (general intelligence and visual-spatial scores) and, to avoid restriction-of-range problems, each sample included a broad range of scores. During fMRI, each participant solved a set of spatial rotation problems along with control problems. The visual-spatial task activated mainly frontal and parietal areas but, contrary to the efficiency hypothesis, there was no basic finding of inverse correlations between intelligence and brain activation during the task. Activation in the posterior cingulate and precuneus was related to intelligence. These are two of the default network areas proposed as additions to the PFIT (Basten et al., 2015). The authors interpreted this as a possible indication that deactivation in areas of the default network might indicate greater task demands for less intelligent participants. They also found that in the females, higher intelligence was related to greater activity in task-related areas for more difficult problems. In short, as predicted by our first law, the results reinforced the complexity of the efficiency concept.

In the second study of efficiency, Basten and colleagues obtained fMRI in 52 participants while they performed a working memory task of increasing difficulty (Basten et al., 2013). They made an important distinction between two kinds of brain areas: task positive, where activation increased during task performance; and task negative, where activation decreased during performance. They correlated intelligence test scores to activation in both kinds of areas separately. In the networks formed by task positive activations, higher intelligence was related to less efficiency. In the task negative networks, higher intelligence was related to greater efficiency. These opposing findings, similar in male and female subsamples, suggest that whole brain analyses of the efficiency hypothesis may be more confusing than regional analyses.

Despite the initial appeal of the simple efficiency hypothesis regarding individual differences in intelligence, subsequent research continues to underscore a complex set of issues. On the one hand, efficiency remains a popular concept for thinking about neural circuit activity and how it relates to complex cognition (Bassett et al., 2015). On the other hand, the concept has been characterized as so vague as to be useless, although it still has potential explanatory power if better defined and measured (Poldrack, 2015).

As mentioned, since the first edition of this book, new connectivity research now illustrates advances in the potential explanatory power that brain efficiency studies can provide. For example, Kirsten Hilger and colleagues have published several well-designed neuroimaging studies of intelligence; we discussed two of them in Section 4.1 regarding network modules. Another one focused on brain efficiency and characterized functional connectivity among key global networks (i.e., whole networks linking anatomically disparate brain areas) and among separate nodes within the networks (i.e., anatomically close areas) based on resting-state fMRIs for 54 individuals (Hilger et al., 2017a). They found that measures of global network efficiency were not correlated to WAIS IQ scores but that there was evidence that node efficiency measures were correlated to IQ in three brain regions. These regions were associated with specific aspects of information processing that could help explain efficient communication among brain areas. After a detailed discussion, the authors concluded that:

> [O]ur analyses imply that with respect to network topology, brain regions that were previously related to salience processing … and the filtering of irrelevant information from further processing … play a crucial role in explaining individual differences in intelligence. We speculate that the observed differences in network integration of these three regions may enable intelligent people to more quickly detect, evaluate, and mark salient new stimuli for further processing and to protect ongoing cognitive processing from interference of irrelevant information, ultimately contributing to higher cognitive performance and high intelligence. (Hilger et al., 2017a: 20)

This kind of research nicely ties intelligence to cognitive neuroscience (Barbey et al., 2021; Euler & McKinney, 2021; Hilger & Sporns, 2021).

But you might note that this 2017 study was based only on resting-state fMRI with a relatively small sample. What would task-dependent fMRI show? Another study compared fMRI during rest and seven different tasks to identify multiple network dynamics that are related to intelligence test scores (Thiele et al., 2022). Findings are reported for a sample of 134 individuals along with a replication in another sample of 184. The central element in this study was assessing how functional networks change from task to task, where change is called reconfiguration. The general finding was that reconfiguration across multiple networks was inversely correlated to a measure of g extracted by the factor analysis of 12 cognitive measures. That is, the individuals with higher g-scores showed less reconfiguration from task to task, suggesting a more general neural reflection of the positive manifold (i.e., all mental ability tests are positively correlated with each other) and a more efficient organization of brain networks. They also argued that the findings revealed "insights into human intelligence as an emergent property from a distributed multitask brain network."

As sophisticated as these connectivity studies are, there are even more direct ways to measure information flow among brain areas. As noted previously, electrophysiological techniques such as EEG and EP methods measure electrical signal changes in the brain millisecond by millisecond, making these techniques invaluable for charting cognitive processes that often last less than a second or two. Despite some technical limitations, they have a long history in intelligence research and the potential for further explicating relevant aspects of information processing (see the comprehensive review in Euler and Schubert, 2021). For example, several recent electrophysiological studies have been reported by Anna-Lena Schubert and colleagues, including ones that find relatively strong associations between intelligence measures and speed of cognitive processing (Frischkorn et al., 2019; Schubert & Frischkorn, 2020; Schubert et al., 2017; Schubert et al., 2019; Schubert et al., 2020). However, a clever experiment using nicotine to increase speed of information processing suggested this may not be a causal relationship. The authors concluded that "structural properties of the brain may affect both the speed of information processing and general intelligence and may thus give rise to the well-established association between mental speed and mental abilities" (Schubert et al., 2018: 66).

Another approach to measuring efficiency uses the noninvasive neuroimaging technique based on the MEG. MEG detects minute magnetic fluctuations created as groups of neurons fire on and off. The spatial resolution of this technique is about a millimeter but, like EEG methods, the time resolution of a millisecond makes this especially appealing for studying information flow in the brain. Magnetic signals also have less distortion than EEG signals as they pass through the skull, an advantage for detecting the spatial localization of activity. When MEG is acquired while a person solves a cognitive problem, millisecond-by-millisecond fluctuations related to neurons firing can be detected and tracked through the entire brain. There are a number of issues surrounding the interpretation of such fluctuations, but they can potentially provide insight into how individual brains process information during problem solving.

One group of researchers, for example, used MEG during a choice reaction time task to assess the timing and sequence of brain activations that might be related to intelligence (Thoma et al., 2006). The task was chosen because choice reaction time is correlated to intelligence (choice reaction time tasks require making a decision about which response is correct; simple reaction time tasks just require a response to a stimulus). Fast reaction times in choice reaction time tasks, reflecting faster information-processing speed, are associated with higher intelligence test scores in many studies; reaction time in simple tasks is not (Jensen, 1998b; Jensen, 2006; Vernon, 1983). The MEG results from 21 young adult males suggested that activation sequences involving a posterior visual processing area and a sensory motor area were related to scores on the RAPM test of abstract reasoning (described in Chapter 1). This was a pioneering use of MEG to study intelligence but it did not have the advantage of a large sample or a model to test so the complex MEG results are necessarily tentative. Another MEG study of 20 university students investigated efficient information flow during a verbal memory task. The results suggested that "an efficient brain organization in the domain of verbal working memory might be related to a lower resting-state functional connectivity across large-scale brain networks possibly involving right prefrontal and left perisylvian areas" (Del Río et al., 2012: 160). The PFIT was not tested directly but the results are an encouraging example of the potential for MEG analyses for detecting the sequence and timing of information processing.

MEG is a tricky, expensive technology and not many MEG machines have been available to researchers. This contrasts with the MRI methods that are now widely available. Many psychology departments have one or more MRI machines under their control along with legions of graduate students familiar with sophisticated image analysis software, developed by mathematical experts specifically for cognitive studies. MEG is still very much in development as a research tool. For example, one research group used MEG and fMRI in the same sample to study optimal methods for revealing network connectivity (Plis et al., 2011) and other groups have used data for graph analysis of connectivity (Maldjian et al., 2014; Pineda-Pardo et al., 2014), but intelligence was not a variable in any of these studies. Another research group studied reading difficulties and found MEG activations in three areas correlated to IQ scores but the sequential timing among areas was not reported (Simos et al., 2014).

After relatively slow adaption for intelligence studies, MEG has finally been used to compare functional brain networks (defined by the frequencies of fast-oscillatory assessments) between individuals with high and average fluid intelligence scores (Bruzzone et al., 2022). Structural connectivity also was assessed with DTI in the 66 young adult participants (both men and women; average age 25 years). They found that the DTI white matter structural connectivity in the high-fluid individuals was stronger than in the average group across brain areas that included some in the PFIT, and could be interpreted as providing more efficient information processing, similar to the findings of Li et al. (2009). The MEG functional data only address the brain efficiency hypothesis indirectly and the complicated results are open to interpretation and not to a simple summary. You, the reader, are not the only one who finds that the more sophisticated brain imaging becomes, the more the results can become less clear, especially in the context of complex methods requiring technical knowledge (the same is true for genetic studies). Please repeat Haier's law #1: No story about the brain is simple. Nonetheless, this study is another important advance for identifying network dynamics that differ according to the level of fluid intelligence.

I'm looking forward to more MEG research to test hypotheses about network dynamics in a more straightforward way since MEG can better localize brain areas below the cortex compared to EEG methods. For example, the PFIT hypothesizes that intelligence is related to a specific sequence of activation across specific areas during problem solving. Generally, the sequence starts in posterior sensory processing areas, travels forward to parietal and temporal association areas where information is integrated, and then moves on to frontal lobe areas for hypothesis testing and decision-making. How often this sequence might be repeated while solving a particular problem could be a key variable related to individual differences in intelligence. The exact areas involved in the sequence could also differ among individuals and so could the timing of the sequence. MEG provides a means to assess the actual sequence and compare it to what the PFIT predicts. For example, individuals with high intelligence test scores might engage a different set of PFIT areas than individuals with average intelligence test scores. Perhaps fewer areas would define a sequence in the high-score group, consistent with efficiency. Or individuals may engage the same set of PFIT areas in the same sequence irrespective of intelligence, with high intelligence related to a faster speed of engaging or repeating the sequence of areas.

Finally, there is emerging evidence associating efficiency of neurons to intelligence test scores. One study used an advanced MRI technique called neurite orientation dispersion and density imaging (Genc et al., 2018). This technique allows quantitative inferences about neuron structure. Using multiple samples, they found that *less* density and *less* arborization of dendrites (the branching ends of neurons) are related to higher intelligence scores. The authors interpret the inverse correlations as suggesting that "the neuronal circuitry associated with higher intelligence is organized in a sparse and efficient manner, fostering more directed information processing and less cortical activity during reasoning ... these results offer a neuroanatomical explanation underlying the neural efficiency hypothesis of intelligence" (Genc et al., 2018: 1).

However, another research group found a *positive* correlation between intelligence and the complexity of dendrites in temporal lobe pyramidal neurons (Goriounova et al., 2018: 1):

[L]arger dendritic trees enable pyramidal neurons to track activity of synaptic inputs with higher temporal precision, due to fast action potential kinetics. Indeed, we find that human pyramidal neurons of individuals with higher IQ scores sustain fast action potential kinetics during repeated firing. These findings provide ... evidence that human intelligence is associated with neuronal complexity, action potential kinetics and efficient information transfer from inputs to output within cortical neurons.

This finding was based on the temporal lobe (the only area studied) and at first glance seems to be in the opposite direction of what Genc et al. found throughout the cerebral cortex. In a follow-up study, the authors provided a more detailed microstructural analysis of cortical layers in the temporal lobe. They showed that a thicker cortex in subjects with higher intelligence did not contain more neurons, but rather similar numbers of larger cells at lower neuronal densities (Heyer et al., 2021). Thus, lower neurite density in the Genc et al study could manifest at a cellular level as lower density of neurons with larger dendrites. There is some evidence that supports this from brain samples taken during surgery (Douw et al., 2021).

Yes, I know this is confusing. The weight of evidence about neuronal efficiency and intelligence is still accumulating; remember law #3: It takes time to resolve inconsistent data. But in the big picture, taking intelligence research to the cellular level is a remarkable advance. Here is another aspect of this progress. It is possible, for instance, to associate GWAS data to types of brain cells and tissue-specific transcriptome (RNA molecules) data from postmortem human brains (Ardlie et al., 2015). This methodology has the potential to associate specific gene expression to proteins that work in cell function and brain development. Even newer methods are becoming available for mapping synapses and neurotransmitter activity along with identifying pathways of gene expression and understanding gene influences on brain development (Bhaduri et al., 2021; Hansen et al., 2021; Makowski et al., 2022; Mountjoy et al., 2021). Much more can be expected. These approaches are taking intelligence research ever deeper into the brain (Goriounova & Mansvelder, 2019) and suggest possible causal links between neurobiology and intelligence. Linking the molecular mechanisms at each level of explanation to each other will be the great challenge for the next decade or more. To meet this challenge, new research programs focused on the molecular genetics and biology of intelligence will rival molecular research programs for learning and memory, which, as we know, are aspects of intelligence. More glimpses of the future await in Chapter 6.

4.3 Predicting IQ from Brain Images

Imagine if colleges and universities gave applicants for admission a choice between submitting either standardized test scores or a brain image. As discussed in Chapter 1, SAT scores are a good estimate of general intelligence and that is an important reason why they are good predictors of academic success, although they and other standardized tests are being eliminated in the USA for college admission for a variety of reasons (Wai and Bailey, 2021; Wai et al., 2019). Can a better estimate of intelligence or predictor of academic success be extracted from a brain image? Can a brain image avoid the criticisms used against standardized tests? The first is an empirical question, and a positive answer is probably far less scary than you might think. In fact, brain images are likely to be more objective, especially structural images, and not sensitive to a host of factors that can potentially influence psychometric test scores, such as motivation or anxiety, although these factors may be far less important than believed (Bates & Gignac, 2022). Whether you are a good test taker or not is irrelevant for getting a scan. Brain images are generally less expensive than SAT preparation courses or formal IQ testing and getting a brain image is far less time-consuming. There is no preparation, you spend about 20–30 minutes in the scanner, and you can have a nap during structural image acquisition. Still not interested in this possibility?

Whether or not there are any practical applications for predicting IQ from brain images, the ability to do so would signal a more advanced understanding of brain–intelligence relationships than we currently have. In fact, predicting IQ from neuroscience measures such as those obtained from neuroimaging is one of two major goals of intelligence research. The other one is the ability to manipulate brain variables to enhance IQ, which is tackled in Chapter 5.

There is a long history of trying to predict IQ from brain measurements that dates back to the early EEG research of the 1950s and 1960s. At least one patent to do so was issued in 1974 (US 3,809,069). In 2004, a group from the University of New Mexico, including my colleague Rex Jung, obtained a patent (US 6,708,053 B1) to measure IQ based on neurochemical signatures in the brain assessed by MRI spectroscopy. This claim was derived from their research correlating IQ to N-aspartate in a single brain area (see Chapter 3) (Jung et al., 1999a; Jung et al., 1999b). In 2006, a group from South Korea filed a patent application to measure IQ from a combination of structural and functional MRI assessments and that patent eventually was issued in 2012 (US 8,301,223 B2). Their patent is supported by previous research, including our MRI work (Haier et al., 2004), and on research from the South Korean group that reports correlations between predicted IQ scores and actual IQ scores for different samples (Choi et al., 2008; Yang et al., 2013). Let me note that no commercial potential for these patents is apparent to me at this time. In my view none of these patents represent an immediate threat to the publishers of the SAT or the WAIS IQ test because their validity has yet to be established in any large-scale, independent replication trials. I am doubtful that such studies will be positive for these specific methods. This is because predicting an individual's IQ based on group average data is quite difficult. Nonetheless, I am optimistic that strong neuroimaging-based IQ prediction is possible, and we have already reviewed examples based on connectivity analyses. As we shall see shortly, there is now good evidence to support my optimism for even stronger predictions. My skepticism and my optimism are both derived from my view about the importance of individual differences. Let me explain.

Conceptually, predicting IQ or any intelligence factor from neuroimaging is straightforward. Success depends on how strongly brain variables, individually or in combination, correlate to intelligence test scores. Recall from Chapter 1 that IQ scores are good estimates of the g-factor because they are a combination of scores, age and sex corrected, on several subtests that tap different cognitive domains. Presumably, different cognitive domains require different brain networks, so various brain measures for different domains might be combined to predict IQ. As we noted in Chapter 3, whole brain size has a modest correlation with IQ. It is not strong enough for brain size alone to substitute for IQ but the correlation is certainly a base to build on.

There are a number of statistical approaches for combining measures to make a prediction. The most basic one applied for predicting IQ scores is the multiple regression equation. This method, and related versions of it, use the correlation between IQ and each measure after removing common relationships between measures. For example, if variable A correlates to IQ and so do variable B and variable C, they cannot be simply combined without first statistically removing the common variance between A and B, A and C, and B and C. The remaining correlations to IQ for each variable are called partial correlations. Regression equations combine partial correlations between each variable and IQ along with computing weights for each variable that maximize the IQ prediction. In the ABC example, A might be weighted more than B and B might be weighted more than C in order to make the strongest IQ prediction. The resulting equation can then be applied to a new person's data and an IQ score predicted. The correlation between a predicted IQ score and the actual IQ score for a large group of individuals must be nearly perfect for the equation to be acceptable as a substitute for an actual IQ score. It is not sufficient if an equation produces a statistically significant correlation between the predicted IQ and the actual IQ. Whenever a regression equation is calculated in a research sample, the exact same equation must be applied to an independent sample so the predicted and actual IQ score correlation can be replicated. This is cross-validation of the equation and this step is required because the original equation may produce a spuriously high correlation by incorporating chance effects, especially in small samples. In our early research, we tried the regression approach on several occasions but each regression equation failed on cross-validation so neither publication or patent was attempted.

So far, to the best of my knowledge, none of the patented methods of predicting IQ from brain measurements has achieved this crucial step. One recent paper attempted to predict IQ scores from structural MRIs collected from different sites (Wang et al., 2015). They reported good correlations with two different regression models that incorporated gray and white matter, but there was no cross-validation in independent samples, participants (N = 164) ranged in age from 6 to 15 years old, sex effects were not investigated, and no clear description of IQ testing was detailed. They identified 15 brain areas that were included in the prediction but there was no attempt to integrate these areas to the PFIT or any other intelligence framework, and the areas are not generally found in other imaging studies of intelligence. At best, without independent replication the findings are tenuous. The regression models are of interest but it is too early to judge whether the analyses will have predictive validity, as the authors hope. Here is their final sentence: "It should be emphasized again that our work paves a new way for research on predicting an infant's future IQ score by using neuroimaging data, which can be a potential indicator for parents to prepare their child's education if needed" (Wang et al., 2015: 15). It's an optimistic view of a potential commercial market, but significantly more caution here would have been wise.

An important contribution to this literature comes from the continuing longitudinal study of children from Scotland described in Chapter 1. These researchers collected structural MRIs on 672 individuals with an average age of 73 years and representing the full range of intelligence (Ritchie et al., 2015). They used structural equation modeling, a form of regression equation, to compare four models for combining several different MRI assessments to determine which structural brain features were most related to individual differences in intelligence based on a g-factor extracted from a battery of cognitive tests. They found that the best model accounted for about 20 percent of the variance in the g-factor. Total brain volume was the single measure that contributed most of the predicted variance in this model. White matter along with cortical and subcortical thickness contributed some additional variance but iron deposits and micro-bleeds did not. The main issue for future study was whether additional measurements of other brain variables such as corpus callosum thickness or functional variables might add predictive variance beyond 20 percent. This project had a large sample and multiple cognitive measures so it is a solid study of older men and women. How the results might differ for children or younger adults is an open question for replication and cross-validation studies.

Why would straightforward prediction approaches such as the ones described fail to cross validate? Correlations between any two variables are based on individual differences for each variable. That is, there must be variance among individuals for correlations to exist. Regression equations generally work on group data where there is variance on all the variables. In the case of intelligence, there may be many combinations of the same set of variables that predict any specific IQ equally well. For example, one set of brain variables might characterize a person with an IQ score of 130 but another person with the identical IQ score of 130 might be characterized by a different set of brain variables. In a group of

100 people all with IQs of 130, how many different sets of brain variables related to intelligence might there be? Compounding the problem, two individuals both with WAIS IQs of 130 may have very different subtest scores indicating different cognitive strengths and weaknesses despite the same overall IQ (Johnson et al., 2008). The same problem may exist independently at several IQ levels so the brain variables that predict high IQ might be different from the ones that predict average or low IQ, even though the relevant genes may be the same across the entire IQ range, as discussed in Chapter 2. Age and sex also could be important factors for identifying optimal sets of variables for predicting IQ.

There is another major source of difficulty. No two brains are the same structurally or functionally, even in identical twins. Most brain image analysis starts with morphing each brain into a standard size and shape, called a template. This step artificially reduces individual differences in brain anatomy by creating an "average" brain. To account for the imprecision of "average" brain anatomy, analyses typically add a step of smoothing in an effort to minimize the imprecision. Nonetheless, forcing all brains into a standard space introduces error into efforts to predict IQ from images. Some template methods create more error than others. Consider a neuroimaging study of male–female differences. Should the males be standardized to a male template and the females to a female template, or should everyone be standardized to the same template? Many neuroimaging studies use a standard template supplied by the analysis software and other studies create a template from only the participants in the study. No one way is always correct but the issue presents a problem for efforts to predict IQ. A study of 100 postmortem brains highlights the issue (Witelson et al., 2006). Their strongest finding was that 36 percent of differences in verbal ability scores was predicted by cerebral volume. However, age, sex, and handedness influenced regression analyses between other anatomical features and different cognitive abilities in complex ways. The authors cautioned neuroimaging researchers to take these factors into account.

When all these issues are considered, the prediction of IQ using regression methods becomes less straightforward. How many different regression equations may be necessary? This is why I have been skeptical of this approach. One alternative approach may involve the use of profile analysis. This is common in personality testing where a profile of scores on different personality scales is used to characterize an individual. Profiles are used extensively to interpret personality tests such as the Minnesota Multiphasic Personality Inventory (MMPI), for example. MMPI scores from the different subscales can be used in regression analyses, but the

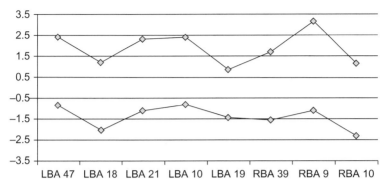

Figure 4.7 Brain profiles of two individuals both with an IQ of 132. The graphs show the amount of gray matter in eight PFIT areas identified by numbered BAs (L = left; R = right). The y-axis is based on standardized gray matter scores so positive numbers show values greater than the group mean; negative numbers show values less than the group mean. Although the profile shapes are similar for these two individuals, one has substantially more gray matter in the eight areas than the other.
(Courtesy Richard Haier)

analysis of individual profiles allows comparisons of groups of people defined by similar profiles across the subscales to determine variables related to the profile type. We illustrated this approach for predicting intelligence by creating profiles of individuals based on the amount of gray matter in several PFIT areas and tried to relate the profile to IQ score (Haier, 2009b). As shown in Figure 4.7, this demonstration did not work so well. The two individuals shown with equally high IQ scores had different gray matter profiles. It may be a promising approach for future study in large samples. In fact, an encouraging report used patterns of fMRI activation (including in some PFIT areas) to predict profiles of cognitive performance while a small sample of 26 participants solved deductive reasoning problems (Reverberi et al., 2012). The analyses show the complexity of the problem but the results illustrate my optimism that individual differences can be a solution to the complexity and not just a nuisance. This is an evolving view (Finn and Rosenberg, 2021).

There is another potential application for predicting IQ. As we discussed in Chapter 1, the g-factor definition of intelligence is sufficient for many empirical research questions, but what if we could define intelligence based on quantifiable brain measures instead of psychometric scores? If brain parameters can predict IQ, then why not define IQ in terms of brain parameters? We do not know if twice the gray matter in a particular part of the cortex, for example, makes one twice as smart. We are now able to explore redefining intelligence in ways that incorporate neuro-metric assessments. In Chapter 5, we explore this idea further when we discuss future research possibilities.

So, after all these considerations, can intelligence be predicted from neuroimaging? For the first edition of this book, my short answer was no. The longer answer was, not yet. The weight of evidence was nascent. But, as mentioned, just as I was finishing the final draft of the first edition, the brain fingerprint study was published (Finn et al., 2015). After reading this study I was ready to change the answer about predicting intelligence from neuroimaging from "not yet" to "looking good." In fact, as I write this second edition, there are now so many studies that predict intelligence from brain connectivity patterns that it is both a joy and a challenge to decide which ones to include.

I already described two connectivity studies that showed sex differences (Fraenz et al., 2021; Jiang et al., 2020). The one from Jiang and colleagues specifically predicted IQ scores; the most variance accounted for after cross-validation in an independent sample was about 12 percent in females and only about 5 percent in males. Here are some other noteworthy studies to illustrate the increasing strength of prediction, which, in my view, looks like inexorable progress.

First is a study based on resting-state fMRI obtained from 884 participants in the Young Adult Human Connectome Project (Dubois et al., 2018). Although technical, this study is well worth reading for its sophisticated research design and image analysis, and a scholarly/insightful discussion. The researchers derived a *g*-factor from ten cognitive tests and correlated it to connectivity patterns correcting for age and sex. Overall, the cross-validated result was an impressive correlation of about 0.46 accounting for almost 20 percent of the variance. Although several brain areas included in the prediction were consistent with the PFIT, no particular brain structure or network was more predictive than others. There were no separate analyses by sex but the use of resting-state connectivity was further refined in a subsequent methodological paper to further explicate relevant networks (Hebling Vieira et al., 2021). In their 2018 discussion, Dubois and colleagues made an interesting speculation: "It would be intriguing to find that humans share with other species a core set of genetically specified constraints on intelligence, but that humans are unique in the extent to which education and learning can modify intelligence through the incorporation of additional variability in brain function" (Dubois et al., 2018: 10). Note that this would be an example of why environmental influences work through biological mechanisms.

A second illustrative study, also using Connectome data, reported an even stronger result based on fMRI activation patterns in individual areas (not connectivity among areas) during seven diverse tasks rather than during a resting state (Sripada et al., 2020). Generally, the more complex the task, the stronger the prediction of a general cognitive ability score with the highest cross-validated correlation of 0.50 indicating 25 percent of variance was predicted. Just using parietal-frontal and default networks increased the correlation to 0.66 (Sripada et al., 2020: figure 6). The authors concluded: "These results suggest a picture analogous to treadmill testing for cardiac function: Placing the brain in a more cognitively demanding task state significantly improves brain-based prediction of General Cognitive Ability" (Sripada et al., 2020: 3186). Another methodological paper using the same databases and measures improved predicted variance to about 40 percent when the analyses were focused on more fine-grained assessments of individual differences in cortical patterns (Feilong et al., 2021). Still better prediction of up to 70 percent was claimed by combining networks to predict the factor structure of 12 subtests from a test of intelligence (Soreq et al., 2021). Too good to be true? The details of these predictions go beyond our interest here except to say that, as we have encountered before, the more sophisticated the analyses, the harder it is to give a meaningful summary. But, stepping back and seeing these reports published since the first edition of this book, it is apparent to me that a credible weight of evidence is emerging that individual differences in intelligence, especially *g*, can be predicted by individual differences in brain structure and function as assessed by neuroimaging. More detailed reviews of these efforts can be found in (Basten & Fiebach, 2021; Cohen & D'Esposito, 2021; Drakulich & Karama, 2021; Hilger et al., 2022; Martinez & Colom, 2021; Vieira et al., 2022). *Here is a key point*: Pending additional independent replications, these brain/intelligence predictions are among the strongest for any variables in psychology. We are a long way from uninformed assertions that intelligence cannot be studied scientifically.

4.4 Are "Intelligence" and "Reasoning" Synonyms?

This may seem an odd question, but there is an anomaly in the research literature that deserves some consideration at this point. There is a specialization within the field of cognitive psychology that studies reasoning. Relational reasoning, inductive reasoning, deductive reasoning, analogical reasoning, and other kinds of reasoning are subjects in a variety of studies, including ones that use neuroimaging to identify brain characteristics and networks related to reasoning. The anomaly is that more than a few of these cognitive neuroscience studies of reasoning do not use the word intelligence and they often fail to cite relevant neuroimaging studies of intelligence. This is problematic because tests of reasoning are highly correlated to the *g*-factor (Jensen, 1998b). In fact, analogy tests have some of the highest *g*-loadings of any mental ability tests. Obviously, this means that findings from intelligence studies are quite relevant to reasoning research and vice versa.

In my view, the artificial preference of "reasoning" over "intelligence" made by some researchers has its origins in a long-held view within cognitive psychology that the word "intelligence" is too loaded with controversy and therefore must be avoided completely. It is not unusual to find that books in the field of cognitive psychology and cognitive neuroscience do not include "intelligence" in the index. Language counts. No one is fooled by substituting "reasoning" for "intelligence," although some granting agencies may think so.

Generally, neuroimaging studies of reasoning show network results consistent with intelligence studies, although reasoning studies tend to differentiate more components of information processing and accompanying subnetworks. This is an important difference and a positive one for identifying the salient brain components for different cognitive processes involved in intelligence factors. An excellent example is a sophisticated fMRI study that compared groups of high school students (N = 40) defined by high and average fluid intelligence scores while they performed problems of different difficulty that required geometric analogical reasoning (Preusse et al., 2011). Hypotheses were based in part on the PFIT and on brain efficiency. The authors concluded that:

[The high-IQ students] display stronger task-related recruitment of parietal brain regions on the one hand and a negative brain activation–intelligence relationship in frontal brain regions on the other hand ... We showed that the relationship between brain activation and fluid intelligence is not mono-directional, but rather, frontal and parietal brain regions are differentially modulated by fluid intelligence when participants carry out the geometric analogical reasoning task. (Preusse et al., 2011: 12)

The integration of reasoning and intelligence findings in this work demonstrates the richness of interpretation possibilities and helps advance the field.

Two other interesting and well-done fMRI papers investigated analogical reasoning although neither one mentioned intelligence. They appeared in a special section on "the neural substrate of analogical reasoning and metaphor comprehension" in the *Journal of Experimental Psychology: Learning, Memory, and Cognition* (only one of the other six papers in this section mentioned intelligence). The first example used an analogy-generation task in a sample of 23 male college students and found corresponding brain activity in a region of the left frontal-polar cortex, as hypothesized (Green et al., 2012). Exploratory analyses revealed more distributed activations (Green et al., 2012: figure 3), which is seemingly consistent with the PFIT framework, but the discussion linked the findings to creativity not intelligence. The left frontal-polar cortex had also been linked to *g* in the lesion study we described earlier (Glascher et al., 2010). Similarly, the second example reported a systematic investigation of analogical mapping during metaphor comprehension in 24 Carnegie Mellon University undergraduates (males and females combined) (Prat et al., 2012). The findings showed activations consistent with the PFIT and brain efficiency, although neither PFIT or intelligence–efficiency findings from other studies were referenced. These two studies are solid contributions to the reasoning research literature but they were mostly overlooked in the intelligence literature.

At minimum, in my view, reasoning research reports should include "intelligence" as a key word for indexing and the relationship of reasoning tests to intelligence should be acknowledged in the discussion of results that show brain–reasoning relationships. Intelligence reports should do the same for reasoning. Moreover, there is a growing recognition that the results of cognitive/imaging experiments might change dramatically depending on whether the participants are selected for high or average IQ scores (Graham et al., 2010; Perfetti et al., 2009; Preusse et al., 2011). New imaging technologies such as MEG may allow even more detailed analysis of information flow during reasoning/problem solving, especially if levels of intelligence are included in the research design of studies. More collaboration between reasoning researchers, with cognitive expertise, and intelligence researchers, with psychometric expertise, is the best way to integrate these two rich empirical traditions.

4.5 Common Genes for Brain Structure and Intelligence

In Chapter 2 we discussed quantitative and molecular genetic findings related to intelligence, but we deferred the presentation of genetic studies of intelligence that included neuroimaging. Now that neuroimaging has been introduced as it is used in studies of intelligence, let us consider the powerful combination of genetic and neuroimaging methods to study intelligence.

As the hunt for specific genes continues, the newest and most compelling quantitative genetic twin studies of intelligence go beyond the simple question of whether there is a genetic component or not. They focus on what the genetic component may do in the brain, even without knowing any specific genes. Paul Thompson and colleagues reported the first twin study that used MRI to assess and map the heritability of cortical gray matter volume and relate it to intelligence (Thompson et al., 2001). They studied a small sample of 10 MZ pairs and 10 DZ pairs. The estimated genetic contribution for gray matter volume based on similarity between twin pairs varied across brain regions, which was a somewhat surprising finding at the time. The highest genetic contributions were in cortical areas of the frontal and parietal lobes. Moreover, the correlation between IQ scores and gray matter in the frontal lobes was statistically significant. Based on the unique combination of IQ testing and neuro-imaging in twins, this study provided unique evidence of what had been suspected by many researchers: Individual differences in intelligence were due, at least in part, to the genetics of brain structure, specifically cortical gray matter volume. Despite the small sample, the importance of this finding was underscored by its publication in the prominent journal *Nature Neuroscience*. An accompanying commentary noted that the high heritability implied that gray matter development was apparently less sensitive to experience than might be expected (Plomin & Kosslyn, 2001).

Researchers in the Netherlands have published a compelling series of findings that draw on larger samples of twins. We introduced some of their findings in Chapter 2 concerning shared and nonshared environmental influences on intelligence. Here we summarize more of their important MRI findings that indicate there are common genes for intelligence and brain structure. In 2002, they published findings about gray and white matter heritability and intelligence, also in *Nature Neuroscience* (Posthuma et al., 2002). Heritability estimates were high for both and whole brain white matter was slightly more heritable than whole brain gray matter. Moreover, by comparing MZ and DZ twins, the authors found that the correlation between gray matter volume and general intelligence was due entirely to genetic factors. No variance was attributable to shared or nonshared environmental factors. Later, they expanded their sample of twins to increase statistical power and examine correlations between gray, white, and cerebellar volumes and different intelligence factors (Posthuma et al., 2003). All three volumes were correlated with working memory capacity and related to a common genetic basis. Processing speed was genetically related to white matter volume. Perceptual organization was related both genetically and environmentally to cerebellar volume. Verbal comprehension was not related to any of the three volumes. This group also showed that gray and white matter in specific brain areas had a common genetic basis with IQ (Hulshoff-Pol et al., 2006).

Similar results from the Netherlands were found in 112 9-year-old twin pairs (van Leeuwen et al., 2008), indicating early genetic influences on intelligence in the maturing brain. A longitudinal MRI study of adult twins examined changes in CT over a five-year period and found the degree of cortical change (also called plasticity) had a strong genetic basis (Brans et al., 2010). Change was related to IQ. Higher IQ scores were related to cortical thinning over time in the frontal lobes and to thickening in the parahippocampus, an important brain structure in the temporal lobes involved with memory. The actual cortical changes were a fraction of a millimeter but even small amounts of brain tissue can be important. In Chapter 3, we noted that an earlier study reported cortical thinning during early childhood was related to higher IQ scores (Shaw et al., 2006). This adult twin study of IQ and brain plasticity in frontal lobes and the parahippocampus concluded that both variables might have some common genetic basis. One novel finding concerned the subcortical parahippocampus, which was unusual because most studies have focused on the cortex. Another MRI study drawing on the Netherlands twin data examined whether the volume of several subcortical areas was related to IQ. Only the volume of the thalamus, an important hub of brain circuit connectivity, was related to IQ and a common genetic component was implicated for both (Bohlken et al., 2014). Although CT has been associated with IQ in several studies, there is also an indication that cortical surface area may show even stronger associations with cognitive abilities and related genes based on a study of 515 middle-aged twins that compared both thickness and surface area measures (Vuoksimaa et al., 2015). This field is quite dynamic as new approaches to data analysis evolve and extend previous findings with increased accuracy. They contribute to the weight of evidence regarding genes and brain structure with mostly consistent findings for intelligence.

White matter integrity is a particular focus of intelligence research given its heritability and the DTI results we have noted here and in Chapter 3. The Thompson team used DTI in a sample of 92 Australian twins (23 MZ and 23 DZ pairs) to quantify a measure of white matter fiber integrity called fractional anisotropy (FA) (see Chapter 3). They mapped the heritability of FA throughout the cortex and found the highest values in frontal and parietal lobes (bilaterally), and the left hemisphere occipital lobe (Chiang et al., 2009). IQ scores (FSIQ, performance

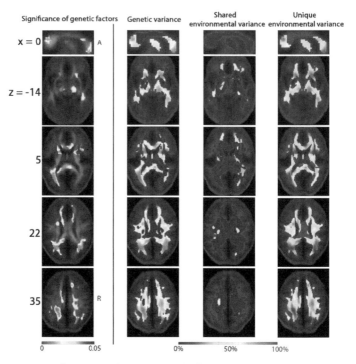

Figure 4.8 Maps of genetic and environmental influences on white matter integrity (measured by FA). Each row shows a different axial brain view (horizontal slice). Red/ yellow shows strongest results. Left column shows significance of genetic influences. Other columns show the strength of the FA measure for genetic, shared, and nonshared environment, respectively. (Adapted with permission (Chiang et al., 2009: figure 4)) Note: A black and white version of this figure will appear in some formats. For the color version, please refer to the plate section.

IQ, verbal IQ) were correlated to specific fiber tracts (higher integrity was associated with higher IQ) and these correlations were also mapped. Based on cross-trait mapping, they concluded that common genetic factors mediated the correlations between IQ and FA, suggesting a common physiological mechanism for both. When the data are displayed as maps, the results are compelling. Figure 4.8 shows the distribution of FA variance for genetic, shared, and nonshared environment. Figure 4.9 shows the cross-trait mapping for FSIQ. In 2011, the Chiang group expanded these findings. Based on a larger sample of 705 twins and their nontwin siblings, they examined the effects of age, sex, SES, and IQ on the hereditability of the FA measure (Chiang et al., 2011b). There were complex interactions for various brain regions, but in general genetic influence was

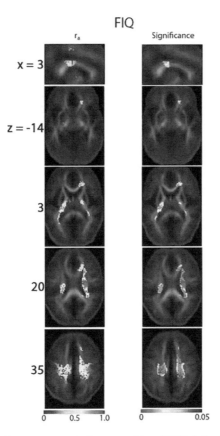

Figure 4.9 Overlap of common genetic factors on FA and FSIQ (left column) based on cross-trait analysis of areas shown in Figure 4.8. Right column shows statistical significance. Each row shows a different axial (horizontal slice) brain slice. (Adapted with permission (Chiang et al., 2009: figure 7))
Note: A black and white version of this figure will appear in some formats. For the color version, please refer to the plate section.

greater in adolescents compared to adults, greater in males than females, greater in those with high SES, and greater in those with higher IQ.

Schmithorst and colleagues, as noted in Chapter 3, had reported age and sex differences in the earliest DTI studies of intelligence. A comprehensive DTI study of 1,070 children aged 6 to 10 years old from the Netherlands supported the Schmithorst findings and further reported FA correlations with nonverbal intelligence and with visual-spatial ability (Muetzel et al., 2015). A three-year longitudinal study of adolescent twins and their siblings used DTI and graph analysis to map the

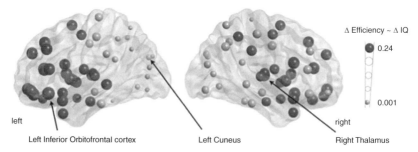

Figure 4.10 Correlations between three-year change in IQ score and change in local brain efficiency measured with FA. The largest purple spheres show the strongest IQ/efficiency change correlations. (Adapted with permission (Koenis et al., 2015))
Note: A black and white version of this figure will appear in some formats. For the color version, please refer to the plate section.

heritability of white matter fiber integrity (Koenis et al., 2015). The efficiency of white matter networks, assessed with FA, was highly heritable with genetic influences accounting for as much as 74 percent of variance. Moreover, there was a provocative finding related to intelligence. For the subgroup of individuals who showed a change in IQ scores over the three-year period, individual increases in scores were correlated to increases in local network efficiency in frontal and temporal lobe areas. These findings are shown in Figure 4.10. The authors speculate that finding ways to promote efficiency of white matter networks may optimize teenage cognitive performance. Most parents of teenagers will try anything. So will most teenagers.

There are now so many studies in this area that it is easy to be confused. Here's the short, unsurprising story. Genes influence both the brain and intelligence. What does influence mean? It can mean many things. For example, one study looked at polygenic score prediction of intelligence and found that it was mediated by CT and brain surface area (Lett et al., 2019). This kind of influence is interesting but until specific genes and their expression are identified, we cannot distinguish directly whether genes influence brain morphometry, which then influences intelligence, or whether genes influence intelligence, which then influences brain morphometry. It is also possible that many genes influence both brain morphometry and intelligence (pleiotropy) and only some of them are common to both. Informative examples and discussions of issues about predicting intelligence with imaging and genetic data can be found in Lee and Willoughby (2021) and Anderson and Holmes (2021).

As you see so far in this chapter, the quantitative analysis of neuro-imaging has become quite sophisticated with complex multivariate statistical methods. The quantitative analysis of genetic data is also quite complex. Teams of researchers that include mathematicians in addition to imaging experts and genetic experts now carry out this research. Intelligence is increasingly a focus of interest and experts in intelligence research are becoming part of such teams. Summarizing results in this chapter from the combination of the neuroimaging and the quantitative genetic domains without oversimplifying is a challenge, but at minimum the progress and excitement in the field should be clear to all readers. We are light years past earlier controversies about whether there is a role for genetics for understanding individual differences in intelligence. The challenge in Section 4.6 is explaining the complexity of neuroimaging analysis combined with the even greater complexity of molecular genetic analysis. The details may be difficult and, to the untrained, the gene nomenclature of letter–number combinations seems random or irrational. But here's the main point. The results are exciting and speak to the optimism that the hunt for genetic and brain mechanisms that effect intelligence is gaining ground as complexities are slowly disentangled.

4.6 Brain Imaging and Molecular Genetics

There are now many studies that combine neuroimaging and genetic analyses so we must choose which ones best illustrate progress. Let's continue with more in the sequence of papers by Chiang and colleagues. In a DTI analysis of 455 twins and their nontwin siblings, they found an association between white matter integrity and the Val/Met polymorphism related to BDNF, the brain growth factor involved in normal neuron function. They suggested that BDNF might be related to some intellectual performance indirectly by modulating white matter development in some fiber tracts (Chiang et al., 2011a). In another paper, this group pursued the idea that the genetics of white matter brain wiring is fundamental to intelligence (Chiang et al., 2012). They focused their imaging efforts in a novel way to help identify specific genes related to brain connectivity and intelligence. As we noted earlier in this chapter, in 2009 they had reported that there were genes in common for the integrity of white mater tracts and intelligence based on cross-trait maps of the respective heritability of each. They expanded this approach using DTI and DNA data from a sample of 472 twins and their nontwin siblings. The basic idea was to use statistical methods of clustering many variables based on their similarity. First, thousands of points within white matter fibers were clustered to find brain systems with common genetic determination. The white matter measure was FA, as described previously. Then they used DNA in a genome-wide scan and network analyses to identify a network of genes that was related to white matter integrity in major tracts. FA in some hubs in the white matter network was related to IQ scores (Chiang et al., 2012: figure 9).

The results of this analysis are complex and include 14 specific genes listed (Chiang et al., 2012: table 5), along with what was known about each one's function in 2012. This kind of study illustrates both the complexity of understanding how genes function and a major neuroscience direction for future intelligence research. There is a long road between observations like these and making any practical use of them. But if replicated, identifying genes related to intelligence and how they function can point to potential mechanisms for enhancing intellectual performance if the cascade of genetic influences on functional molecular events can be manipulated at the right stage of brain development. This includes manipulating more general genetic effects on brain structures such as white matter integrity that may influence intelligence indirectly (Kohannim et al., 2012a; Kohannim et al., 2012b). We discuss enhancement more in Chapter 5.

In the rapid evolution of molecular genetic studies, the next major advance was the aggregation of huge samples in worldwide multicenter collaborations formed to investigate brain diseases and normal cognition (Khundrakpam et al., 2021). We noted some of these in Section 2.6. These groups, and the resulting publications, are a triumph of logistical, political, and scientific achievement. One of the largest is named ENIGMA (Enhancing Neuro Imaging Genetics through Meta-Analysis). One of their papers included a finding that related individual differences in intelligence to a specific variant in a gene called HMGA2 that is related to brain size (Stein et al., 2012). Their discovery and replication samples included thousands of participants who had completed neuroimaging in addition to cognitive and DNA testing. As in other studies, the finding explained only a tiny fraction of variance in intelligence, but it illustrates a definitive victory for finding a gene needle in a haystack of DNA.

Another group used part of the ENIGMA data set and data from the European Union IMAGEN consortium to investigate gene associations with CT assessed with MRI and intelligence in 1,583 14-year-old adolescents (Desrivieres et al., 2015). Measures of CT have some technical advantages over VBM measures of gray matter volume and, as mentioned, CT has been associated with intelligence in large representative samples of children and adolescents (Karama et al., 2009a; Karama et al., 2011). This IMAGEN report started by examining the relationship between CT and nearly 55,000 SNPs. One variant (rs7171755) was associated with thinner cortex in the left frontal and temporal lobes, and in a small subsample with WAIS scores. They further found that this variant affected expression of the NPTN gene, implicated in glycoprotein encoding required for synaptic health. This finding illustrates how synaptic events may influence brain maturation of CT, which in turn may influence intelligence. The cascade of intervening steps is complex but it is also a finite problem to solve. The methodology and complexity of analyses that led to these findings require an advanced technical background. But even this summary, along with the studies summarized in Section 2.6, illustrates how sophisticated the hunt for specific genes has become.

All these studies demonstrate that identifying the many genes with small effects thought to be involved in intelligence is not an insurmountable challenge. Once there is more progress, epigenetic effects can be investigated for individual genes, but as of yet there are not enough data to test specific epigenetic hypotheses. I repeat: At minimum, these very large samples showing DNA associations with measures of intelligence should put to rest any doubts about a genetic role for intelligence. Recall our three laws once again: No story about the brain is simple; no one study is definitive; it takes many years to sort out conflicting and inconsistent findings and establish a weight of evidence. The studies summarized in this section illustrate the rapid pace of progress for understanding the complexities surrounding the genetic aspects of intelligence.

We opened Chapter 3 by referring to our 1988 PET report with a question: "Where in the brain is intelligence?" Over 30 years later, neuroimaging is providing informative data on this question, and the question itself has become more refined. Genetic studies in combination with neuroimaging are beginning to suggest specific brain mechanisms involved in individual differences in intelligence. Neuroscience research aimed at understanding intelligence has a firm basis and is progressing at a fast pace based on evolving neuroimaging and genetic technologies and methods. There is now a context for thinking about how brain parameters can be used to predict or even define intelligence. There is also a developing empirical context for thinking about how brain mechanisms might be manipulated to enhance intelligence, and these are the subjects of Chapter 5.

Chapter 4 Summary

- New neuroimaging methods, especially graph analysis, have revealed structural and functional brain networks related to intelligence test scores.
- Overall, the brain networks identified in many intelligence studies are consistent with the PFIT framework with possible modifications for consideration.
- Many studies find correlations between brain measures and IQ scores and predicting intelligence for individuals from brain images now seems possible.
- In general, neuroimaging studies of reasoning report results that are consistent with studies of intelligence, although many studies of reasoning avoid discussing any overlap. More collaboration between reasoning researchers, with cognitive expertise, and intelligence researchers, with psychometric expertise, is the best way to integrate these two rich empirical traditions.
- The combination of quantitative genetics and neuroimaging indicates that individual differences in brain measures and intelligence have genes in common.
- The combination of molecular genetics and neuroimaging might identify specific genes and related brain mechanisms that influence individual differences in intelligence.

Review Questions

1. What is the strongest evidence that intelligence depends on multiple areas distributed throughout the brain?
2. What are different ways that brain efficiency can be measured?
3. What brain measure shows the strongest correlation to IQ scores and why is this not sufficient for predicting IQ from brain images?
4. Why are studies of analogical reasoning related to studies of intelligence?
5. Describe how the combination of quantitative genetics and neuroimaging has advanced intelligence research.
6. What are the implications of combining molecular genetics and neuroimaging for understanding brain mechanisms related to intelligence?

Further Reading

What Does a Smart Brain Look Like? (Haier, 2009b). Written for a lay audience, this is an overview of the PFIT of intelligence and what imaging studies may mean for education.

"Where Smart Brains Are Different: A Quantitative Meta-analysis of Functional and Structural Brain Imaging Studies on Intelligence" (Basten et al., 2015). And a newer review (Basten and Fiebach, 2021). Both are excellent reviews of neuroimaging studies of intelligence for advanced readers.

The Cambridge Handbook of Intelligence and Cognitive Neuroscience (Barbey et al., 2021). This edited volume has 23 chapters written by experts on topics that include imaging and other aspects of cognitive research. Mostly for advanced readers.

"The Biological Basis of Intelligence: Benchmark Findings" (Hilger et al., 2022). An excellent overview of this field.

"Genetic Variation, Brain, and Intelligence Differences" (Deary et al., 2022). Another great review paper.

CHAPTER FIVE

The Holy Grail

Can Neuroscience Boost Intelligence?

We who have worked on this project at Beekman University have the satisfaction of knowing we have taken one of nature's mistakes and by our new techniques created a superior human being.

Professor Nemur's fictional character in *Flowers for Algernon* delivers this remark during his presentation at a psychology conference describing how he increased the IQ of a mentally retarded man to super genius level. Daniel Keyes (1966)

I know Kung Fu.

Keanu Reeves as Neo in the movie The Matrix (1999), several seconds after a learning program on fighting is uploaded directly into his brain

A tablet a day and I was limitless… I wasn't high. I wasn't wired. Just clear. I knew what I needed to do and how to do it… I read the Elegant Universe by Brian Greene in 45 minutes, and I understood all of it!

Two characters in the movie *Limitless* (2011) after taking an IQ pill

Learning Objectives

- What are important features for any study claiming to show an increase intelligence that make the results credible?
- Why are people drawn to studies that claim relatively quick and easy ways to increase intelligence?
- What is the concept of transfer and why it matters for studies that claim intelligence can be increased?
- What methods can be used to stimulate the brain for experimentally testing possible causes of increased intelligence?
- Are there ethical issues about enhancing intelligence with drugs or brain stimulation that differ from issues about changing the environment to increase intelligence?

Introduction

Why would the then governor of Georgia ask the state legislature to buy a classical music CD for every newborn infant in the state? Why learn to memorize longer and longer strings of random numbers? Why do school systems with limited funding purchase expensive computer games for children to play during class? What exactly are the best 5 or 7 or 10 tips for increasing your IQ by 17 to 40 points?

This chapter is about sense and nonsense regarding the possibility of increasing intelligence. The good news is that neuroscience may some-day offer the possibility of increasing intelligence based on an under-standing of the brain mechanisms involved, including mechanisms that can be influenced by a variety of means. The bad news is that claims that we already know how to do this are naïve, wrong, or misrepresentations. These claims are not just on the internet or in books written by authors with no particular scientific expertise. Some are found in research reports published after peer review in highly respected scientific journals. How could this happen?

Higher intelligence is better than lower intelligence. Few people seri-ously disagree, although this is debated endlessly in some circles even though, if given a choice, no one chooses to have lower intelligence. But this assertion is often misunderstood to mean smarter people are better people. Look around – they are not. Based on my assertion, it is my view that all intelligence research speaks to the goal of enhancement, either directly or indirectly. This is a worthy goal; just ask the parents of a child with low IQ or a cognitive disability. It is also a primary goal of most parents for their children whether articulated so bluntly or not. There may be some people who do not care to be smarter, but I do not know any of them. And, as we discuss in Chapter 6, based on the normal dis-tribution, 16 percent of the population have IQ scores under 85 – that's about 53,000,000 people in the United States alone. Is this not a reason to think about enhancement as a means to make everyday life a bit easier?

Achieving the goal of increasing intelligence will require an understand-ing of what intelligence factors are, how best to measure them, how they develop, how they relate to specific brain mechanisms, and how those mechanisms may be malleable. There is a long history of trying to increase intelligence. I cannot document it, but I suspect this was a subject of inter-est to the alchemists, ancient builders, and even earlier mystics. So far as modern scientific efforts go, there is no appreciable success when success is defined by independent replication of empirical research results that last over time based on sophisticated assessments of intelligence in well-designed studies. This is my view based on the weight of evidence regard-ing increasing g, but not everyone agrees with my assessment. There are some examples of small, temporary increases that likely are non-g aspects of intelligence and these will be noted. But, for me, increasing g is key.

In Chapter 1, we noted the critical measurement problem that IQ scores are not a ratio scale, making change scores before and after an intervention nearly impossible to interpret. To repeat, IQ points are not measures like inches or pounds. This key problem, at the heart of claims about increasing intelligence, is all but ignored in the studies we review in this chapter. In Chapter 2 we reviewed the failures of earnest compensatory and early childhood education programs to boost IQ. These programs have other positive results but the weight of evidence does not support any claims concerning increased intelligence as assessed with IQ or other psychometric tests. One hypothesis is that this failure may be due in large part to the genetic influences on intelligence demonstrated in the many studies we discussed in Chapters 2 and 4.

Nonetheless, apparently undaunted by past failures and inherent measurement problems, or ignorant of them, there are newer reports in the scientific literature that claim to raise IQ scores dramatically in children and adults. We examine three of these specific claims under the implied heading: "Don't let this happen to you." These claims are based on the use of classical music, memory training, and computer games to raise IQ. By showing how such claims should be evaluated skeptically, I hope to inoculate you against future declarations of alleged breakthroughs or landmark results. Following these cautionary case studies, we then examine equally dubious claims about drugs that increase IQ. After that we move on to the exciting possibility of increasing intelligence by neuroscience means that test the boundary between science and science fiction.

5.1 Case 1: Mozart and the Brain

Mozart died in 1791, but 202 years later Mozart became the focus of a craze. But it wasn't a musical one. It started with a brief letter published in the respected scientific journal *Nature*. The letter claimed that listening to a particular Mozart sonata for 10 minutes temporarily increased IQ by eight points (Rauscher et al., 1993). Eight points is about half a standard deviation, which is quite a large effect for a mere 10-minute intervention. Intelligence researchers recognized immediately that this sounded too good to be true. It was not true but, amazingly, it took six years to dampen popular enthusiasm after a critical review (Chabris, 1999). Another 11 years were needed to finally put this claim to rest with the publication of a comprehensive review article titled, "Mozart Effect – Shmozart Effect: A Meta-analysis" (Pietschnig et al., 2010). The title speaks for itself. Over the 17 years this popular myth endured, an uncountable number of Mozart and other classical music CDs were

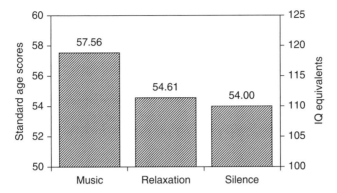

Figure 5.1 The bar graphs that launched the Mozart effect. Spatial intelligence test scores and IQ equivalents (*y*-axis) are shown after listening for 10 minutes to Mozart, a relaxation tape, or silence. A different test and different participants were used for each condition. (Reprinted with permission (Rauscher et al., 1993))

purchased with the expectation of increasing IQ just by listening. School music programs gained new support and music lessons gained a new rationale. An uncountable number of high school science fair projects investigated various aspects of the Mozart effect, tested mostly on friends and family. In fairness, these were hardly terrible consequences of a wrong idea. Possibly with the exception of a few accordion lessons, no one was harmed but no one's IQ increased either.

The original 1993 report was based on 36 college students who were tested on measures of abstract reasoning after a 10-minute exposure to three different conditions. The reasoning measures were three different spatial reasoning tests from the Stanford–Binet intelligence test battery. The three experimental conditions were listening to Mozart's sonata for two pianos in D major, listening to a relaxation tape, and listening to silence (I know you cannot listen to silence but the sentence requires parallel construction). After each condition, one of the three tests was given. For the Mozart condition, the standard test score was 57.56. This was statistically higher than 54.61 for the relaxation condition and 54.00 for the silence condition. These standard test scores were "translated" to spatial IQ scores of 119, 111, and 110, respectively. These findings are shown in Figure 5.1 from the original report. The authors stated: "Thus, the IQs of subjects participating in the music condition were 8–9 points above their IQ scores in the other conditions" (Rauscher et al., 1993: 611). They also noted that the enhancing effect only lasted for the 10–15 minutes of testing. They encouraged additional research on the duration of the time between listening and testing, on variations in the time of listening, on effects on other measures of intelligence and memory, on other compositions and styles of music, and on possible differences between musicians and nonmusicians. And so, the Mozart effect for increasing IQ was born along with countless science fair projects.

Whoever the peer reviewers were for *Nature*, they were apparently unaware that treating the three different reasoning tests as equal measures of abstract reasoning, based on the fact they were correlated to each other, was a terrible psychometric procedure. They also failed to require information about the participants with respect to IQ or musical experience and ability. And most distressing, the translation of individual test scores to spatial IQ scores and the claim of an eight-point increase was psychometrically naïve, as we discussed in Chapter 1. A case could be made that this report of a single experiment in a small sample with an extraordinary finding was not based on extraordinary evidence and therefore was not ready for publication, especially in *Nature*.

Although the authors focused on spatial IQ, the resulting media coverage was not so specific and the Mozart effect was widely understood to enhance IQ in general. In addition to the media coverage, some intelligence researchers at the time seized on any findings that suggested IQ was highly malleable as evidence against a strong role for genetic influences on intelligence. The *Nature* report also spoke to a desire for an easy way to acquire greater intelligence at a relatively small cost and no risk. According to an article in the *New York Times* (January 15, 1998), Governor Zell Miller of Georgia proposed spending $105,000 a year from the state budget to purchase classical music tapes or CDs (compact disks for you younger readers) for the approximately 100,000 children born in Georgia each year (iPods and streaming were not yet a reality). During his budget address to lawmakers, the governor played part of Beethoven's "Ode to Joy" and asked: "Now don't you feel smarter already?" He also remarked that from his experience growing up, "[m]usicians were folks that not only could play a fiddle but they also were good mechanics." The article also quoted a skeptic: "I'm familiar with those findings and, at the moment," said Sandra Trehaub, a professor of psychology at the University of Toronto who studies infants' perception of music, "I don't think we have the evidence to make that statement unambiguously. If we really think you can swallow a pill, buy a record or a particular book or have any one experience and that that's going to be the thing that gets you into Harvard or Princeton, then that's an illusion." Notably she did not mention Yale.

Independent replication of findings is a cornerstone of the scientific method. Additional research was catching up on Mozart. Five years after the original publication, two critical letters and a response from the original paper's primary author appeared in *Nature* (Chabris 1999). The first critical letter from Dr. Christopher Chabris was a meta-analysis of 16 studies of the Mozart effect that included 714 participants and several different reasoning tests. This kind of analysis combines all the results across studies. The overall Mozart effect was tiny for general intelligence and a bit larger for spatial-temporal tests. He too converted individual test scores to IQ equivalents and found about 1.4 points for general intelligence and about 2.1 points for spatial-temporal reasoning. These conversions are always suspect, as we have discussed, but they serve a narrow purpose of showing how small the effect looks since such small fluctuations are likely due to standard measurement errors found in all mental tests. Chabris concluded that these small effects were likely due to the effect of positive mood invoked by an enjoyable experience such as listening to Mozart. In his formulation, enjoyable experience increases arousal, especially in the right hemisphere, which processes spatial-temporal information. A second letter in this citation from Dr. Kenneth Steele and colleagues reported a complete failure to replicate the findings from the original 1993 experiment. Some of their results indicated poorer test performance after listening to Mozart.

In response, Dr. Rauscher noted that the original report did not claim an increase in intelligence. She contended that the claim was limited to spatial-temporal tasks involving mental imagery and temporal ordering. She pointed out that the smaller number of studies in Chabris' meta-analysis that used only tests of spatial-temporal reasoning did show an increase after Mozart listening, and she critiqued his enjoyment arousal hypothesis. She also critiqued the studies from the Steele group because they were not exact replications. She acknowledged that there were inconsistent results from independent studies and concluded: "Because some people cannot get bread to rise does not negate the existence of a 'yeast effect'" (Chabris 1999: 827).

Dr. Rauscher's key clarification regarding the claim that increased general intelligence was an erroneous inference from the original report is correct but confusion arose in part from the labeling of their figure axis (reproduced here as Figure 5.1) of "IQ equivalents." Moreover, the university press release that accompanied the original *Nature* publication contributed to the confusion. My copy of that release (embargoed until 6.00 pm Eastern Daylight Time, October 13, 1993) begins with the finding about "spatial intelligence" but then quotes the researchers as saying: "Thus, the IQs of subjects participating in the music condition were 8–9 points above their scores in the other two conditions." The distinction between spatial and general intelligence could have been clearer. After the 1999 exchange in *Nature*, controversy about a Mozart effect on any mental ability tests persisted. More studies were published with conflicting and inconsistent results.

Another, larger meta-analysis was published in 2010 and this one is widely regarded as the final blow (Pietschnig et al., 2010). This comprehensive analysis included nearly 40 studies and over 3,000 participants. An important feature of this analysis was the inclusion of unpublished studies because studies with negative results often fail to be published. This can bias the literature toward positive studies. Another feature was separate analyses for studies done by the original authors of the 1993 report for comparison with studies done by other researchers. Overall, this meta-analysis showed a small effect for Mozart on spatial task performance and a nearly identical small effect for other music conditions. Including the unpublished studies corrected this result to an even smaller effect. The results from only those studies done by the original researchers generally showed greater effects than from studies by other investigators, indicating a confounding influence of lab affiliation. The meta-analysis authors concluded that, "[o]n the whole, there is little left that would support the notion of a specific enhancement of spatial task performance through exposure to the Mozart sonata KV 448" (Pietschnig et al., 2010: 322). The title of this meta-analysis paper said it all: Mozart effect – Shmozart effect.

Whether the newborns of Georgia got their CDs or not, it is clear that the public understanding of the 1993 *Nature* report went far beyond what the study's authors intended. The original study had been conducted at my university under the auspices of the prestigious Center for Learning and Memory, and I knew the senior author, Dr. Gordon Shaw. Although I was unaware of the study before it was published, I subsequently had a number of affable conversations with him. A physicist by training, he was interested in the brain and problem solving and, before he passed away, he was developing a theory relating the complexity of music composition to cognition. He regretted the widespread misunderstandings about the original finding and general intelligence, but he remained convinced that music and cognition were linked in a positive way. His work with Dr. Rauscher helped stimulate interest in this important area. Whatever the many rich benefits of music exposure and training are, increased intelligence, general or spatial, is not one of them. The Mozart effect should be a cautionary tale for any researcher who claims dramatic increases in IQ after an intervention. Unfortunately, the lessons have not been taken to heart and such claims continue.

5.2 Case 2: You Must Remember This, and This, and This ...

Another extraordinary claim about increasing intelligence was published as the cover article in the *Proceedings of the National Academy of Sciences* (*PNAS*) (Jaeggi et al., 2008). Mozart was not mentioned, but the report claimed that training on a difficult task of working memory resulted in a "dramatic" improvement on a test of fluid intelligence. As noted in Chapter 1, fluid intelligence (often expressed with the notation Gf) is highly correlated to the *g*-factor and many intelligence researchers regard them as synonymous terms. Moreover, this surprising finding was augmented by two important observations: The effect increased with more training, suggesting a kind of dose response, and the effect transferred from the memory-training task to an "entirely different" test of abstract reasoning. The authors concluded: "Thus, in contrast to many previous studies, we conclude that it is possible to improve Gf without practicing the testing tasks themselves, opening a wide range of applications" (Jaeggi et al., 2008: 6829).

This pronouncement was a bombshell. It received wide media coverage and public attention. Like the original Mozart report, this report also was seized upon by researchers intent on showing that general intelligence could not be something fixed or genetic since it could be increased dramatically with a memory-training exercise. For most experienced intelligence researchers, however, this claim was immediately reminiscent of the 1989 cold fusion claim of an astonishing breakthrough thought to be impossible by most physicists. The cold fusion result turned out to be a heat measurement error made by eminent researchers in one field who were inexperienced in the measurement technicalities required in another field. Could a measurement error of fluid intelligence possibly be a factor in the *PNAS* report? You know where this is going.

The rationale for the *PNAS* memory-training study was simple. Memory is a well-established component of intelligence, so improving memory by training could improve intelligence. Ignoring that both could be related to a third underlying factor, a critical component for testing this simple train of thought would be that the training task must be independent of the intelligence test. In other words, the memory-training effect should transfer to a completely different test that did not require memory. For example, training a person to memorize the order of cards in a deck might transfer to their ability to remember a sequence of 52 random numbers because both are similar tests of memory. It would be more impressive if training to memorize cards resulted in better scores on a test of analogies (analogy tests usually have high g-loadings). It would be even more impressive if four weeks of training to memorize cards resulted in twice the improvement on analogy tests than did two weeks of card training.

For the *PNAS* memory-training experiment, 35 university students were randomly assigned to one of four training programs (one student dropped out for a total of 34 participants) and 35 other students were assigned to four control groups that received no training. Of the 70 students, the male–female ratio was about 50–50. Thus, each group had a very small number of about eight male or female participants. Each participant was tested before and after training (or the same control intervals) on either the RAPM test (discussed in previous chapters and used only for one training group) or the similar Bochumer Matrizen-Test (BOMAT) of abstract reasoning used for three training groups. There was no explanation for using two tests instead of one. Each test had two forms: one for the pre-test and one for the post-test. The four training programs differed on the number of training sessions between the pre- and post-testing: 8, 12, 17, and 19 days. These same intervals defined pre- and post-testing for the four control groups who received no training.

Memory training for all four groups used a well-known task in cognitive psychology. It's called the n-back test, where n stands for any integer. The idea is that a long series of random numbers or letters or other elements are presented one at a time on a computer screen to the participant. In the 1-back version with letters, for example, whenever any letter is repeated twice in a row, that is, the same letter is 1 back in the series, the participant presses a button. This is quite easy because it requires keeping only one letter in working memory until the next letter appears. If the next letter is not the same, no button is pressed and the new letter must now be remembered until the next letter appears. In the 2-back version, the button is pressed if the same letter was presented 2 letters before. This requires keeping two letters in working memory. The 3-back version is more difficult and 4-, 5-, and 6-back become considerably more difficult. Note that the letters (or numbers or whatever elements are used) can be presented visually or through earphones. For this study, participants were trained to do both a visual and an auditory version *simultaneously*. I'm not kidding. To understate the obvious, this is quite difficult and it is surprising that only one person dropped out. A more detailed description of the task is provided in Textbox 5.1.

Textbox 5.1: The dual n-back test
In Figure 5.2, the 2-back version is shown where spatial positions and letters are the elements used. The person presses a button whenever the same element is repeated with one intervening element. The spatial position elements are presented visually one at a time and the letters are presented one at a time through headphones. In the dual version, both the spatial and the letter elements are presented simultaneously for 500 milliseconds each with 2,500 milliseconds between elements. This is illustrated in the top row of Figure 5.2. It shows a sequence of spatial positions (white squares) in elements presented one at a time starting on the left. The middle element should trigger a button press because it is a repeat of the identical element 2-back

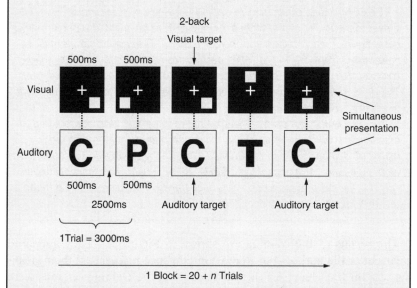

Figure 5.2 Illustration of the dual n-back memory task. This is a 2-back example. Two versions are run simultaneously. The top row shows the visual-spatial version. The location of the white box in each presentation must be remembered. If the same location is repeated after one intervening presentation, a button is pressed because the same location is repeated two presentations back. The bottom row shows the auditory letter version. Each letter presentation is made through earphones. When the same letter is repeated 2-back, a button is pressed. After training on each version separately, both versions are presented simultaneously and people practice until they can perform 3-back, 4-back, or more better than chance. In this illustration the order of presentation is from left to right, one presentation at a time.
(Reprinted with permission (Jaeggi et al., 2008))

(the element on the left end of the row). The bottom row shows the letter version. The middle element "C" should trigger a button press because it is an identical repeat of the "C" 2-back (at the very left end of the row). The "C" at the right end of the row is also a trigger because it is an identical repeat of the middle "C" 2-back. Once a person learns to do this difficult memory task better than chance, they move on to the harder 3-back version, which in turn progresses to 4-back and 5-back versions, and so on until performance cannot be learned better than chance. This is all a bit tricky to understand the first time you read it but once you get how the n-back works, you'll appreciate how difficult the training becomes. Remember that the claim is that training on this task increases your fluid intelligence (without giving you headaches or damaging your self-worth).

The BOMAT test of abstract reasoning is based on visual analogies and is similar to the RAPM test described in Chapter 1. In the BOMAT a 5 × 3 matrix has a figure in each cell except one cell is blank. The missing figure must be determined from the logical rules derived from other components (shape, color, pattern, number, spatial arrangement of the elements of the figure). The person taking the test must recognize the structure of the matrix and select from six possible answers one that allows the logical completion of the matrix. The RAPM test used a 3 × 3 matrix, so fewer elements are required to maintain in working memory while solving each item compared to the 5 × 3 matrix of elements used in the BOMAT. This is why the BOMAT is more of a working memory test and why it is similar to the n-back. This similarity under-cuts the claim that training on the n-back transfers to a completely different test of fluid intelligence (Moody, 2009).

The results of the training are shown in Figure 5.3. They appeared clear-cut to the authors but to most intelligence researchers their meaning was far less clear. All the participants in the training sessions were combined into one group (N = 34) and all the controls into another group (N = 35). Average n-back difficulty increased for the training group from about 3-back at the start to about 5-back at the end. The groups did not differ on the pre-test of abstract reasoning. Both groups showed average increased abstract reasoning scores at post-test. This was about a 1-point increase for the control group and about 2 points for the training group. Note these are not IQ points; they are the number of correctly answered items on the test. This small change was statistically significant and described as "substantially superior." When the intelligence test score increase was graphed against days of practice, the group with eight days showed less than a 1-point increase whereas the group with 19 days of practice showed nearly a 5-point gain.

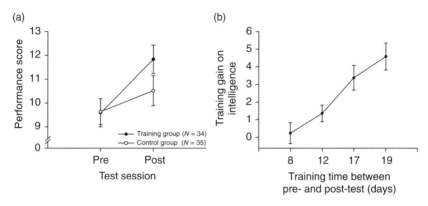

Figure 5.3 The line graphs that claimed a "landmark" result for memory training. Panel (a) shows pre and post n-back training fluid intelligence test scores (*y*-axis) for training and control groups. Panel (b) shows the gain on intelligence test scores (*y*-axis) plotted against the number of training days. (Reprinted with permission (Jaeggi et al., 2008))

The authors boldly concluded:

The finding that cognitive training can improve Gf is a landmark result because this form of intelligence has been claimed to be largely immutable. Instead of regarding Gf as an immutable trait, our data provide evidence that, with appropriate training, there is potential to improve Gf. Moreover, we provide evidence that the amount of Gf-gain critically depends on the amount of training time. Considering the fundamental importance of Gf in everyday life and its predictive power for a large variety of intellectual tasks and professional success, we believe that our findings may be highly relevant to applications in education. (Jaeggi et al., 2008: 6832)

I do not know whether they contacted the governor of Georgia, or any other state, with this newsflash but they ignited a memory-training frenzy.

The first devastating critique came quickly (Moody, 2009). Dr. Moody pointed out several serious flaws in the *PNAS* cover article that rendered the results uninterpretable. The most important was that the BOMAT test used to assess fluid reasoning was administered in a flawed manner. The test items are arranged from easy ones to very difficult ones. Normally, the test taker is given 45 minutes to complete as many of the 29 problems as possible. This important fact was omitted from the *PNAS* report. The *PNAS* study allowed only 10 minutes to complete the test so any improvement was limited to relatively easy items since the time limit precluded getting to the harder items that are most predictive of Gf, especially in a sample of college students with restricted range. This nonstandard administration of the test transformed the BOMAT from a test of fluid intelligence to a test of easy visual analogies with, at best, an unknown relationship to fluid intelligence. Interestingly, the one training group that was tested on the RAPM showed no improvement. A crucial difference between the two tests is that the BOMAT requires the test taker to keep 14 visual figures in working memory to solve each problem whereas the RAPM requires holding only eight in working memory (one element in each matrix is blank until the problem is solved). Thus, performance on the BOMAT is more heavily dependent on working memory. This is the exact nature of the n-back task, especially since the version used for training included the spatial position of matrix elements quite similar to the format used in the BOMAT problems (see Textbox 5.1). As noted by Dr. Moody: "Rather than being 'entirely different' from the test items on the BOMAT, this [n-back] task seems well-designed to facilitate performance on that test." When this flaw is considered along with the small samples and issues surrounding small change scores of single tests, it is hard to understand the peer review and editorial processes that led to a featured publication in *PNAS* that claimed an extraordinary finding that was contrary to the weight of evidence from hundreds of previous reports.

Subsequent n-back/intelligence research has progressed in stages similar to those in the Mozart effect story. Dr. Jaeggi and colleagues published a series of papers addressing some of the key design flaws of the original study and reported results consistent with their original report (Jaeggi et al., 2010; Jaeggi et al., 2011; Jaeggi et al., 2013; Jaeggi et al., 2014), as did some other researchers. Far more studies by other investigators failed to replicate the original claim of increased Gf, especially when they used more sophisticated research designs that included larger samples and multiple cognitive tests to estimate Gf as a latent variable along with other intelligence factors to determine whether improved n-back performance transferred to increased intelligence scores (Chooi and Thompson, 2012; Colom et al., 2013; Harrison et al., 2013; Melby-Lervåg and Hulme, 2013; Redick et al., 2013; Shipstead et al., 2012; Thompson et al., 2013; Tidwell et al., 2013, von Bastian and Oberauer, 2013, 2014).

Undaunted by these independent failures to replicate, Dr. Jaeggi's group published their own meta-analysis, including the negative studies. Their analysis supported a four-point IQ increase due to n-back training (Au et al., 2015). They ignored warnings about IQ conversions and change scores, and they failed to note that four points is the estimated standard error of IQ tests. Other researchers quickly reanalyzed this meta-analysis (Bogg & Lasecki, 2015). They concluded the small effect reported by Au and colleagues likely resulted from the small sample sizes of most studies included in the meta-analysis because they were statistically underpowered and biased toward a spurious result. Therefore, they cautioned that the small training effects on Gf could be artifacts. Another comprehensive independent meta-analysis of 47 studies concluded that there were no sustainable transfer effects for memory training (Schwaighofer et al., 2015), although the authors encouraged more research with better study designs. Finally, there is also some evidence that small apparent increases in test scores after memory training can be due to improved task strategies rather than to increased intelligence (Hayes et al., 2015).

Eight years after the initial *PNAS* report, the weight of evidence from independent studies found essentially no transfer effects from memory training to intelligence scores that are truly independent of the training method (Redick, 2015). At this stage, most positive results about n-back training and intelligence came from Dr. Jaeggi and her colleagues. Most researchers remained highly skeptical and most have moved on to other projects despite some earlier enthusiasm for the possibility of increasing Gf with memory training (Sternberg, 2008). The Shmozart paper effectively ended most research on the Mozart effect. Would the compelling reports by Bogg and Lasecki and by Redick have the same impact on the n-back – shman-back intelligence story?

Well, it did not. As I write the second edition of this book, the n-back story remains alive. On the one hand, another meta-analysis of 87 well-designed studies reinforced the skepticism. The authors concluded that "working memory training programs appear to produce short-term, specific training effects that do not generalize to measures of 'real-world' cognitive skills. These results seriously question the practical and theoretical importance of current computerized working memory programs as methods of training working memory skills" (Melby-Lervåg et al., 2016: 512). On the other hand, in the best scientific tradition, Jaeggi's group responded with a detailed critique of this meta-analysis. These authors of the original claim came to a different conclusion: "There still seems to be an overall small but significant effect size of n-back training on improving Gf test performance" (Au et al., 2016: 336). They argued that even a small improvement is proof of concept that justifies more research. And the Jaeggi group has published new studies including work on methodological issues (Au et al., 2020) and one study that emphasizes the importance of investigating n-back effects on the individual differences level since some people show a training effect while others do not (Pahor et al., 2022).

There are additional negative results and doubts remain strong (Gobet & Sala, 2022; Redick, 2019; Watrin et al., 2022; Wiemers et al., 2019). I remain skeptical. In my view, the weight of evidence still favors no appreciable effect on fluid intelligence; small effects are most likely due to ever-present measurement noise in test change scores (Haier, 2014). Nonetheless, we likely are in for at least a few more years of investigating the original claims.

An interesting coincidence is that Dr. Jaeggi relocated to my university a few years ago to the School of Education and we became friends despite a complete disagreement about whether memory training increases intelligence. Based on the history of similar claims in the past, I suspect memory-training research will become less directed at improving intelligence and more directed at other cognitive and education variables (Wang et al., 2019). In fact, there is growing interest in broader cognitive training using computer games to increase school achievement, as we see in the next case.

5.3 Case 3: Can Computer Games for Children Raise IQ?

There is a large research literature and considerable controversy about whether computer games may have any beneficial cognitive effects. Whatever effects computer games may have on learning, attention, or memory (Bejjanki et al., 2014; Cardoso-Leite & Bavelier, 2014; Gozli et al., 2014), our focus here is on the narrow question of whether computer game training demonstrably increases intelligence. One research group from the University of California, Berkeley claimed a 10-point increase in performance IQ following computer game training of basic cognitive skills involved in reasoning and processing speed in a study of children from low-socioeconomic backgrounds (Mackey et al., 2011). Reminiscent of the 2008 *PNAS* n-back study, the Berkeley researchers bluntly concluded that, "[c]ounter to widespread belief, these results indicate that both fluid reasoning and processing speed are modifiable by training" (Mackey et al., 2011: 582). Let's see.

The study involved 28 children aged 7–10 years old. The students were randomly assigned to one of two training groups. One group (n = 17) trained on commercial computer games thought to foster fluid reasoning (i.e., fluid intelligence or the *g*-factor) and the other group (n = 11) trained on commercial computer games thought to foster brain processing speed. Each training intervention occurred during school for an hour on two days a week for eight weeks, although the average number of training days was about 12 for each group. On training days (two per

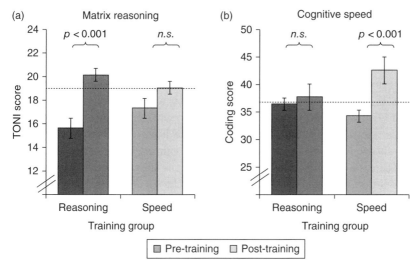

Figure 5.4 The findings that countered "widespread belief" and were the basis for optimism for closing cognitive gaps for disadvantaged children. Panel (a) shows that computer game training on matrix reasoning (n = 17) increased reasoning scores but not speed of processing scores. Panel (b) shows that cognitive speed training (n = 11) increased coding scores but not reasoning scores.
(Reprinted with permission (Mackey et al., 2011))

week) each group worked on four different computer games for about 15 minutes each. Pre- and post-training assessments for fluid reasoning (FR) were based on the Test of Nonverbal Intelligence (TONI-version 3) and for processing speed two tests were used: Cross Out from the Woodcock–Johnson Revised test battery and Coding B from the Wechsler Intelligence Scale for Children IV. The test details are not necessary to understand the results.

The group trained on FR showed about a 4.5-point increase in TONI nonverbal intelligence scores on the post-test and no significant increase on the processing speed tests. For the group trained on processing speed, the opposite was found. There was a significant increase in the coding score but no change in the FR score. The authors translated the raw score 4.5-point increase to an increase of 9.9 IQ points, more than half a standard deviation. Four of the children apparently increased by over 20 IQ points. They concluded that the main message was hope that cognitive gaps in disadvantaged kids, especially any related to FR, could be closed with a "mere 8 weeks of playing commercially available games" (Mackey et al., 2011: 587). News coverage followed. So did grant funding.

The key finding is shown in Figure 5.4. There are several problems that by now should be familiar to you. The sample sizes are very small and IQ scores at this age often fluctuate by several points. The apparent IQ increases could easily be due to chance effects with undue influence in small samples, as noted in Section 5.2 (Bogg and Lasecki, 2015). This is more likely given that the children who improved the most on some training tasks were not the children who showed the greatest FR gains. Actually, the children who had the lowest FR before the training showed the greatest increase after training, suggesting the effect was due, at least in part, to regression to the mean (statistically, repeat scores on average tend to move back to the group mean). Overall, the results are interesting but trusting they indicate a new finding "counter to widespread belief" is a dubious conclusion, especially when the widespread belief is based on the weight of evidence from hundreds of other studies.

This study may be the basis of some generic commercial claims that computer games can increase IQ (without specific attribution to this research study). I am unaware of any replication studies of the University of California, Berkeley findings of such large IQ increases, positive or negative, either by the original authors or by other researchers. This is odd given the claimed potential for these findings to overturn widely held beliefs. Most intelligence researchers remain highly skeptical of a 10-point IQ increase attributed to general cognitive training.

Let's pause for a moment and talk about whether there is any relationship between video gaming and intelligence. For the first edition of this book, I noted that a comprehensive study found virtually no relationship between video game experience and fluid intelligence in a large sample of young adults (Unsworth et al., 2015). Interestingly, newer work suggests that video games might be used to measure intelligence (Kokkinakis et al., 2017) and this was demonstrated by extracting a general factor from performance on a battery of video games that was correlated to a g-factor extracted from a battery of cognitive tests (Quiroga et al., 2015; Quiroga et al., 2019; see also a comprehensive summary of gaming–intelligence relationships in Quiroga & Colom, 2020). This research implies that video games can be used to assess intelligence but, even if that ever comes to pass, it is a separate question from whether gaming can increase intelligence.

But, at the time of this second edition, there is a new claim about computer game playing and increasing IQ scores published in another top journal (Sauce et al., 2022). Like the other examples in this section, this study was widely reported in popular media with sensationalized headlines like, "Video Games Can Boost Children's Intelligence." Here we go again, but this time the study is well designed and has a large sample. Do the results, however, justify such bold headlines? I think not.

The study started with a database of 9,855 US children aged 9–10 years old. They had completed several cognitive measures and 5,169 were retested on the same measures two years later (the same items were used at both testings). Three primary independent variables – screen time watching digital media, gaming screen time, and socializing screen time (all self-reported) – were correlated to a common factor labeled intelligence extracted from the cognitive tests. From the database, genetic data from PGSs and family SES were used as control variables, which are wonderful additions to the research design. There were many sophisticated statistical analyses but only one set of results are of interest to us; they address how the independent variables predict intelligence change scores. From the paper's results section:

The follow-up after two years showed that screen time Socializing had no independent effect on the change in intelligence (... $p = 0.220$). Interestingly, we found a positive effect on the change in intelligence from screen time Watching (... $p = 0.047$; or 1.8 IQ points) as well as from screen time Gaming (... $p = 0.002$; or 2.55 IQ points), with more time watching digital videos or playing video games leading to greater gains in intelligence. (Sauce et al., 2022: 3)

That's it. 2.55 IQ points which, if accurate, is easily explained as measurement error or by practice since the same items were used at both testing times. Just screen time watching apparently gained 1.8 IQ points so the added benefit of gaming would be even smaller.

Also, note that intelligence was not assessed by standard IQ tests, so how they converted their general factor scores to IQ points is not clear and such conversions include error variance. In my view, this study – perhaps the best designed and analyzed so far – adds to the weight of evidence that gaming does *not* increase intelligence in children. Apparently encouraged by the authors, the media coverage took the opposite view.

Some commercial companies market computer-based training programs to parents and to school systems with the explicit or implied goal of closing cognitive gaps, especially for students from disadvantaged backgrounds (see Chapter 6 for more about SES and intelligence). Most reputable companies are careful to avoid making explicit claims about increasing intelligence. One company, however, claimed in their 2014 report (downloaded from the internet) that their brain-training program raises IQ by an average of 15 points for their clients. Their clients who start the program with "severe cognitive weakness" show average gains of 22 IQ points. The report has many pages of impressive-looking statistical analyses, tables, and graphs that show apparently amazing results for users of their program, but the report does not list a single publication where the statistics and claims have undergone independent peer review. Other companies sometimes cite individual published research reports, especially with small samples, as evidence for the validity of computer training programs in increasing mental performance. This kind of cherry picking is quite common where only studies that support a claim are noted while ignoring other studies that do not.

In the first edition, I concluded that neuro-education and brain-based learning are attractive concepts for educators but, in my view, there is not yet a compelling weight of evidence of successful applications so considerable caution is required (Geake, 2008, 2011; Howard-Jones, 2014). Not much has changed, as noted by an editorial introducing an informative special issue of 14 papers on the topic published in *Frontiers of Human Neuroscience* (Arevalo et al., 2022). It stated:

In several countries, the search for (and almost parallel supply of) neuroscience-related courses has grown exponentially, but the rate at which this has happened almost guarantees that quality does not match quantity. Another area of rapid growth, especially in North America and Europe, has been the industry of brain-based products, mostly pseudo-scientific endeavors that target parents, teachers, schools, and even local governments. (Arevalo et al., 2022: 1)

The authors went on to say: "The gap between cognitive neuroscience and learning is still very conspicuous. And one of the consequences of this distance is the appearance and propagation of myths that in many cases have some scientific support" (Arevalo et al., 2022: 1). Potential buyers of such programs, especially those claiming increases in intelligence, are advised to keep three words in mind before signing a contract or making a purchase: independent replication required.

Speaking of independent replication, none of the three case studies discussed so far (the Mozart effect, n-back training, and computer training) included any replication attempt in the original reports. There are other interesting commonalities among these studies. Each claimed a finding that overturned long-standing conclusions from many previous studies. Each study was based on small samples. Each study measured putative cognitive gains with single test scores rather than extracting a latent factor like g from multiple measures. Each study's primary author was a young investigator and the more senior authors had few previous publications that depended on the psychometric assessment of intelligence. In retrospect, is it surprising that numerous subsequent studies by independent, experienced investigators failed to replicate the original claims? There is a certain eagerness about showing that intelligence is malleable and can be increased with relatively simple interventions. This eagerness requires researchers to be extra-cautious. The peer-reviewed publication of extraordinary claims requires extraordinary evidence, which is not apparent in Figures 5.1, 5.3, and 5.4. In my view, the basic requirements for the publication of "landmark" findings would start with replication data included along with original findings. This would save many years of effort and expense trying to replicate provocative claims based on fundamentally flawed studies and weak results. It is a modest proposal, but probably unrealistic given academic pressures to publish and obtain grant funding.

Before leaving this section on increasing intelligence in children, there is another interesting report to consider. Whereas the three cases discussed so far are presented as cautionary examples, this one is a positive illustration of how progress in the field might be advanced more prudently. This report is based on meta-analyses of "nearly every available intervention involving children from birth to kindergarten" to increase intelligence (Protzko et al., 2013). These researchers from New York University (NYU) maintain the Database of Raising Intelligence. This database includes studies designed to increase intelligence that have the following components: a sample drawn from a general, nonclinical population; a pure randomized controlled experimental design; a sustained intervention; and a widely accepted, standardized measure of intelligence as an outcome variable. Four meta-analyses are reported on the effects of dietary supplementation to pregnant mothers and neonates, early educational interventions, interactive reading, and sending a child to preschool. Here is a summary of the main results for each of these four analyses.

The nutrition research was limited mostly to studies of the long-chain fatty acid called PUFA (don't ask why this name), an ingredient in breast milk necessary for normal brain development and function. This analysis was inspired by earlier evidence of higher IQ in breast-fed children compared to bottle-fed children (Anderson et al., 1999). The 2013 meta-analysis included 10 other studies of 844 total participants. The analysis suggested that a 3.5 IQ point increase was associated with long-chain PUFA when it was used as a dietary supplement. However, a review of 84 related studies suggested several possible confounding factors, including that parents with higher IQs tended to breastfeed more.

The conclusion was that the small IQ increase in children attributed to breastfeeding may actually be due to confounding factors, including the genetics of IQ (Walfisch et al., 2013). This was also the conclusion of an earlier prospective study of sibling pairs where one was breast fed and the other was not (Der et al., 2006). Thus, the weight of this evidence does not support breastfeeding as a way to increase a child's IQ. Similar analyses for iron, zinc, B6, and multivitamin supplements were less encouraging for increasing IQ based on the available evidence. There are many newer studies of breastfeeding and intelligence. Nonetheless, the question of breast feeding and intelligence is still unsettled according to a comprehensive review by Stuart Ritchie. He concludes: "At the risk of sounding like a broken record, it's another area where – in the morass of confusion and overstatement, studies that vary wildly in quality, and failure to use the optimal research designs – science has let us down" (Ritchie, 2022).

The second meta-analysis focused on early education. In Chapter 2 we described a few key intervention studies that failed to show lasting IQ increases. The NYU analyses incorporated 19 studies going as far back as 1968. Some had interventions that went on for more than three years. Although some individual studies did show IQ increases for some infants, altogether the meta-analysis indicated that there was no appreciable effect on IQ. The third meta-analysis focused on interactive reading and incorporated 10 studies totaling 499 participants. For children under four years old, the meta-analysis indicated about a six-point increase in IQ when the child was an active participant in the reading. The authors speculate that this intervention may influence language development which than indirectly influences IQ. Active reading is now widely recommended to parents. The fourth meta-analysis focused on preschool and included 16 studies of 7,370 participants, mostly with low family income backgrounds. Altogether the analysis indicated a four-point increase in IQ but up to a seven-point increase for the subset of programs that included a specific emphasis on language development. Interestingly, longer preschool attendance was not related to greater increase in IQ points. How long any of the putative increases may last and the brain mechanisms that might be relevant are not yet known.

Here is a key point about claims of increased intelligence: Even if statistically significant, the reported IQ increases are still mostly about the size of the standard error of IQ tests (2–4 points), especially given that intelligence test scores in this young age range are less reliable and often fluctuate for many reasons over short periods of time. Many of the studies included in the four meta-analyses have the same small sample issues that characterized the three case studies and the n-back meta-analysis done by Au and colleagues on studies of memory training. Continued skepticism is warranted for any effect these interventions may have on intelligence.

As a final point in this section, I have noted the eagerness that underlies many of the claims about fairly simple interventions alleged to increase fluid intelligence and demonstrate malleability by environmental means. As we discussed in Chapter 2, the evidence for environmental influences is actually not that strong and more caution is warranted generally when talking about intelligence or success (Moreau et al., 2019). I agree with David Moreau's summary from his latest review paper about malleability of cognitive abilities (Moreau, 2022: 418):

Given the promise of significant change with little investment or resources, cognitive interventions are appealing to a wide range of individuals and institutions. In many respects, however, robust scientific evidence to confirm these benefits is still lacking, and the underlying mechanisms of improvement remain poorly understood … For now, caution is thus required, especially when these findings are used to support large-scale policies. Individual differences in human ability are profoundly complex – a view that recognizes these differences, rather than stigmatize them or further a rhetoric of extreme cognitive malleability, is one that is not only more accurate, but also provides the foundations for a fair and just society.

5.4 Where Are the IQ Pills?

The genetic and neuroimaging studies described in Chapters 2, 3, and 4 provide compelling evidence that intelligence has a strong basis in neurobiology, neurochemistry, and neurodevelopment. Actual brain mechanisms that influence or control brain structures and functions related to intelligence are not understood to any significant degree. If certain neurotransmitters, for example, are found to play a central role in relevant cognitive mechanisms (say working memory), then drugs that increase or decrease the activity of those neurotransmitters may show effects on intelligence test scores. Synaptic events regulated by neurotransmitters may be the place for interventions. These include changing the level of the neurotransmitter or how fast it is replenished, or changing the sensitivity of the receptors that respond to the neurotransmitters. On the other hand, if drugs are accidently found to increase scores on IQ tests, inferences about how those drugs work on neurotransmitters in the synapse can generate new hypotheses about what brain mechanisms might be most relevant to intelligence. This logic for drug effects is the same as applied in the intervention studies we discussed earlier in this chapter. Drugs influence brain mechanisms more directly than memory training, for instance, so drugs may have greater intelligence-boosting potential. The study criteria for showing an effect on intelligence for any drug is also the same: a sample that includes a range of normal IQ scores, multiple measures of intelligence to extract a latent g-factor, double-blind placebo-controlled trials with random assignment, a dose-dependent response for any short-term effect (greater dose shows greater enhancement), a follow-up period to determine any lasting effects, and independent replication. And, of course, a ratio scale of intelligence would make an increase most convincing although none yet exists (Haier, 2014); see Textbox 6.1 for a possible way to define a ratio scale for intelligence.

The internet has countless entries for IQ-boosting drugs, and there are many peer-reviewed studies of cognitive-enhancing effects on learning, memory, and attention for drugs such as nicotine (Heishman et al., 2010). Psycho-stimulant drugs used to treat attention deficit hyperactivity disorder and other clinical disorders of the brain are particularly favorite candidates for use by students in high school, college, and university and by adults without clinical conditions who desire cognitive enhancement for academic or vocational achievement. Many surveys show that drugs are already widely used to enhance aspects of cognition and a number of surrounding ethical issues have been discussed (Dresler et al., 2019; Knafo & Venero, 2015; Sharif et al., 2021). Some of these issues are presented in Textbox 5.2, which includes one of the first surveys of cognitive enhancement among scientists. Overall, well-designed research studies do not strongly support such use (Bagot & Kaminer, 2014; Farah et al., 2014; Husain & Mehta, 2011; Ilieva & Farah, 2013; Smith & Farah, 2011). Even fewer studies are designed specifically to investigate drug effects directly on intelligence test scores in samples of people who do not have clinical problems. I could find no relevant meta-analysis that might support such use. In short, there is no compelling scientific evidence yet for an IQ pill. As we learn more about brain mechanisms and intelligence, however, there is every reason to believe that it will be possible to enhance the relevant brain mechanisms with drugs, perhaps existing ones or new ones. Research on treating Alzheimer's disease, for example, may reveal specific brain mechanisms related to learning and memory that can be enhanced with new drugs significantly better than existing drugs. This prospect fuels intense research at many multinational pharmaceutical companies. If such drugs become available to enhance learning and memory in patients with Alzheimer's disease, surely the effect of those drugs will be studied in nonpatients to boost cognition.

Because there is a paucity of empirical evidence for raising intelligence, and because psychoactive drugs often have serious side effects, especially when a physician does not monitor their use, no list of drugs claimed to increase intelligence appears in this book. In my view, there are none to list, even for this second edition. The potential for drugs to boost intelligence, however, is directly correlated to the extent to which the biological bases of intelligence are revealed, and as described in previous chapters, the pace of discovery is increasing. Drugs, however, may not be the only way to tweak neurobiological processes. There are fascinating hints at other methods. We now turn to what may sound like science fiction efforts to enhance intelligence and related cognition. They are not fiction and they are mind blowing, almost literally.

Textbox 5.2: Cognitive-enhancing drugs

*The journal **Nature** published a commentary about ethical issues concerning putative cognitive-enhancing (CE) drugs in 2007 (Sahakian and Morein-Zamir, 2007) and a 2008 commentary (Greely et al., 2008) based on an informal survey of 1,400 scientists from 60 countries about their use of such drugs (Maher, 2008). Ethical questions raised in the 2007 commentary included whether the use of CE drugs in healthy individuals without neurological or psychiatric disorders was cheating or fair, whether CE drugs should be available without medical supervision, and whether someone might feel undue pressure to use CE drugs for themselves or for their children if they knew others were doing so at school or work. The 2008 survey of scientists revealed that 20 percent already used drugs to improve concentration; 70 percent would risk mild side effects to boost brainpower; 80 percent defended the right to take such boosters; and greater than 33 percent would feel pressure to give their kids brain boosters if other kids used them. The survey report included four specific comments from responders that illustrated basic ethical issues. **Safety**: "The mild side effects will add up to be profound in due course and may even require stronger therapy to control the addiction" wrote a young man from Nigeria. **Erosion of character**: "I wouldn't use cognitive enhancing drugs because I think it would be dishonest to myself and all the people who look to me as a role model" wrote a young person from Guyana. **Distributive justice**: "Morally puts a disadvantage to people without access" wrote a middle-aged person from the United States. **Peer pressure**: "As a professional, it is my duty to use my resources to the greatest benefit of humanity. If 'enhancers' can contribute to this humane service, it is my duty to do so" wrote a senior citizen from the United States.*

The authors of the 2008 **Nature** commentary argued that "[s]ociety must respond to the growing demand for cognitive enhancement. That response must start by rejecting the idea that 'enhancement' is a dirty word." Since these three commentaries were published, there are many new surveys about CE use around the world (Dietz et al., 2013; Dietz et al., 2016; Franke et al., 2014). Different survey methodologies and different samples make it hard to fix rates of regular versus occasional use or the motivations for use. Nonetheless, there is general agreement that the trends for CE use are increasing, especially for high school and college students in America despite limited evidence of efficacy (Smith & Farah, 2011; also see Sahakian & Kramer, 2015). A comprehensive meta-analysis of surveys confirms that rates of students using CE drugs are increasing around the world (Sharif et al., 2021).

One presentation of ethical considerations surrounding pharmacological cognitive enhancement (PCE) formulated six principal issues:

(1) The medical safety-profile of PCEs justifies restricting or permitting their elective or required use. (2) The enhanced mind can be an "authentic" mind. (3) Individuals might be coerced into using PCEs. (4) There is a meaningful distinction to be made between the treatment vs. enhancement effect of the same PCE. (5) Unequal access to PCEs would have implications for distributive justice. (6) PCE use constitutes cheating in competitive contexts. (Maslen et al., 2014: 1)

The discussions about enhancement and these issues are mostly limited to cognitive elements of attention, learning, and memory. Specifically enhancing intelligence for all individuals is not yet a primary focus of ethical discussions although the ethics of researching any group differences is another matter (Carl, 2018; Cofnas, 2020). Generally, there is a positive ethical view in the narrow context of increasing IQ scores through nonpharmacological interventions to minimize school achievement gaps (like the ones in the three case studies). One area is not as clear. As we noted in Chapter 2, polygenic scores are being used in some cases of embryo selection to screen for potentially low intelligence. There is emerging discussion about using such scores to select for potentially high intelligence (Pagnaer et al., 2021; Tellier et al., 2021; Turley et al., 2021). If, as I believe, more intelligence is better than less, is there not a moral obligation in favor of such selection by DNA or enhancement by pharmacological means? What do you think about these issues? The 1996 movie **Gattaca** and the 2011 movie **Limitless** may provoke some ideas.

5.5 Magnetic Fields, Electric Shocks, and Cold Lasers Target Brain Processes

This section briefly introduces five odd-sounding technologies that alter brain processes and may have implications for enhancing cognition and intelligence. Most importantly, these techniques allow for the experimental manipulation of cortical and subcortical activity to study the impact on cognition. This ability provides exciting and important opportunities for determining cause and effect relationships that go beyond studies that report correlations between brain variables and performance on mental tests. They herald the beginning of a new phase of intelligence/ brain research.

The first technique is transcranial magnetic stimulation (TMS). TMS uses a device containing a metal coil to produce strong magnetic field pulses when electricity is applied in short bursts. When the coil is placed over a part of the scalp, the magnetic field fluctuations pass through the scalp and skull undistorted into the brain. The fluctuations induce electrical currents that depolarize neurons in the underlying brain cortex. The rate of pulses and their intensity can be varied to increase or decrease cortical excitation. As a research tool, TMS can be used to test whether a particular region of cortex is involved in a cognitive task. For example, inducing cortical deactivation might result in poorer performance and inducing activation might result in better performance, or in the case of efficiency, vice versa. A review of over 60 TMS studies done over the last 15 years (Luber & Lisanby, 2014) concluded that this technique has promise for enhancing a range of cognitive tasks, although intelligence is not specifically discussed and this review is not a quantitative meta-analysis. According to the authors, TMS may affect brain mechanisms to increase task performance in at least two general ways: either by direct impact on neurons that increases the efficiency of task-relevant processing or by disrupting processing that is task-irrelevant and distracting to performance. Some enhancement effects attributed to the first category are for tasks involving nonverbal working memory, visual analogic reasoning, mental rotation, and spatial working memory, among others (Luber & Lisanby, 2014: table 1). Enhancement effects attributed to the second category include tasks of verbal working memory, spatial attention, and sequential item memory (Luber & Lisanby, 2014: table 2). In addition to laboratory experiments, the authors also discuss some real-world applications for TMS, including cognitive rehabilitation after brain injury.

Since the first edition of this book, TMS has been demonstrated to alleviate clinical depression and is now widely available as a possible treatment (George et al., 2000; Loo and Mitchell, 2005; Sonmez et al., 2019). There is also some evidence that TMS enhances aspects of cognition in impaired psychiatric patients (Kim et al., 2019) and in patients with mild cognitive impairment (Jiang et al., 2020). The brain mechanisms for TMS effects appear to be nonspecific although much remains to be learned (Siebner et al., 2022). As far as I can determine, there are no reports of TMS effects specifically on measures of intelligence, but this is still an experimental method that may be useful for intelligence research. For example, TMS to specific areas might disrupt problem-solving performance to give hints about relevant brain mechanisms that could be targeted with other means to enhance performance.

The second technique is transcranial direct current stimulation (tDCS). In other words, this delivers electric shocks directly to the head. The currents are quite mild and the shocks barely noticeable. They are not used as punishment shocks for wrong responses to a cognitive task or like electric convulsive therapy that induces seizures that ameliorate deep clinical depression. tDCS currents are generated by a nine-volt battery and pass between electrodes on the scalp. Depending on the parameters used, this current can increase or decrease neuronal excitability under the electrode locations similar to the effect of TMS. Early tDCS studies were encouraging (Clark et al., 2012; Utz et al., 2010). Writing about enhancement effects for both TMS and tDCS, one research group (McKinley et al., 2012: 130) noted:

These techniques are perhaps best suited for career fields where certain cognitive skills such as vigilance and threat detection are essential to preserving human life. Because such jobs are plentiful in the military, it is no surprise that the US Air Force has recently begun investing in noninvasive brain stimulation for its efficacy in benefiting human cognitive performance.

Another group reviewed tDCS enhancement effects from many studies of attention, learning, and memory in healthy adults (Coffman et al., 2014). This qualitative review concluded that "battery powered thought" had considerable potential for certain cognitive tasks, although the review did not address intelligence directly. This review included some research on how tDCS might influence the brain mechanisms underlying cognitive enhancements. They noted possible roles for aspects of glutamate, GABA, NAA, NMDA, and BDNF regulation and function (see similar findings from molecular genetic studies noted in Chapter 4). Predictably, however, a newer, comprehensive quantitative analysis of tDCS and cognition in healthy adults was more discouraging (Horvath et al., 2015b). They found essentially no effects for outcome measures of executive function, language, or memory. They also found no reliable neurophysiological effects (Horvath et al., 2015a). Further doubts about any enhancing effects on intelligence were expressed in a study that showed decreased performance on the WAIS-IV intelligence test battery following tDCS stimulation (Sellers et al., 2015). These authors reported two studies (total of 41 adults), both double-blind, between subjects design (i.e., the same individuals tested before and after tDCS), using a sham stimulation control condition (fake connections to look like the real device but there is no current). tDCS was applied bilaterally to frontal lobe areas in one study and unilaterally in the other study. In both studies, the tDCS condition was compared to the sham condition. In both studies tDCS was associated with degraded performance on certain WAIS subscales. No improved performance was observed. This is the lesson learned for virtually all claims of cognitive enhancement so far: early promising findings must be reliably reproducible by independent investigators and survive comprehensive quantitative analyses.

There is not much new data to report about tDCS and intelligence for this second edition. One intriguing study looked at tDCS and three other forms of brain stimulation combined with executive function training (tasks of working memory, inhibition, and cognitive flexibility) in 82 participants divided into four groups (Brem et al., 2018). A fluid intelligence measure was based on three different matrices reasoning tests. Compared to a control condition, this measure showed small increases when training sessions were combined with tDCS and another stimulation sessions (but not for tACS – see next paragraph). This is another example of potential importance, but so far, in my opinion, the reported effects do not add up to a compelling weight of evidence for tDCS effects on intelligence.

There are other potentially informative stimulation studies at early stages to report in the context of this chapter, albeit with the warning that replication studies remain to be done. Instead of direct stimulation, a variant of this technique uses a mild alternating current, called transcranial alternating current stimulation (tACS). This is our third technique. Whereas tDCS produces a general stimulation to the brain, tACS can be targeted to specific areas. Of interest here, two studies have reported tACS-induced enhancement specifically on fluid intelligence tests. In the first study, tACS was used experimentally to alter natural oscillation frequencies that are generated by neuronal activity (Santarnecchi et al., 2013). Oscillation frequencies in the brain are related to mental task performance but whether they are cause or consequence is an open question. The participants were 20 young adults and the "imperceptible" tACS was applied by scalp electrode over the left middle frontal lobe. Compared to sham stimulation (a control condition), rhythmic stimulation within the gamma band (a particular frequency) induced by tACS resulted in faster solution times for only more difficult items such as those on the RAPM test. This suggested a causal relationship between the oscillations and the influence on the test. Note that the enhancement effect was assessed by shorter times to solution, where time is a ratio scale. The authors concluded that their finding "supports a direct involvement of gamma oscillatory activity in the mechanisms underlying higher order human cognition" (Santarnecchi et al., 2013: 1449).

Another study of intelligence in 28 young adults compared tACS in the theta band (a different frequency) applied to either the left frontal lobe or to the left parietal lobe; sham stimulation was also used as a control (Pahor & Jausovec, 2014). tACS was given for 15 minutes prior to completing two tests of fluid intelligence. The tests were a modified version of the RAPM and the paper folding and cutting (PF&C) test of spatial ability from the Stanford–Binet IQ test battery. EEG was also obtained during both tests. The authors concluded that:

Left parietal tACS increased performance on the difficult test items of both tests (RAPM and PF&C) whereas left frontal tACS increased performance only on the easy test items of one test (RAPM). The observed behavioral tACS influences were also accompanied by changes in neuroelectric activity. The behavioral and neuroelectric data tentatively support the P-FIT neurobiological model of intelligence. (Pahor & Jausovec, 2014: 322)

There are some inconsistencies and contradictions in the findings of these two independent tACS studies of fluid intelligence, and in the studies using tDCS, but they may provide hints about the salient brain mechanisms. They further demonstrate the potential of brain stimulation techniques for the systematic manipulation of neuron activity in humans to determine effects on cognitive performance. Surely more research will be forthcoming, with refined experimental designs that include larger samples and an emphasis on individual difference variables such as age, sex, and pre-existing brain excitability (Krause & Cohen Kadosh, 2014; Santarnecchi et al., 2015; Santarnecchi et al., 2016; also see review by Antal et al., 2022). Brain stimulation with these techniques is experimental but the mechanics of building tDCS and tACS devices are fairly simple. There are reports of homemade "brain shock" devices used by gamers and others looking for enhanced cognition. Some commercial companies sell such devices for a range of self-uses. Independent replication research supporting their claims, if any, would be important to evaluate. Applying homemade or commercial electrical devices to your brain might have unintended consequences. Please do not compete for a Darwin Award by trying this at home.

Deep brain stimulation (DBS), the fourth technique, is the conceptual equivalent of a heart pacemaker. DBS applies mild electrical stimulation to microelectrodes surgically implanted into specific brain areas below the cortex by a team of medical specialists. It is a major invasive procedure not readily accomplished at home. The stimulation can be constant or applied when needed. DBS has demonstrable clinical applications for alleviating the symptoms of Parkinson's disease and clinical depression, and it is under study for other brain disorders. There are also a number of studies of DBS on learning and memory that suggest possible enhancement under some conditions (Suthana & Fried, 2014; Widge et al., 2019; see review of effects on memory impairment in Hescham et al., 2020). As far as I know, there are not yet DBS studies of intelligence. There is an interesting question as to whether brain areas related to cognitive enhancement effects identified with TMS, tDCS, or tACS can be more accurately localized with neuroimaging specific to an individual person and then be targeted with the precise localization of DBS electrodes. Could constant DBS in multiple areas enhance the g-factor, especially in individuals with low IQ, or could on-demand DBS in a specific area enhance specific mental abilities in any of us? This is a long way from listening to Mozart or n-back training. Do these possibilities sound more enticing than compensatory education? Such rank speculation is offered only in this section to stimulate your imagination about the importance and potential of neuroscience approaches to intelligence research.

While you are thinking about this, here is one more noninvasive brain stimulation technique that invites speculation. Our fifth technique is based on lasers. Light from low-power "cold" lasers in the near-infrared range penetrates the scalp and skull and can affect brain function. One group of researchers reported preliminary evidence that this technique can enhance some kinds of cognition when aimed at different brain areas. They describe how laser light affects the brain this way: "Photoneuromodulation involves the absorption of photons by specific molecules in neurons that activate bioenergetic signaling pathways after exposure to red-to-near-infrared light" (Gonzalez-Lima & Barrett, 2014: 1). Imagine this special laser light aimed from a distance at an unsuspecting person's brain to either enhance or disrupt cognition. Sounds like a screenplay idea (or even something from current news reports regarding mysterious attacks on government employees called the Havana Syndrome that cause brain problems). Enough speculation.

The data are preliminary and lasers can be quite dangerous. Do not try this at home either.

5.6 The Missing Weight of Evidence for Enhancement

In Chapters 1, 2, 3, and 4 the weight of empirical evidence supported, respectively: the *g*-factor concept; an important role for genetics in explaining individual differences in intelligence; intelligence-related networks distributed throughout the brain; and, to a lesser extent, the idea that efficient information flow around the brain was related to intelligence. In this chapter, despite many provocative claims and intriguing findings, no weight of evidence yet supports any means or methods for enhancing intelligence, fluid or otherwise.

From time to time, I am asked by magazine health writers to offer tips on increasing IQ. My answer is always the same and usually induces a long silence from the writer. There are no such tips – not even one that is supported by the weight of evidence. Eat better? Exercise? Engage in mentally challenging activity? All are good suggestions for general health and well-being but no specific effects for boosting intelligence can be substantiated. Not surprisingly, these writers never quote me, although science writers sometimes do in more substantial articles. I am happy to be the voice of reasonable skepticism to help stop the spread of bad information. One online magazine listed 10 tips for boosting IQ including listening to classical music, memory training, playing computer games, and learning a new language. For each tip, they listed the putative IQ point increase claimed by someone, and then they added up all the points to support the nonsensical headline promising ways of boosting your IQ 17–40 points. Really.

Although enhancement of intelligence is an important goal for neuroscience research, the weight of evidence to date indicates there is a long and winding road ahead for meeting this goal with drugs, genetics, electric or magnetic stimulation, or lasers. The road appears no shorter for education and cognitive training approaches. These roads have no posted speed limits or guardrails so crashes are inevitable. Moreover, my assertion that enhancement is an important goal is not universally recognized. If it were, considerably more federal and foundation funding would be directed toward achieving it and not just for disadvantaged children. After all, many national challenges from technological and economic innovation to cybercrime and cyber warfare pit the smartest against the smartest. This is a serious business. Silly magazine tips are not helpful.

What are the most likely research pathways toward enhancing intelligence, especially *g*? If I had to bet, two pathways are the most likely. The first is from neuroimaging studies that aim to identify specific networks and circuits in the brain. At some point, once the relevant ones are known, experiments to modify them could yield increases in reasoning and problem-solving ability. The other pathway would be derived from genetics research that gives clues about brain mechanisms at the level of neurons and synapses. This would open the door to experiments that modify the systems and their influence on reasoning and problem solving. Both pathways are complex and long-term projects. The neuroimaging path may be shorter than genetic interventions given current technology. But, in the long run, genetics may hold the key for the most dramatic enhancements possible.

In Chapter 2 we discussed Doogie mice, a strain bred to learn maze problem-solving faster than other mice. In Section 4.6 we reviewed some possible ways genes might influence the brain. Even if hundreds of intelligence-relevant genes are discovered, each with a small influence, the best case for enhancement would be if many of the genes worked on the same neurobiological system. In other words, many genes may exert their influence through a final common neurobiological pathway. That pathway would be the target for enhancement efforts (see, e.g., the Zhao et al. 2014 paper summarized in Section 2.6). Similar approaches are taken in genetic research on disorders such as autism and schizophrenia and many other complex behavioral traits that are polygenetic. Finding genes, as difficult as it is, is only a first step. Learning how those genes function in complex neurobiological systems is even more challenging. But once there is some understanding at the functional system level, then ways to intervene can be tested. This is the step where epigenetic influences can best be explicated. If you think the hunt for intelligence genes is slow and complex, the hunt for the functional expression of those genes is a nightmare. Nonetheless, we are getting better at investigations at the genetic and molecular functional levels (Chen et al., 2022; Makowski et al., 2022) and I am optimistic that, sooner or later, this kind of research applied to intelligence will pay off with actionable enhancement possibilities. The nightmares of neuroscientists are the driving forces of progress.

None of the findings reported so far are advanced enough to consider actual genetic engineering to produce highly intelligent children. There is a major development in genetic engineering technology, however, with implications for enhancement possibilities. A method for editing the human genome is called CRISPR/Cas9 (Clustered Regularly Interspaced Short Palindromic Repeats/Cas genes). I don't understand the name either, but this method uses bacteria to edit the genome of living cells by making changes to targeted genes (Sander & Joung, 2014). It is noteworthy because many researchers can apply this method routinely so that editing the entire human genome is possible as a mainstream activity. Two developers of this technique earned Nobel Prizes in 2020. This applied technology is already alleviating genetic diseases. Once genes relevant for intelligence and how they function are identified, this kind of technology might provide the means for enhancement, although the enormous complexity of gene–gene interactions and other combinations of factors may limit the application of gene-editing techniques for complex traits such as intelligence. Perhaps that is why the name of the method was chosen to be incomprehensible to most of us.

Most of this chapter is about what does not work to enhance intelligence. It is fair to say that education and cognitive approaches have made little demonstrable progress after many years of concerted efforts and neuroscience approaches are relatively nascent. We should not be discouraged, just as we are not discouraged that the hunt for genes that influence intelligence has progressed slowly. The brain is complex and its secrets are not easily revealed. All science is technology driven and intelligence research is no exception. As discussed in this chapter and previous ones, there are exciting possibilities for enhancing intelligence based on new neuroscience technologies and new information about brain structure, function, and development. From my perspective of nearly 50 years in the field, the pace of discovery is quickening. There is no clear roadmap for the future, but Chapter 6 presents some neuroscience perspectives on emerging approaches for learning even more about intelligence and the brain.

Chapter 5 Summary

- Despite many claims, there is no way yet to increase any intelligence factor that survives independent replication and creates a compelling weight of evidence.
- Studies that have made claims of enhancement have serious flaws including "teaching to the test," generalizing from small samples, and treating small score changes on single tests as indications of large changes in underlying intelligence factors.
- Psychoactive drugs and various nondrug methods of stimulating the brain may have potential for the cognitive enhancement of attention, learning, and memory but there is no weight of evidence yet that these methods enhance intelligence.

- Ultimately, enhancement may depend on not only finding specific genes related to intelligence but also on the harder problem of understanding how those genes function on a molecular level, including epigenetic influences.

Review Questions

- Why is the "weight of evidence" concept especially important for claims about enhancing intelligence?
- What are three examples of research findings that claimed sizable increases in IQ that proved incorrect?
- Explain the concepts of "transfer" and "independent replication."
- What are five methods of brain stimulation that may influence cognition?
- What are six ethical issues concerning the use of drugs for cognitive enhancement?

Further Reading

"Cognitive Enhancement" (Farah et al., 2014). This is a comprehensive discussion of enhancement issues.

"Increased Intelligence Is a Myth (So Far)" (Haier, 2014). Explains why intelligence test score increases do not mean intelligence has increased.

"Hacking the Brain: Dimensions of Cognitive Enhancement" (Dresler et al., 2019). This is a concise overview of different dimensions of cognitive enhancement.

"The Use and Impact of Cognitive Enhancers among University Students: A Systematic Review" (Sharif et al., 2021). This is an in-depth meta-analysis of worldwide student use of CE drugs.

As Neuroscience Advances, What's Next for Intelligence Research?

We choose to go to the moon and do the other things … not because they are easy, but because they are hard.

> President John F. Kennedy, speech at Rice University, September 12, 1962

The remarkable thing is that although basic research does not begin with a particular practical goal, when you look at the results over the years, it ends up being one of the most practical things government does.

> President Ronald Reagan, radio address, April 1, 1988

Without a doubt, this is the most important, most wondrous map ever produced by human kind.

> President Bill Clinton, remarks on completion of the first survey of the entire Human Genome Project, June 26, 2000

As humans we can identify galaxies light years away, we can study particles smaller than an atom, but we still haven't unlocked the mystery of the three pounds of matter between our ears.

> President Barack Obama, statement introducing the Federal Human Brain Initiative, April 2, 2013

The CHIPS and Science Act is going to inspire a whole new generation of Americans to answer [Steve Job's] question, "What next?"

> President Joe Biden on signing the CHIPS and Science Act of 2022, August 9, 2022

Learning Objectives

- What is chronometrics and why is it an advance over psychometrics?
- How does the study of memory and super-memory inform research on intelligence?
- How does research on animals provide insights about neurons and intelligence?
- How does neuroscience's understanding of brain circuits advance building intelligent machines?
- Given problems of definition, how can there be a neuroscience of consciousness and creativity?
- Why might SES and intelligence be confounded on the neural level?

Introduction

Paradoxically, in any area of scientific inquiry, the more we learn, the more we do not understand. An answer to one question often leads to a new question never before formulated. Advances depend on our intellect and imagination to make sense of new empirical observations obtained from creative methods and technologies that are constantly improving to provide new kinds of data. Think about the experimental validation of the standard model in particle physics as a result of observations made with multibillion-dollar accelerators. These huge, worldwide efforts also revealed new mysteries such as dark energy that cannot be solved by existing methods so new ones must be invented. Each generation of researchers builds on the recent past and extends into the near future. The early researchers who studied rudimentary language and perceptual deficits in patients with brain damage could not imagine the neuroscience tools now available to address questions about intelligence and the brain. All the advances in genetic and imaging methods and the potential for understanding and perhaps enhancing intelligence discussed so far in this book are just the beginning. There is more to come, but the pace depends on whether there is a commitment for generous funding that is dispensed wisely. A focus on research that will likely yield practical results is not necessarily wise, as history shows many examples of unforeseen major benefits from seemingly arcane basic research. It is nearly impossible to imagine, but what if a country ignored space exploration and announced that its major scientific goal was to achieve the capability to increase every citizen's g-factor by a standard deviation? By the end of this chapter, you might not think this is so impossible.

In every area of science, each stage of progress becomes more expensive and complex to conduct logistically and it becomes more complex to interpret results. For neuroimaging, CAT scans are more complex than X-rays; structural MRI is more complex than CAT. PET is more complex than EEG; functional MRI is more complex than structural MRI; MRI spectroscopy and DTI are more complex than structural or fMRI; MEG is more complex than MRI. Each new technology provides better spatial and temporal resolutions and amasses bigger and bigger data files that require more advanced computing power for processing and analyses. MRI shows brain tissue in millimeters but even this is far too big for showing individual neurons or synapses. MEG shows brain activity changes every millisecond but this is far too slow to show nanosecond neurochemical events in the synapse. Neuroscience techniques are available to study the brain at the level of single neurons and synapses so it is not beyond imagination that these techniques can be applied to questions about intelligence. Advances in the intelligence field will likely come from the integration of findings from basic research on clinical brain disorders, aging, and normal cognitive processes such as learning, memory, and attention from both animal and human studies that expose events that are smaller and smaller, faster and faster, and deeper and deeper in the brain.

This chapter highlights six developing lines of inquiry relevant to intelligence. Before discussing them, here is a brief recap of three main points developed in the previous chapters. (1) Based on the weight of evidence, intelligence is something that can be defined, measured, and studied scientifically, especially the g-factor, which correlates to many real-world outcomes, and brain structure and function, and has a strong genetic basis. (2) Neuroimaging research is beginning to identify specific brain characteristics related to intelligence differences among people, and genetic research is beginning to identify specific genes that influence intelligence. These advances, driven by technology, are moving intelligence research in a more neuroscience direction. (3) How brain characteristics related to intelligence develop from genetic, biological, and environmental factors, and their combinations, is not yet understood. But once we have a better understanding of how these factors work in the brain, we should be able to manipulate them to increase intelligence either to close any gaps among groups or to raise everyone, perhaps dramatically. Building on these three points and moving forward, here are six exciting areas to watch for progress, each in its own section.

6.1 From Psychometrics to Chronometrics to "Gamification"

On one side of the equation that links genetic and neuroimaging data to intelligence, we have the most up-to-date multimillion-dollar equipment and teams of specialists to collect and analyze complex data sets. On the other side of the equation, we have a psychometric test score, often from a single test that costs a few dollars. This is quite a mismatch, or more accurately a chasm. Decades ago, the earliest imaging studies of intelligence and the earliest quantitative and molecular genetic studies of intelligence used the same intelligence tests still used today. To advance the field, the study of intelligence can no longer be limited to psychometric test scores. As noted in Chapter 1, a sophisticated measurement of intelligence is badly needed to match the sophisticated genetic and neuroimaging assessments that are widely available. At minimum, a latent variable approach that extracts a factor from a battery of tests is required. The optimal assessment of intelligence will require a ratio scale, as noted in Chapter 1.

Let's review why this is so using a new example. Suppose you have an intervention that is designed to increase happiness (pick whatever intervention you like). You measure happiness by asking participants to rate their happiness on a scale from 1 to 10, where a 10 represents the most happiness. You find an average happiness score of 4 for a group before the intervention. After the intervention, the group average has increased to 8. If you are naïve about measuring constructs such as happiness, you might conclude that your intervention resulted in people becoming twice as happy based on the change from 4 to 8. You would be wrong to conclude this. Your happiness scale is an interval scale where points are not equivalent and each person has a subjective idea of what 4 or 8 means. Eight on an interval scale is not literally 2×4. Eight pounds, however, is literally 2×4 pounds because pounds are points on a ratio scale that is bounded by an actual zero point of no weight. A pound of bricks weighs the same as a pound of feathers. A pound is a pound irrespective of what is being measured.

Intelligence test scores, like all measures of happiness, are all on interval scales. Your score has meaning only relative to other people, typically expressed as a percentile. If you are at the 95th percentile, how much more intelligent are you than someone at the 90th percentile? You are not 5 percent more intelligent. We do not have a measure of intelligence as a quantity. In Chapter 4, we discussed whether intelligence could be defined by quantifying brain characteristics such as amount of gray matter, thickness of the cortex, connectivity of networks, or the integrity of white matter. These are all potential ratio scales but imaging is not a practical basis for wide use as an intelligence test in most settings.

Another way to create a ratio measurement of intelligence depends on the measurement of time (eight seconds is literally twice four seconds). The concept is to create a standard battery of mental tests where the time it takes to arrive at an answer is the basis for the measurement rather than the number of correct answers. Intelligence could then be defined as speed of information processing during a standard set of test items. A person who had an average time of four seconds on a battery of information-processing test items would be literally twice as fast as a person with an average of eight seconds. The validity of information-processing speed as an alternative definition of intelligence would require research establishing what this metric might predict in terms of academic or other achievements. In fact, a considerable body of research like this already exists.

In his last book before he died, Arthur Jensen summarized this research and considered the technical obstacles to overcome for developing a new kind of intelligence test based on information-processing time (Jensen, 2006). He called this approach "chronometrics" and examples are given in Textbox 6.1. If research supports the validity and reliability of the chronometric approach by establishing correlations with intelligence test scores, its use in future genetic and neuroimaging studies could narrow the sophistication-of-measurement gap. Jensen expressed the optimistic view that chronometric approaches could elevate intelligence research to a natural science. Combined with other neuroscience approaches, the pace of discovery would surely increase with this kind of measurement.

Textbox 6.1: Chronometric assessment of intelligence

As proposed by Jensen (2006), mental chronometry is based on two fundamental concepts. The first is that the time it takes to make a decision is a measure of brain processing speed. This is often referred to as reaction time (RT) or response time. RT studies have a long history in psychology going back more than 100 years. Many cognitive tasks have been used in RT studies. Often, they are called elementary cognitive tasks (ECTs). One of the most replicated findings is that RT increases with task complexity. Another core finding is that people with faster RTs generally have higher IQ scores. Therefore, the measurement of RT can potentially be used to measure intelligence and RT is an especially attractive metric because time is a ratio scale. RT for most ECTs is measured in milliseconds or seconds, depending on the task. The second fundamental concept is standardization. Different researchers often do RT studies using different devices. This lack of uniformity introduces method variance that confounds the RT assessment of individuals and makes it difficult to compare studies or to combine data from different studies into a large data set. Jensen proposed building a standard device to test RT to a standard set of diverse ECTs. Jensen believed that the combination of RT measures with a standardized method for testing ECTs would advance the study of intelligence beyond the limitations of psychometric tests such as the WAIS

For example, one ECT involves eight buttons arranged in a semi-circle. To start a trial, the person being tested presses a finger on a home button below the semi-circle, holding it down. Three of the eight buttons then light up simultaneously. One of them will be further away from the other two. As quickly as possible, the person releases the home button and presses the lighted button furthest away from the other two. This is called an "odd man out" task. After a series of such trials, a person's average RT is computed. A different ECT requires the person to memorize a string of numbers (or letters or shapes) after seeing them for a brief period on the display screen. Then a target number (or letter or shape) appears and if the target was in the string memorized, the designated "yes" button is pressed. If the target is not in the string, the "no" button is pressed. As the trials continue, the string gets longer so more memory is scanned in order to decide yes or no. This increases RT and higher-IQ people generally scan memory faster than lower-IQ people. Another ECT shows two words on the screen simultaneously. If they are synonyms, one button is pressed; if not, another button is pressed. There are many variations of these tasks and many other ECTs. Research will establish which ECTs will generate RTs that, in combination, make a good battery to assess intelligence. There are many technical issues to resolve. There is a long road ahead for this research before mental chronometry might replace psychometric tests of intelligence. At the end of his book, Jensen concluded that "chronometry provides the behavioral and brain sciences with a universal absolute [ratio] scale for obtaining highly sensitive and frequently repeatable measurements of an individual's performance on specially devised cognitive tasks. Its time has come. Let us get to work!" (2006: 246). This method of assessing intelligence could establish actual changes due to any kind of proposed enhancement in a before and after research design. The sophistication of this method for measuring intelligence would diminish the gap with sophisticated genetic and neuroimaging methods.

It may even be possible to define intelligence by brain characteristics such as speed of information processing among areas or the amount of gray matter tissue in certain areas. The advantage of such definitions would be that they are quantitative on a ratio scale. Imagine that your information-processing speed on a standard test battery is twice as fast as someone else. Whether this might predict something about your future academic success or other variables better than an IQ score is an open question for empirical study.

Unfortunately, we are not likely to see this question addressed any time soon. It turns out that the ideal of chronometric testing to attain a ratio scale is no longer a priority with any research group, as far as I know.

Instead, many multicenter collaborative projects used educational attainments (years of education) as an easily available substitute to estimate IQ when thousands (or millions) of participants also completed DNA testing for GWAS. Now there is more emphasis on computerized testing that can be done at home over the internet.

Several research groups have developed their own computerized cognitive batteries from which g can be extracted, but the most recent effort exemplifies solid psychometric properties along with an element of engagement achieved by presenting five tasks in a "gamified" format called Pathfinder (Malanchini et al., 2021). The five include a vocabulary test, a missing letter test, a verbal analogies test, a visual puzzle test, and a matrix reasoning test. The development of this battery included twin studies of heritability and PGS prediction of Pathfinder g-scores. Pathfinder is a comprehensive and impressive project that may represent the epidemy of computerization of traditional psychometric tests, and it can likely achieve the goal of being completed by millions of research participants. This is a major advance, but it is not the ratio scale envisioned by chronometrics (although response times to Pathfinder items are recorded so a speed measure is possible for additional research). It is possible to assess millisecond reaction times on home computers in a standardized way, so perhaps a new generation of researchers will take up the chronometric challenge.

6.2 Cognitive Neuroscience of Memory and Super-Memory

In Chapter 1, we noted that one definition of intelligence is individual differences in the cognitive processes of learning, memory, and attention. Most cognitive neuroscience research does not include any assessment of intelligence as either an independent or dependent variable. As we reviewed in Chapter 4, results from any study of learning, memory, language, or attention may differ if participants are selected on the independent variable of high or low IQ or g-factor scores. As noted in Chapter 5, when intelligence is included as a dependent variable, as in the n-back training studies, the assessment is typically based on a single test score rather than on a latent variable extracted from a test battery.

All this is the old bad news. The more recent good news is that cognitive psychologists are becoming more interested in the relationships among language, memory, attention, and intelligence (Barbey et al., 2021). One area with considerable research is the relationship between working memory and the *g*-factor. In some psychometric studies they are empirically virtually identical (Colom et al., 2004; Kane & Engle, 2002; Kyllonen & Christal, 1990). In other studies, they are overlapping but separate constructs (Ackerman et al., 2005; Conway et al., 2003; Kane et al., 2005). Imaging research suggests some overlap in brain areas for both (Colom et al., 2007) and both may have genes in common (Luciano et al., 2001; Posthuma et al., 2003) but these issues are not yet settled (Burgaleta & Colom, 2008; Colom et al., 2008; Knowles et al., 2014; Thomas et al., 2015). The ultimate goal is to understand how intelligence may integrate fundamental cognitive processes such as memory and attention and the way they influence language and learning (Cowan, 2014). This will require cooperation among different research groups with access to many samples of individuals across the full range of intelligence that have completed a large, diverse battery of cognitive tests, DNA analysis, and neuroimaging with structural and functional methods. We are just beginning to see such comprehensive projects, as noted in Chapters 2 and 4.

Super-memory cases are also of increasing interest. In Chapter 1, we mentioned Daniel Tammet's recitation from memory of 22,514 digits of pi. According to the Guinness World Records, however, the current record for reciting pi from memory is an amazing 70,000 digits. A previous record of 67,890 was held by a person (CL) who is not a savant. He uses mnemonic methods (i.e., memory tricks – see Textbox 6.2) that allow the storage and retrieval of large amounts of information. One fMRI study recruited several participants in the World Memory Championship and found several brain areas were activated when mnemonic procedures were used (Maguire et al., 2003). Unfortunately, each participant used a different mnemonic strategy so the imaging results were not easily interpretable. At age 28 years old, CL, the holder of the Guinness record for memorizing 67,890 digits of pi, was studied with fMRI while he used his strategy and a strategy designed by the researchers as a control condition (Yin et al., 2015). CL has many years of training on his mnemonic method, which the authors described this way: "CL used a digit-image mnemonic in studying and recalling lists of digits, namely associating 2-digit groups of '00' to '99' with images and generating vivid stories out of them." An example of this method is created in Textbox 6.2. Eleven male graduate student controls were also scanned and tested in the same strategy conditions. According to the authors, the results suggested that CL relied on brain

Textbox 6.2: A memory trick

Here's how you can train yourself to memorize a long string of numbers such as pi just in case you decide to do so. Before memorizing the digits, create a list of words to represent 100 pairs of sequential digits. For example, if the sequence contained 0,0 that would be remembered as a dog. If the sequence contained 0,1 that would be remembered as a fish. Assign a word to the combination of 0,3 and another word for 0,4 and so on for 0,5 ... 5,0, 5,1 ... 9,9. The words can be, for example, animals, tools, famous historical figures, or anything you choose. As you begin to memorize the long string of numbers, convert sequential pairs to your pre-memorized list of 100 words and create a story linking each consecutive word. The more outrageous the story, the easier it is to remember. Let's say this is your standard list of words for each two-digit combination (showing only 8 of 100 pairs):

> *00 dog*
> *01 fish*
> *02 Lincoln*
> *03 hammer*
> *... 29 robin*
> *... 51 airplane*
> *... 86 shoe*
> *... 99 bank*

Now here is a string of 18 numbers to memorize: 860229000299000151. Convert this string to pairs and then convert the pairs to your pre-assigned word for each pair: 86 is shoe; 02 is Lincoln; 29 is robin; 00 is dog; 02 is Lincoln; 99 is bank; 00 is dog; 01 is fish; 51 is airplane. Then you memorize a story you create that is rich in visual imagery, such as: My shoe fits Lincoln and he kicks a robin that is eaten by a dog but Lincoln takes it to the bank where a dog is eating a fish on an airplane. With practice, this sentence is easier to remember, and after you memorize it you can convert the words back to two-digit pairs. It may seem quite awkward, but this kind of mnemonic strategy can be used effectively to remember many things from numbers to names. It requires considerable practice and imagination, but some people get remarkably good at it and a few people are extraordinary. This is how CL memorized 67,890 digits of pi. Unlike some of the electrical or drug-enhancement techniques discussed in Chapter 5, this is one thing you can try at home. There is no evidence, however, that increasing your ability to memorize like this increases your intelligence (see Chapter 5). It is an open question as to whether people who learn to excel at memorizing using this method are people who already start with high intelligence scores.

areas related to episodic memory rather than verbal rehearsal. The imaging results are actually quite complex and open to interpretation (Sigala, 2015).

PET was used in a similar study of a mental calculation prodigy (Pesenti et al., 2001). Mental calculators are exceptionally accurate and fast at solving complex calculations in their heads. Whereas some savants apparently have this ability without training, the person studied in this report, 26-year-old R. Gamm, is a healthy individual and not a savant. He had, however, "trained his memory for arithmetic facts and calculation algorithms several hours each day for about six years" starting at age 20. For example, he could calculate two-digit numbers to various powers (e.g., 99^5 equals 9,509,900,499 or 53^9 equals 3,299,763,591,802,133). He could also do roots, sines, and divisions of prime numbers, and apply an algorithm to perform calendar calculations to name the day of the week for any date (another ability found in savants). Gamm was compared to six nonexpert male students scanned as controls performing the same tasks during PET determination of regional blood flow. PET scans were obtained during a calculation task and a memory retrieval task. The results showed brain activations in several areas common to both Gamm and the controls, but Gamm also showed unique activations when the two task conditions were contrasted. Gamm showed more activation in medial frontal and the parahippocampal gyri, the upper part of the anterior cingulate gyrus, the occipito–temporal junction in the right hemisphere, and the left paracentral lobule (see where these areas are in Figure 6.1). The authors concluded that:

[C]alculation expertise was not due to increased activity of processes that exist in non-experts; rather, the expert and the non-experts used different brain areas for calculation. We found that the expert could switch between short-term effort requiring storage strategies and highly efficient episodic memory encoding and retrieval, a process that was sustained by right prefrontal and medial temporal areas. (Pesenti et al., 2001: 103)

In other words, Gamm's brain worked differently.

Imaging such rare individuals with exceptional mental ability achieved by training may provide insights about the effects of intensive strategy training over years on brain networks or insights about unusual brain connections that apparently result by chance or unknown factors in the case of savants. There is no indication that either CL or R. Gamm showed an increase in g related to their respective intensive memory training.

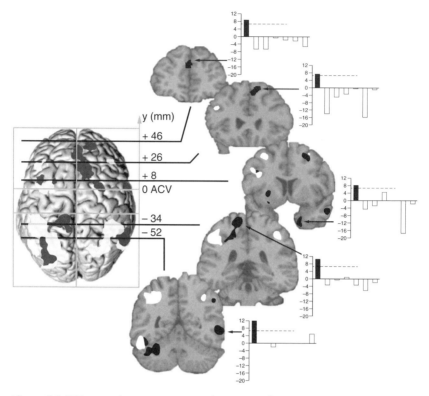

Figure 6.1 PET scans of an expert memory champion performing complex mental calculations compared to six nonexpert controls. Brain areas uniquely activated by the expert are shown in green; areas activated both by the expert and nonexperts are shown in red. Bar graphs show activations in each area for each person (red bar is the expert). (Reprinted with permission (Pesenti et al., 2001))
Note: A black and white version of this figure will appear in some formats. For the color version, please refer to the plate section.

6.3 Bridging Human and Animal Research with New Tools Neuron by Neuron

On the much smaller spatial scales of neurons and synapses, intelligence is not a major focus of interest for most neuroscience researchers. There are some attempts to relate neurotransmitters and other aspects of synaptic function to intelligence in molecular genetic studies, as we noted in Chapters 2 and 4 (Douw et al., 2021; Goriounova et al., 2018; Goriounova & Mansvelder, 2019; Heyer et al., 2021). Many questions remain ripe for examination. For example, does the number or type of mitochondria inside neurons (from any particular brain area) have any relationship to individual differences in the g-factor or other mental abilities? There are contrasting views about this possibility (Burgoyne and Engle, 2020; Geary, 2018, 2019; Matzel et al., 2020). An older postmortem human study suggested a relationship between the complexity of dendrites and education level (an indirect measure of intelligence) (Jacobs et al., 1993), but the direction of the relationship could go either way and replication is required (Genc et al., 2018). There are many possibilities to study intelligence on this level, especially if technology ever advances to the point where noninvasive measurements of single neurons and synapses can be made in humans.

Until such a time, animal studies provide some intriguing observations that suggest a tentative bridge to human studies. A systematic lesion study in rats, for example, found that several discrete brain areas were related to general problem-solving ability because lesions to those areas degraded performance on several different problem-solving tasks (Thompson et al., 1990). Lesions to other areas degraded performance only on specific tasks. The areas implicated in this study were compared to early PET studies of reasoning in humans but showed only limited overlap (Haier et al., 1993). Nonetheless, this combination of problem-solving tasks and lesions provided an animal model for brain–intelligence relationships that expanded the pioneering work of Karl Lashley (Lashley, 1964) and indicated that the g-factor is not unique to humans. Studies of genetically diverse (outbred) mice learning a variety of tasks also indicate a g-factor. The results from one study of mice sound strikingly like those from human studies:

Indicative of a common source of variance, positive correlations were found between individuals' performance on all tasks. When tested on multiple test batteries, the overall performance ranks of individuals were found to be highly reliable and were "normally" distributed. Factor analysis of learning performance variables determined that a single factor accounted for 38 percent of the total variance across animals. Animals' levels of native activity and body weights accounted for little of the variability in learning, although animals' propensity for exploration loaded strongly (and was positively correlated) with learning abilities. These results indicate that diverse learning abilities of laboratory mice are influenced by a common source of variance and, moreover, that the general learning abilities of individual mice can be specified relative to a sample of peers. (Matzel et al., 2003: 6423)

They also demonstrate the importance of an individual differences approach, even in mice (Sauce and Matzel, 2013).

Continuing this line of research, Matzel and Kolata summarized human imaging studies of memory/intelligence and mice experiments that tested causal relationships between aspects of selective attention, working memory, and general cognitive ability. They concluded that the data suggested "that common brain structures (e.g., prefrontal cortex) mediate the efficacy of selective attention and the performance of individuals on intelligence test batteries. In total, this evidence suggests an evolutionary conservation of the processes that co-vary with and/or regulate 'intelligence' and provides a framework for promoting these abilities in both young and old animals" (Matzel and Kolata, 2010: 23). Having such potent animal models of intelligence can help drive future neuroscience experiments, especially moving down the spatial and temporal scales from accumulated brain activity in specific areas to more precise measurements in neurons and synapses using methods not applicable to humans. There is some suggestion, for example, that physical and mental training in mice may increase the number of neurons and their survivability in specific brain areas (Curlik et al., 2013; Curlik and Shors, 2013). There is also some suggestion in mice that the signaling efficiency of the dopamine D1 receptor in the prefrontal cortex may relate to both memory tasks and intelligence tests (Kolata et al., 2010; Matzel et al., 2013; see also Alavash et al., 2018; Kaminski et al., 2018; Karalija et al., 2021; Lee et al., 2021). And there are informative mice studies of gene–environmental interactions (Sauce et al., 2018; Sauce & Matzel, 2018). It is too early to evaluate whether these findings represent a weight of evidence, but these studies demonstrate how an animal model of intelligence will help direct neuroscience progress down to the level of neurons and synapses.

Another illuminating example is the use of fluorescent proteins that literally light up neurons and synapses. The first fluorescent protein was discovered in jellyfish decades ago and that discovery has evolved into amazing techniques that create new fluorescent proteins and remarkable ways to introduce them into cells. Once inside a neuron, fluorescent proteins can track electrical activity and map neural circuits in the brain. Different fluorescent proteins attach to different neurochemicals and produce different colors. This means that the distribution of individual neurotransmitters can be mapped. In fact, individual neurons can be made to have a unique color so individual neuron pathways and their neurochemical signals can be mapped. Fluorescent studies of intelligence in mice would be fascinating. Doogie meets Mickey in a new *Fantasia* movie.

In Chapter 5, we briefly introduced one photo-neuro-modulation method that used red laser light to activate or deactivate neurons. Newer optogenetic and chemogenetic methods are more specific and based on modifying synaptic receptors so neurons react to special light-sensitive chemicals. Both methods have been used to modify mouse behavior by turning on light. In the process, experimental studies reveal neuro-circuits involved in complex behaviors and suggest ways to modify them.

The optogenetic method basically works like this. Normally, neurons fire when they receive a brief electrical pulse across synapses from neighboring neurons. This pulse changes the neurochemistry of the receiving neuron to create another pulse that travels to neighboring neurons in the circuit. The key neurochemical change involves proteins within the neuron. The electrical pulse stimulates the protein to create a new pulse to fire the next neurons in the circuit, and this cascade of firing continues until inhibitory signals diminish or stop the firing. Optogenetic techniques create light-sensitive proteins in specific populations of neurons. These neuron clusters can be induced to fire by applying light in controlled experiments. Light is delivered directly into neurons using hair-thin fiber optic thread after light-sensitive proteins have been introduced genetically into neurons of interest. In a mouse model of depression, for example, light stimulation of neurons in the medial frontal cortex relieves symptoms (Covington et al., 2010). Symptoms of cocaine addiction in mice can be reversed by light stimulation to neurons that project to the nucleus accumbens (Pascoli et al., 2012). Aggressive or sexual behavior in mice can be activated when a burst of light stimulates different neurons in the hypothalamus (Anderson, 2012). Optogenetic methods can be combined with the CRISPR-Cas9 method of gene editing (described in Chapter 5) so specific gene expression (turning genes on and off) can be targeted with light (Nihongaki et al., 2015) and can be used to explicate the activation details of specific brain circuits (Li et al., 2022). This incredible field is growing rapidly and there are many examples of experiments that could eventually lead to therapies for brain disorders, and perhaps to the enhancement of mental abilities (Adamczyk & Zawadzki, 2020; Aston-Jones & Deisseroth, 2013; Wolff et al., 2014). This sounds like science fiction, but it is happening now in a laboratory near you. Screenwriters, pay attention.

The chemogenetic method is a complementary approach for turning neurons on and off. This technique is based on creating "designer receptors exclusively activated by designer drugs" known by the acronym DREADD (Urban and Roth, 2015); how do they find these names? Recently, researchers developed a variation of DREADD that allowed neurons to be turned on *and* off, rather than the previous limitation of on *or* off techniques (Vardy et al., 2015). This allowed these researchers to turn hunger and activity levels on or off in mice for periods of time longer than can be done with optogenetic methods.

Another example uses DREADD methods to turn off specific neurons in the locus coeruleus (LC) in monkeys to study effects on vigilance and attention (Perez et al., 2022). The LC is the source of the neurotransmitter norepinephrine which is involved with learning, memory, and attention. Both the LC and norepinephrine are particularly interesting for fluid intelligence, as detailed in a comprehensive research review (Tsukahara & Engle, 2021). They conclude:

Various brain theories of intelligence have been proposed that range from lower-order metabolic functions … to higher-order functions, such as the functional organization of large-scale brain networks … The properties and functions of the locus coeruleus–norepinephrine system connect the lower-order and higher-order brain functions, and therefore it has explanatory power to bridge various brain theories of intelligence. In general, subcortical and brainstem structures are at the intersection of sensory, cortical, and motor brain functions, and they need to be considered for a more complete picture of the biological basis of intelligence and cognitive abilities to emerge. (Tsukahara & Engle, 2021: 10)

The neuroimaging methods described in Chapters 3 and 4 give researchers a view of the brain like the view of a city from a high-flying airplane: a unique and informative view not possible before the invention of the airplane. These new neuroscience techniques give researchers experimental control over individual neurons. This is like an aerial view that allows individual cars to be seen on a city street and possibly who is in the car and how fast their heart is beating. We can only imagine further refinements, new DREADDs, and new experiments (Ozawa & Arakawa, 2021). There is breathtaking potential for elucidating intelligence brain circuits if these techniques are applied to animal models of intelligence such as those described in this section by Matzel and colleagues. If such methods are available in humans, the potential for neuroscience/intelligence research staggers the imagination. Are you ready to change your major or thesis topic?

6.4 Bridging Human and Machine Intelligence Circuit by Circuit

The goal of AI research is to create computer software and hardware that mimics human intelligence. There are many wildly successful applications of "smart" technology that continue to change everyday life throughout the world. There are computer programs that beat chess grand masters, *Jeopardy* champions, and poker players. Engineers have developed most advances in AI with limited input from neuroscientists, mostly related to methods based on computational models of neural networks. But, an even more ambitious goal is to create intelligent machines with algorithms based on how neurons communicate in actual brain circuits explicated by basic neuroscience research. This is "real" intelligence.

A popular book by computer engineer and entrepreneur Jeff Hawkins makes a compelling case for building an intelligent machine using this neuroscience-based approach (Hawkins & Blakeslee, 2004). A key point is that computers and brains work on entirely different principles. For example, computers must be programmed and brains are self-learning. His core idea is that the cerebral cortex works fundamentally as a hierarchical system for storing and applying memory, especially memory of sequences, to make predictions about the world, and that this system is the essence of intelligence. One key insight is that the elements of this system are integrated by a single, all-purpose cortical learning algorithm (CLA). Therefore, Hawkins believes that the AI approach of designing separate elements of the system for machines is inherently limited. He believes that it is possible to design machines based on an all-purpose CLA, and that such machines might exceed human mental abilities. Here is how he states the challenge:

> For half a century we've been bringing the full force of our species' considerable cleverness to trying to program intelligence into computers. In the process we've come up with word processors, databases, video games, the Internet, mobile phones, and convincing computer-animated dinosaurs. But intelligent machines still aren't anywhere in the picture. To succeed, we will need to crib heavily from nature's engine of intelligence, the neocortex. We have to extract intelligence from within the brain. No other road will get us there. (Hawkins & Blakeslee, 2004: 39)

Hawkins has created the Redwood Neurosciences Institute and a company called Numenta to make brain-informed intelligent machines a reality. Numenta markets software based on algorithms that identify patterns, trends, and anomalies in large data sets. It is too early to evaluate how the hierarchical CLA concept might relate to the hierarchical g-factor, but it clearly fits the theme of this chapter on neuroscience approaches to intelligence and where they might lead (see Hawkins's new book that continues his theorizing about human and AI: Hawkins, 2021).

AI research is vast and a continued main effort is to develop artificial general intelligence (AGI), which seems to be similar in concept to human g. An influential paper by an AI researcher makes a compelling case for assessing how intelligent a machine may be, not by assessing the specific skills it can perform but rather by its ability to generalize (Chollet, 2019).

It is beyond my expertise to discuss these efforts with any insights other than to hope research on the human g-factor might be helpful conceptually. Three papers written by intelligence researchers addressed such possibilities in a special issue of *Intelligence* that explored future research directions (Haier, 2021). First, Aljoscha Neubauer discussed some broad philosophical and practical implications of intelligence research for AI and the possibility of "super intelligence" (Neubauer, 2021). Among other points, he noted what could be a key question: Will AGI ultimately mimic (or enhance) human g or be something different? There is not yet an answer, but he took an optimistic view that humans would prevail. Second, another thought-provoking paper reviewed what is basically known about human g, and then provided a model of cognition to consider for AI research. Its core conceptual aspect in humans included a meaning-making mechanism called a "noetron" (Demetriou et al., 2021). These authors speculated that an AI version of a noetron would be an important step toward AGI. Recognizing the difference in meaning between raising your index finger and your middle finger requires a noetron (my example, not theirs). Finally, the third paper reviewed AI research and asked the critical question: "How much intelligence is there in artificial intelligence?"(van der Maas et al., 2021). They find much overlap and anticipate even more interplay between human intelligence and AI in the future. Here is a long passage in their own words:

We expect that the recent progress in AI will change the way we think about intelligence. AI forces us to rethink the definition of intelligence. Definitions that center on just information processing and problem solving are perhaps insufficient. [The] observation that intelligence is all about generalization has, so far, withstood the test of time. Many information processing problems, from processing speech to playing chess, appear to be less difficult than perhaps expected. The really hard problem is to deal with completely novel cases. One requirement for solving this hard problem is the ability to learn invariant and thus generalizable patterns. And especially with regard to learning, the progress in AI has been spectacular. The main difference between AI systems of the past, such as expert systems, and modern AI is the fact that they learn. That deep learning and reinforcement learning, the core techniques in current AI, have deep roots in psychology is remarkable and promising for studying how artificial and human intelligence are related. (van der Maas et al., 2021: 7)

They continue in their summary:

AI is relevant to intelligence research because it enhances our understanding of the core mechanisms of human cognition. How the immense neural systems in our brain are able to process extremely complicated information such as speech and produce logical thinking is an extremely difficult question. Having an artificial system that performs such tasks using the same basic principles is extremely useful. Classic questions regarding the modularity of the mind, the origin of creativity, and the organization of long-term memory spring to mind. In addition, we argued that the psychological relevance of AI extends to unexpected areas such as the understanding of individual differences and the development of cognition. It is relatively easy to create a population of AI systems with minor differences in architecture and training regime. Modern AI provides us with a new playing field for individual differences research. (van der Maas et al., 2021: 7)

I find the last two sentences especially energizing for future research.

The concept of building machines based on the way the brain works is also informing the design of microchips. A number of companies are building chips based on a rudimentary understanding of neural circuits. Other companies and research groups are working on building microchips to perform brain functions, especially those related to perception, based on neural circuitry data. The general effort is known as neuromorphic chip technology. Some of these chips are designed to have a direct interface with the brain. Already chips are available to help enhance hearing and vision. These efforts may one day expand to cognitive processes, but I am unaware of any neuromorphic successes related to specific mental abilities let alone general intelligence. Nonetheless, this kind of research fuels speculation about the possibility of "super intelligence" and what it might mean for humanity (Neubauer, 2021). This too is an area ripe for fertile imagination – remember, today's science fiction may be tomorrow's reality and there are already university programs in neuro-engineering.

In Chapters 2 and 4 we introduced some multicenter consortia that are pooling genetic data to create very large samples for statistical analyses that maximize the discovery of small effects related to intelligence that are hard to detect in smaller samples typical of individual studies. There are also other large collaborative research programs that share data from many sources with the goal of mapping the structure and function of the human brain and how it develops. Current technology can produce maps at the neural circuit level. These maps can inform studies of aging, brain disorders, and brain diseases. They may also inform questions about learning, memory, and other cognitive processes. Such studies would be a prelude to addressing how individual differences in mental abilities, including intelligence factors, arise from differences among brains.

A bold goal was announced in 2005 by a group of scientists working in Switzerland. They undertook to create an artificial brain by building biologically realistic models of neurons and networks. Working with an IBM Big Blue supercomputer, they simulated brain activity starting with about 10,000 virtual neurons. This ambitious "Blue Brain" project expanded dramatically in 2009 when the European Union provided additional funding of 1.3 billion dollars and many additional collaborators joined the endeavor, renamed the Human Brain Project (HBP) in 2013. The stated goal was to simulate a human brain, all 80–100 billion neurons with 100 trillion connections. No neuroscience project has ever received this level of support. There is no shortage of controversy about every aspect of this project (Fregnac & Laurent, 2014). Blue Brain is now one aspect of the HBP with the stated goal focused on creating a whole mouse brain.[1] The larger HBP includes cognitive neuroscience, but intelligence per se is not explicitly on the research agenda. Hopefully, at some point someone with access to a simulated brain will wonder about just how smart the virtual brain may be, and what design features may be incorporated to increase its performance.

In the United States, there are more modest initiatives with similar goals of building simulated brains. The Defense Advanced Research Projects Agency funded the SyNAPSE Program (Systems of Neuromorphic Adaptive Plastic Scalable Electronics) in 2008 through 2016 (it sounds like someone badly wanted to call this SYNAPSE and worked backward with a committee). The ultimate aim is to build a microprocessor system that emulates a mammalian brain. In 2013, the White House announced the BRAIN Initiative (Brain Research through Advancing Innovative Neuro-technologies) that provided funding for projects that will lead to detailed functional and structural maps of the brain. This initiative builds on other collaborations that have already been funded, such as the Human Connectome Project, which is one of the few to include cognitive tests from which a g-factor can be derived.

All these major funding initiatives, and many other government–private collaborations, speak well for the future of neuroscience research. They have accelerated the enthusiasm generated more than 30 years ago when President George H. Bush declared that the 1990s would be

---

[1] Blue Brain's Scientific Milestones, www.epfl.ch/research/domains/bluebrain/blue-brains-scientific-milestones/.

the "Decade of the Brain." Unfortunately, at that time intelligence was not mentioned among the targets for research (Haier, 1990). There is an understandable general justification for basic research that may have practical implications for understanding brain diseases and disorders. At some point, these newest twenty-first-century efforts to map and simulate brains may also recognize that attention to intelligence is equally worthy. We now apparently have brain fingerprints that predict intelligence. Once a realistic virtual human brain exists, can creating real intelligence be far behind?

6.5 Consciousness and Creativity

This is a good place to comment briefly on consciousness and creativity. Like intelligence, both are among the highest-order functions of the human brain. If intelligence can be simulated, why not simulate creativity or consciousness? The idea that consciousness has a neuro-scientific basis has become mainstream, in large part based on the popularity of Francis Crick's book, *The Astonishing Hypothesis* (Crick, 1994); Crick shared the Nobel Prize for discovering the molecular structure of DNA. Some of the research efforts to understand the neural basis of consciousness include neuroimaging studies of humans in varying degrees of consciousness induced by different anesthetic drugs. My friend and colleague Michael Alkire, an anesthesiologist, and I published some of the earliest PET imaging studies that investigated this (Alkire & Haier, 2001; Alkire et al., 1995; Alkire et al., 2000; Alkire et al., 1999). We were trying to establish which brain circuits were the last to deactivate as the participant became unconscious. From these studies, we hoped to infer the mechanisms of action for different anesthetic drugs and pinpoint the brain mechanisms responsible for consciousness. There has been no luck so far, but this ambitious goal remains the greatest unsolved mystery of neuroscience (Zhao et al., 2019).

I raise the topic here briefly because during our early PET experiments, we wondered if there might be a link between consciousness and intelligence. We tend to regard everyone who is awake as conscious, but are there degrees of "awakeness"? Are some people more conscious (aware) than others and could such differences be related to intelligence? We have no clear way to assess individual differences in consciousness in individuals who are awake. One hypothesis that could be tested is whether high-IQ people need more (or less) of an anesthetic drug to render them unconscious for surgery, assuming there is a valid measure of depth of anesthesia. We have not pursued this question, but it seems reasonable to suspect that the two highest-order activities of the human brain may have circuits in common. The mechanisms of action of anesthetic drugs remain unclear, but if there are common circuits between consciousness and intelligence, we might speculate that new drugs that work in opposite ways than anesthetic drugs may produce hyper-consciousness or hyper-awareness, which are possible aspects of higher intelligence.

Similarly, I want to discuss briefly neuroscience studies of creativity as they relate to intelligence. My friend and colleague Rex Jung is a neuropsychologist who specializes in neuroimaging studies of creativity. We have pursued whether intelligence and creativity may have common neuro-circuits. There is some overlap between creativity and intelligence (Haier & Jung, 2008; Jung, 2014). Creativity and the creative process are even more difficult to define and assess for empirical research than intelligence and reasoning. The same general approach, however, is applicable. A battery of tests that assesses different aspects of creativity is given and a creativity index is derived either by summing the scores of individual tests (like IQ scores) or by extracting a latent creativity variable such as the g-factor. Aspects of creativity include, for example, measures of originality, fluency of ideas, and divergent thinking. However, whether there is a g-like factor of general creativity that transcends different specialty fields is still an open question. Creative artistic ability in dance, painting, or music might have quite different neural substrates and they might not overlap at all with the neural aspects of creativity in the fields of science, literature, or architecture. There is also the question of genius, a concept equally challenging to define for research. Can a creative genius have lower than average IQ, as it seems in some cases of savants? Can an intellectual genius have no creativity? Does the rare "true" genius require both high intelligence and high creativity? There are not yet clear empirical answers but neuroscience approaches may help resolve these basic issues. Creativity research is a large field and emerging neuroscience research on creativity has expanded rapidly, as shown in a collection of creativity research papers on topics including the role of neurotransmitters, the lack of strong evidence for a left brain–right brain laterality hypothesis, and the relationship between creativity and psychopathology, among others (Jung & Vartanian, 2018). Here we limit our focus to a small number of illustrative studies.

I am not aware of any verified cases where brain damage or illness resulted in increased intellectual ability. However, there are apparently rare cases where people have demonstrated a dramatic new creative ability, often artistic, after they develop frontotemporal dementia (FTD), a degenerating illness similar to Alzheimer's disease. This observation is not typical for FTD patients (Miller et al., 1998; Miller et al., 2000; Rankin et al., 2007). This is intriguing because it raises the possibility that creativity might be unleashed in more people if only certain brain conditions changed, although dementia is hardly a positive change. The general idea, however, is that the disinhibition (i.e., deactivation) of neural circuits and networks caused by the disease process is a key factor because disinhibition allows more associations among brain areas that do not routinely communicate. There are many ways to disinhibit the brain in general, such as drinking alcohol or developing FTD, but disinhibition targeted to particular neural networks related to creativity may be possible without affecting other networks necessary for balance, coordination, memory, and judgment. Are there creativity networks in the brain?

Functional neuroimaging studies have tried to capture brain activity during the creative process. There are now many studies, for example, that have imaged people with fMRI while they performed musical improvisation as an expression of creativity (Bengtsson et al., 2007; Berkowitz & Ansari, 2010; Donnay et al., 2014; Limb & Braun, 2008; Liu et al., 2012; Pinho et al., 2014; Villarreal et al., 2013). Music improvisation is a manageable paradigm in experimental studies whereas imaging studies of the creative process in dance, architecture, and other domains is not as practical. One early study, for example, scanned six male professional jazz pianists with fMRI while they performed two tasks that required either improvisation or overlearned musical sequences (Limb & Braun, 2008). The results of this small study suggested that compared to the overlearned sequence, improvisation was associated with a combination of bilateral deactivation in some areas, especially parts of the prefrontal cortex (including BAs 8, 9, and 46) along with bilateral activation in other areas distributed across the brain including in the frontal lobe (BA 10). These findings are shown in Figure 6.2. A similar fMRI study of 12 male freestyle rap musicians compared the spontaneous creation of rap lyrics to previously memorized sequences, both conditions using the same musical background (Liu et al., 2012). The results also suggested a pattern of deactivations and activations, as shown in Figure 6.3, which is mostly consistent with the results of the Limb and Braun study of jazz pianists. Another study of 39 pianists with varying degrees of improvisation experience reported an association between length of experience and connectivity among brain areas that suggested more efficient information flow during creative expression for the more experienced musicians (Pinho et al., 2014). Less activity in frontal and parietal areas was associated with more improvisation experience as

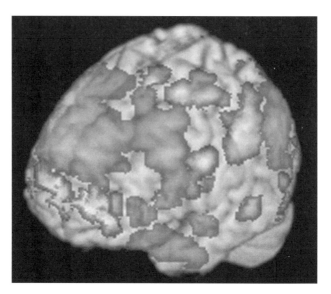

Figure 6.2 Brain activations (red/yellow) and deactivations (blue/green) in jazz pianists during improvisation. (Adapted from Limb and Braun (2008), open access)
Note: A black and white version of this figure will appear in some formats. For the color version, please refer to the plate section.

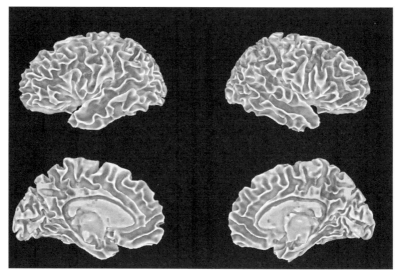

Figure 6.3 fMRI comparison of rappers during improvised and conventional conditions. Yellow represents significant increases in fMRI blood flow during improvisation; blue represents significant decreases. Top row shows cortical surface; bottom row shows medial (inside) surface. (Adapted from Liu et al. (2012) open access)
Note: A black and white version of this figure will appear in some formats. For the color version, please refer to the plate section.

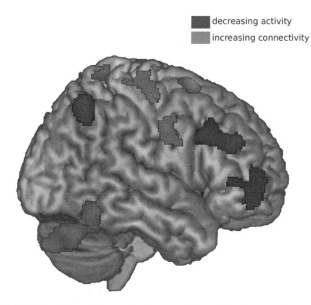

decreasing activity

increasing connectivity

Figure 6.4 fMRI in pianists with varying degrees of improvisation experience. More training is related to less brain activity (blue) during creative expression and to increased functional connectivity among other areas (red). (Reprinted with permission from Pinho et al. (2014: figure 3), free access)

Note: A black and white version of this figure will appear in some formats. For the color version, please refer to the plate section.

shown in Figure 6.4. A meta-analysis of musical improvisation studies like these tried to integrate findings and explain inconsistencies among studies (Beaty, 2015), but one recent review sees some progress despite the complexities inherent in having different kinds of musical creativity (Bashwiner, 2018).

An earlier comprehensive review of 45 functional and structural neuroimaging studies of creative cognition (not limited to musical improvisation) also noted inconsistent findings (Arden et al., 2010). There was a range of different creativity measures across the studies, often only one test score per study, and different imaging methods were used. Perhaps not surprisingly, the results showed disappointedly little overlap among the studies. Figure 6.5, for example, shows the inconsistencies among seven fMRI studies. The authors concluded that a more standardized approach to creativity assessment was necessary for any progress. They proposed eight suggested goals and actions to accomplish them:

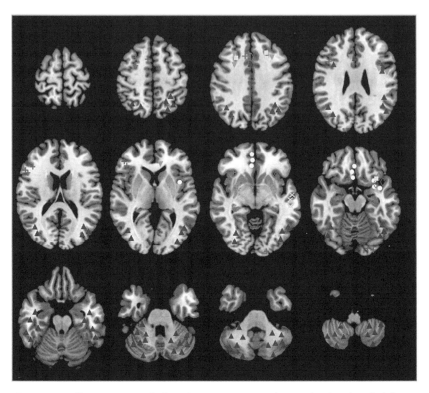

Figure 6.5 Different creativity findings from seven MRI studies. Each colored symbol shows activated brain areas related to creativity from a different study. There is little overlap of areas across studies. Arden et al. (Reprinted with permission from (2010: figure 1))
Note: A black and white version of this figure will appear in some formats. For the color version, please refer to the plate section.

(1) Goal: discover whether creative cognition is domain-specific. Action: test people phenotypically across many domains of creative production to quantify the common variance. (2) Goal: increase reliability of the measure. Action: use exploratory factor analysis – administer diverse creative cognition test batteries to large samples (N > 2000). (3) Goal: improve discriminant validity. Action: include intelligence (indexed by a reliable IQ-type test) and openness to experience (assessed by a reliable personality test) as covariates. (4) Goal: improve ecological validity of the criterion. Action: use evolutionary theory to inform or guide test development. (5) Goal: explore the aetiology of creative cognition. Action: administer creative cognition tests to genetically informative samples such as twins. (6) Goal: improve confidence in our results. Action: increase sample sizes.

(7) Goal: increase comparability across studies. Action: converge on a common brain nomenclature. (8) Goal: increase power of detecting effects. Action: move to study designs that use continuous measures rather than dichotomies such as case-control. (Arden et al., 2010: 152)

Another contemporaneous comprehensive review (Dietrich & Kanso, 2010) noted some general consistencies among creativity studies, including a pattern of activations and deactivations involving frontal areas as well as other areas distributed across the brain in both hemispheres (contrary to the popular idea that creativity is principally a right-hemisphere function). A subsequent critical review came to similar conclusions and listed suggestions for future research that emphasized the important role for collaboration between creativity researchers and cognitive neuroscientists (Sawyer, 2011).

Rex Jung and I attempted to integrate neuroimaging findings from intelligence studies and creativity studies and try to relate them to genius (Jung & Haier, 2013). We focused on consistencies from structural imaging and lesion studies of creativity because they avoid problems of task-specific results that confound functional imaging studies and are a major source of inconsistent results. One study of 40 lesion patients who completed creativity tests was particularly informative because lesions in some areas were associated with deficits on different aspects of creativity (Shamay-Tsoory et al., 2011). Studies of frontal temporal dementia, noted previously in this chapter, were also informative. Based on a combination of these studies, we proposed the "Frontal Dis-inhibition Model" (F-DIM) of creativity (Jung & Haier, 2013). Figure 6.6 shows this model, which is designed for easy comparison to the intelligence PFIT model (see Figure 3.7). Only four F-DIM areas overlap with PFIT (BAs 18/19, 39, and 32), suggesting mostly independent networks for intelligence and creativity. In comparison to the 37 studies that were reviewed for the PFIT, the F-DIM is more tentative since it is based on a smaller number of structural-only imaging studies. The essence of the F-DIM is that networks related to creativity are mostly disinhibitory, especially in frontal and temporal areas that affect other parts of the frontal lobes, the basal ganglia (part of the dopamine system), and the thalamus (an important relay station for information flow) through white matter connections.

With respect to how the F-DIM and the PFIT might relate to genius, we speculated that:

[W]e must look not only to increased neural tissue or activity in key brain regions (e.g., frontal lobes), but perhaps also to some mismatch between mutually excitatory and inhibitory brain regions (e.g., temporal lobes) that

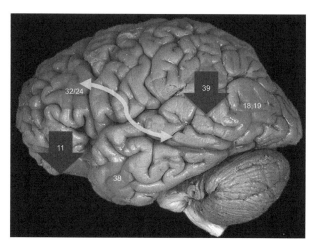

Figure 6.6 F-DIM of creativity. Numbers indicate BAs associated with increased (up arrows) or decreased (down arrows) brain activity based on a review of studies. Blue is left lateralized; green is medial; purple is bilateral; the yellow arrow is anterior thalamic radiation white matter tract. (Reprinted with permission (Jung & Haier, 2013)) Note: A black and white version of this figure will appear in some formats. For the color version, please refer to the plate section.

form a network sub-serving such complex human behaviors as creativity (e.g., planning, insight, inspiration). This notion of a delicate interplay of both increases and decreases in neural mass, white-matter organization, biochemical composition, and even functional activations within and between brain lobes and hemispheres is an important concept. Indeed, it is the rare brain that has highly developed networks of brain regions sub-serving intelligence and (concurrently) the somewhat underdeveloped network of brain regions associated with dis-inhibitory brain processes associated with creative cognition. Such a finely tuned seesaw of complex higher and lower brain fidelity, balanced in dynamic opposition, would almost guarantee the rare occurrence of genius. (Jung & Haier, 2013: 245–246)

Or as we say privately, we really don't know how intelligence and creativity are related to genius on the brain level.

Another comprehensive review of creativity research was based on a meta-analysis of 34 functional neuroimaging studies that included 622 healthy adults (Gonen-Yaacovi et al., 2013). A main analysis examined whether there were brain areas that were consistently activated despite the diversity of creativity tasks performed during the imaging. The analysis, however, was limited because it did not include areas of deactivation. The activation results for all studies together indicate some consistency, as shown in Figure 6.7. The resulting creativity map is consistent with the

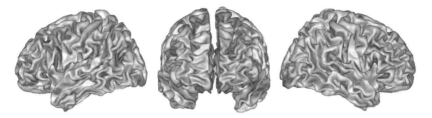

Figure 6.7 Summary findings from 34 functional imaging studies of creativity. Common brain areas of activation are shown revealing distributed networks related to creativity. From Gonen-Yaacovi et al. (2013: figure 1), open access)
Note: A black and white version of this figure will appear in some formats. For the color version, please refer to the plate section.

F-DIM and other studies showing a distribution of salient areas including frontal and parieto-temporal regions, especially the lateral prefrontal cortex. Some of the creativity tasks required the generation of ideas and other tasks required a combination of elements. Separate analyses for both kinds of tasks suggested anterior regions were involved in combining ideas creatively and more posterior regions were involved in freely generating novel ideas. There were also some differences between verbal and nonverbal tasks. In the case of both the shared creativity map (Figure 6.7) and the findings of the two kinds of tasks (not shown), areas in both the right and left hemispheres are associated with creativity, providing additional evidence that creativity is not an exclusive function of the right-sided brain. A reanalysis that includes areas of deactivation would be informative for a more complete picture given other findings related to disinhibition.

In fact, a meta-analysis of 10 small-sample fMRI studies of divergent thinking shows widespread areas of deactivation although, inexplicably, the Gonen-Yaacovi analysis is not cited (Wu et al., 2015). Also, a structural MRI study in 135 adults reported correlations between gray matter and a test of creative fluency and a test of creative originality. Each test was correlated with gray matter in different areas and there was an interaction with intelligence only for fluency (Jauk et al., 2015; also see Bashwiner et al., 2016 and a review by Jauk, 2018). Neuroscience studies of creativity are not yet as advanced as studies of intelligence, but there is increasing interest in how the two may be related in the brain.

Here is a final speculation for this section. If a deep level of disinhibition in certain brain circuits results in becoming unconscious, perhaps a bit less disinhibition may increase creativity. The perception of increased creativity is often a subjective response to "mind expanding" drugs such as LSD (Fox et al., 2018). Disinhibition of the frontal cortex is also associated with dreaming during sleep (Muzur et al., 2002). Obviously, sleep is an unconscious state and dreams are frequently quite creative in content and narrative. Tying creativity and consciousness research together based on neuro-circuits would further demonstrate there is a neural basis for creativity. There also is some genetic evidence regarding creativity (Ukkola-Vuoti et al., 2013), suggesting there might be a potential for enhancing creativity by affecting brain mechanisms. Many drugs have been described subjectively as creativity enhancers, but I am unaware of compelling empirical research that substantiates such observations. Increased creativity has been reported in a few studies that manipulate the brain without drugs (Fink et al., 2010) and there is a small tDCS study (Mayseless and Shamay-Tsoory, 2015), but so far there is no weight of evidence to support these preliminary reports. How intelligence may be related to creativity and consciousness on a neural level is an intriguing question that raises opportunities for imaginative research designs and innovative neuroscientists. Students, that means you.

6.6 Neuro-poverty and Neuro-SES: Policy Implications based on the Neuroscience of Intelligence

The confounding of SES with intelligence was introduced in Section 2.1. Now we consider it further because it remains an important problem that often results in misleading conclusions from research studies. Here is a common train of thought about the importance of SES:

Higher income allows upward mobility, especially the ability to move from poor environments to better ones. Better neighborhoods typically include better schools and more resources to foster children's development so that children now have many advantages. If the children have high intelligence and greater academic and economic success, it could be concluded that higher SES was the key factor driving this chain of events.

Here is an alternative train of thought:

Generally, people with higher intelligence get jobs that require more of the *g*-factor and these jobs tend to pay more money. There are many factors involved, but empirical research shows that *g* is the single strongest predictive factor for obtaining high-paying jobs that require complex thinking. Higher income allows upward mobility, especially the ability to move from poor environments to better ones. This often includes better schools and more resources to foster children's development so that children now have many advantages. If the children have high intelligence and greater academic and economic success, it could be concluded that higher parental intelligence was the key factor driving this chain of events, due in large part to the strong genetic influences on intelligence.

The latter train of thought is hardly new. It was made clear more than 40 years ago in a controversial book mentioned earlier in Chapters 1 and 2, *IQ in the Meritocracy* (Herrnstein, 1973). The argument was reduced to its simplest form in a syllogism: "(1) If differences in mental abilities are inherited, and (2) if success requires those abilities, and (3) if earnings and prestige depend on success, (4) then social standing (which reflects earnings and prestige) will be based *to some extent* on inherited differences among people" (Herrnstein, 1973: 197–198, emphasis added). When this was published in 1973, the evidence for a genetic role in intelligence was strong but not overwhelming and there was room for skepticism; today the evidence is overwhelming and compelling (see Sections 2.5, 2.6, 4.5, and 4.6).

Dr. David Lubinski has written a comprehensive review of the SES–intelligence confounding issue (Lubinski, 2009). Although the context for his paper is cognitive epidemiology, the argument applies to all research using SES as a variable. Essentially, if a study incorporates measures of both SES and intelligence, statistical methods can help disentangle their respective effects. The interpretation of results from any study of SES cannot disentangle which factor is driving the result unless a measure of intelligence is included in the study. Studies of intelligence without considering SES are also problematic. When both variables are included in multivariate studies in large samples, the results typically show that general cognitive ability measures correlate with a particular variable of interest even after the effects of SES are statistically removed. For example, in a study of 641 Brazilian school children, SES did not predict scholastic achievement but intelligence test scores did (Colom & Flores-Mendoza, 2007). An even larger classic study had data on 155,191 students from 41 American colleges and universities. Their analyses showed that SAT scores predicted academic performance with about the same accuracy even after SES was controlled; that is, SES added no additional predictive power (Sackett et al., 2009). Another study of 3,233 adolescents in Portugal found that parents' level of education predicted intelligence in the children regardless of family income. These researchers stated their conclusion straightforwardly: "Adolescents from more affluent families tend to be brighter because their parents are brighter, not because they enjoy better family environments" (Lemos et al., 2011: 1062). The strongest case so far for the predictive dominance of cognitive ability over SES comes from a comprehensive study based on both the 1979 and the 1997 National Longitudinal Surveys of Youth (Marks, 2022). Essentially, the findings showed that the predictions of life success measures by cognitive ability measures were strong and hardly improved after adding SES measures. By contrast, the predictions of success by SES measures decreased appreciably when cognitive measures were added. The author concluded that, "contrary to dominant narratives, cognitive ability is important to a range of social stratification outcomes, and its effects cannot be attributed to socioeconomic background or educational attainment" (Marks, 2022: 1).

Studies with large samples showing that SES effects remain after removing the effects of intelligence are less frequent, although one older meta-analysis suggested that SES independently predicts economic success about as well as intelligence (Strenze, 2007). Another older illustrative example of using both SES and IQ is a study of 110 disadvantaged middle school children. It included maternal IQ along with composite measures of parental nurturance and environmental stimulation (Farah et al., 2008). In the main analysis, parental nurturance was related to memory and environmental stimulation was related to language, after any effects of maternal IQ were statistically removed. The range of maternal IQ, however, was restricted to the lower end of the normal distribution (mean = 83, standard deviation = 9), possibly explaining the lack of an IQ finding, but this study does illustrate why it is important to include IQ measures when investigating specific SES factors. Replication in another sample of disadvantaged children would be important along with obtaining the father's IQ. Replication in a sample of children at higher SES levels would also be informative, as would studies of children at different ages, since the effects of SES on the heritability of intelligence may vary with age (Hanscombe et al., 2012). It is particularly interesting that there is emerging evidence that the SES itself has a strong genetic component (Trzaskowski et al., 2014). Obviously, there are many questions to pursue for establishing a weight of evidence regarding how SES and IQ relate to each other, but the Marks study of National Longitudinal Surveys of Youth samples is the most compelling (Marks, 2022).

One common view in cognitive psychology is that SES–cognitive relationships are mediated by how SES variables influence brain development during early childhood. Other researchers see such relationships as more related to neuroscience, especially when trying to relate such findings to education (Sigman et al., 2014). As you might imagine, the line between cognitive psychology and neurobiology is permeable (Hackman et al., 2010; Neville et al., 2013). The term "cognitive neuroscience" refers to both. Nothing about a major genetic component to intelligence and related neurobiological mechanisms negates or minimizes the importance of SES influences on cognitive psychology variables. Surely, SES is a consequence of many factors, but let's consider just the portion of SES that is confounded with the genetic portion of intelligence. I designate this portion by the term "neuro-SES," and in my view it should be recognized as a matter for research and discussion.

To repeat the main point, studies that make claims about SES variables without including measures of intelligence are difficult to interpret and need to at least acknowledge the "confound" problem before concluding or implying that SES has a causal role. This was a primary point made two decades ago in *The Bell Curve*. Nonetheless, bias toward SES-only explanations remains prevalent. Two high-profile examples illustrate the issue. Both studies use neuroimaging with structural MRI. The first paper is from MIT, reported by Dr. Mackey and colleagues (Mackey et al., 2015) (Dr. Mackey also reported a 10-point IQ increase in disadvantaged children following brief computer game playing in school; see Section 5.3). These researchers set out to study the neuroanatomical correlates of the academic achievement gap between higher- and lower-income students (n = 35 and 23, respectively). The higher-group average yearly family income was $145,465 (95 percent confidence interval between $122,461 to $168,470). The lower-group average yearly family income was $46,353 (95 percent confidence interval between $22,665 to $70,041). It is arguable whether family incomes of over $50,000 constitute a disadvantaged household, but the key finding is still of interest. Structural MRIs showed greater CT in several areas for the high-income group, although other brain measures did not (e.g., cortical surface area, cortical white matter volume). CT differences between the groups in some areas were related to standard test score differences. The authors concluded: "Future studies will show how effective educational practices support academic gains and whether these practices alter cortical anatomy" (Mackey et al., 2015: 8). This is fair enough and certainly supports a commonly held view. However, without assessing the cognitive ability of the parents, we cannot be sure whether the CT difference is related to family income or to the genetics of intelligence. The results from this study would be far more compelling had some estimate or measure of parental intelligence been included to help disentangle SES effects from intelligence effects.

The second paper is a multicenter collaboration reported in *Nature Neuroscience* by Dr. Noble and colleagues (Noble et al., 2015). This MRI study had a large sample of 1,099 children and adolescents. Data included family income, parental education, and genetic ancestry. Income was related to brain surface area even after controlling for parental education. Parental education related to other structural brain characteristics even after controlling for income. These associations were found irrespective of genetic ancestry. The authors state that:

[I]n our correlational, non-experimental results, it is unclear what is driving the links between SES and brain structure. Such associations could stem from ongoing disparities in postnatal experience or exposures, such as family stress, cognitive stimulation, environmental toxins or nutrition, or from corresponding differences in the prenatal environment. If this correlational evidence reflects a possible underlying causal relationship, then policies targeting families at the low end of the income distribution may be most likely to lead to observable differences in children's brain and cognitive development. (Noble et al., 2015: 777)

This is not an unreasonable statement, but one implication of this train of thought might be an experiment that provided modest or large monthly payments to low-income families to improve everyday life with the expectation that the resulting life changes might have subsequent effects on their children's brain and cognitive development; intelligence was not mentioned in the discussion of results. Some recognition and discussion of the neuroscience aspects of intelligence and its intertwining with SES would be important considerations if such an experiment was undertaken.

Well, this experiment is underway and, not surprisingly, my suggestion is not included. A preliminary report of results has received considerable positive attention in the media and less positive attention among many researchers. The report in *PNAS* by Dr. Noble and colleagues (Troller-Renfree et al., 2022: 1) states its significance:

This study demonstrates the causal impact of a poverty reduction intervention on early childhood brain activity. Data from the Baby's First Years study, a randomized control trial, show that a predictable, monthly unconditional cash transfer given to low-income families may have a causal impact on infant brain activity. In the context of greater economic resources, children's experiences changed, and their brain activity adapted to those experiences. The resultant brain activity patterns have been shown to be associated with the development of subsequent cognitive skills.

The reasons for positive media attention are obvious, especially in the presence of the word "causal." But why have many researchers been underwhelmed? Let's unpack some details. The overall research project is called Baby's First Years (BFY) and it is funded by federal grants and private foundations. The goal (from their website) is the "study of the causal impact of monthly, unconditional cash gifts to low-income mothers and their children in the first four years of the child's life." Family income in the year before birth averaged in the low $20,000s. The project has enrolled 1,000 infants from low-income families from four diverse geographic sites around the country. Mothers in one randomly assigned group received a monthly unconditional cash gift of $333; mothers in the other group, acting as a control, received $20 per month unconditionally. Both groups will receive payments for the first 52 months of their child's life. Periodic follow-ups will assess cognition and other developmental variables.

The results in the preliminary report were based on evaluation of EEG brain activity in a subsample of 435 infants when they were 12 months old. The EEG testing was done in the home with mobile equipment. Some of the pre-registered analyses (i.e., stated before analysis to prevent looking at many variables until any differences were found) generally showed the hypothesized EEG differences in infants from the two groups. Statistically, the differences were quite small, but some were consistent with other research that had suggested this kind of brain activity was correlated to better cognition in later childhood. The group EEG differences were the basis for the claim that the higher cash payments caused better brain activity and the presumption that better cognition would follow.

So, why were other researchers underwhelmed? For one thing, the sample of 435 infants in the analyses started with 605 eligible infants. The loss of 170 infants was due to several understandable factors but this is a high attrition rate. Moreover, not all the 1,000 infants enrolled in the study could be contacted for the 12-month evaluation due to Covid restrictions. Together, these losses decrease statistical power and may introduce biases that limit the generalization and replication of any results, especially those based on small group average differences. All the findings in this preliminary report are small and many were not statistically significant. Moreover, there was no information about how the cash was spent so it was not possible to intuit any specific factors that could be associated with the findings. In fairness, the authors recognized these and other weaknesses and limitations, and they cautioned to not overinterpret the results.

Nonetheless, critics wondered why such a preliminary study with obvious shortcomings could be published in a high-profile journal that would be sure to get considerable media attention (some of the authors were not shy about promoting this report). Cynics might say the main purpose was to generate public enthusiasm for the project and continued grant support. My view is that any poverty reduction program based on research deserves a trial as long as there is rigorous independent evaluation that includes analyses to disentangle the well-known SES–intelligence confounding. So far, I am in the underwhelmed group, but I am open to change my mind as new studies become available, especially from research designs that are genetically informative.

The bigger question to consider, however, is whether any neuroscience findings can inform social policies, especially regarding interventions aimed at reducing achievement and other gaps reported for disadvantaged groups. On one side of this question, there are projects such as the BFY that are justified, in part, by previous neuroimaging studies that suggest that poverty results in different brain characteristics that may be related to cognition (Farah et al., 2006; Noble et al., 2006; Noble et al., 2015; Noble and Giebler, 2020). Critics, however, argue that neuroscience studies cannot inform intervention policies because of the confounding of genetic and environmental influences. This argument is detailed by Professor Amy Wax in a comprehensive review of the issues that does not mince words (Wax, 2017). Here is a long passage:

The policy payoff of neuroscience is further undercut by the failure, and indeed the inability, of the research to disentangle the role of environmental and innate factors in generating the detriments of disadvantage. As already discussed, reducing poverty is most likely to make a difference if the experience of poverty, as opposed to innate characteristics, is the main or sole engine of the ill effects associated with low SES. Although traits with a significant genetic component are not necessarily fixed and impervious to interventional improvements, they might be. In any event, because the nature of gene/environmental interaction is as yet poorly understood, the payoff from any proposal must be assessed on a case by case basis. But this constraint, even when acknowledged, is often minimized by scientists working in the neurodeprivation field and scholars commenting on their work. Implicitly or by implication, environmental factors are assumed to dominate, especially when it comes to neuroscience's potential to point to effective solutions.

She continues:

Is science close to sorting out the importance of genes and environment in accounting for the patterns associated with lower SES? The answer is no, and further research faces practical and ethical impediments. As noted, prospective randomized trials, with well-defined treatment and control groups, provide the most promising avenue for disentangling causal mechanisms and identifying effective interventions. But the practical ambit of such trials is limited. At the end of the day, there is no substitute for seeing what works. What matters is whether specific forms of assistance improve target behavior. Visualizing or measuring something about the brain is not an essential component of the inquiry. Brain science, although intrinsically interesting, is inessential clutter. (Wax, 2017: 49–50)

It is still the case that the "Blank Slate" belief, discussed in Chapter 2, promotes SES and other social/cultural influences as critical to intelligence and its development. As noted throughout this book, the weight of evidence does not support the primacy of this view over a genetic one. There is also growing recognition that this view has failed to invigorate successful public policies aimed at closing widely acknowledged gaps in education achievement and cognitive skills shown by many disadvantaged children. A main implication of this book is that the empirical evidence overwhelmingly supports paying more attention to neurobiology as a foundation for changing the status quo. Professor Wax may be correct in the short run but, as argued in previous chapters, neurobiology can be modified, even if there are strong genetic components involved. This simple fact combined with advances in neuroscience research, such as the ones discussed in this chapter, provide some optimism for addressing serious problems that have persisted for decades. This is my long-term optimistic view about the possible policy implications of introducing neurobiology perspectives to research on these problems. But I know it will not be easy.

Not all individuals have a pattern of cognitive strengths that allow barely minimum success in a modern, complex society. This is evident with respect to g and other factors of intelligence. To the extent that different patterns of cognitive strengths and weaknesses are rooted more in neurobiology and genetics than in childhood experience, it is incorrect to blame lack of economic or educational success entirely on poor motivation, poor education, or other social factors. All these things matter but with respect to intelligence, they do not appear to matter that much, as the weight of evidence indicates.

Here is my political bias. I believe government has a proper role, and a moral imperative, to provide resources for people who lack the cognitive capabilities required for education, jobs, and other opportunities that lead to economic success and increased SES. This goes beyond providing economic opportunities that might be unrealistic for individuals lacking the requisite mental abilities. It goes beyond demanding more complex thinking and higher expectations for every student irrespective of their capabilities (a demand that is likely to accentuate cognitive gaps). It even goes beyond supporting programs for early childhood education, jobs training, affordable childcare, food assistance, and access to higher education. There is no compelling evidence that any of these things increase intelligence, but I support all these efforts because they will help many people advance in other ways and because they are the right thing to do. But, even if this support becomes widely available, there will be many people at the lower end of the *g*-distribution who do not benefit very much, despite best efforts. Recall from Chapter 1 that the normal distribution of IQ scores with a mean of 100 and a standard deviation of 15 estimates that 16 percent of people will score below an IQ of 85 (the current minimum for military service in the USA is about 85). In the United States, about 54 million people have IQs lower than 85 through no fault of their own (based on 16 percent of the normal distribution discussed in Chapter 1). There are useful, affirming jobs available for these individuals, usually at low wages, but generally they are not strong candidates for college or for technical training in many vocational areas. Sometimes they are referred to as a permanent underclass, although this term is hardly ever explicitly defined by low intelligence. *Poverty and near-poverty for them is a condition that may have some roots in the neurobiology of intelligence beyond anyone's control.*

The sentence you just read is the most provocative sentence in this book. It may be a profoundly inconvenient truth or profoundly wrong. But if scientific data support the concept, is that not a jarring reason to fund supportive programs that do not stigmatize people as lazy or unworthy? Is that not a reason to prioritize neuroscience research on intelligence and how to enhance it? The term "neuro-poverty" is meant to focus on those aspects of poverty that result mostly from the genetic aspects of intelligence. The term may overstate the case, especially if we learn how to influence genetic probabilities. It is a hard and uncomfortable concept but I hope it gets your attention. This book argues that intelligence is strongly rooted in neurobiology. To the extent that intelligence is a major contributing factor for managing daily life and increasing the probability of life success, neuro-poverty is a concept to consider when thinking about how to ameliorate the serious problems associated with the tangible cognitive limitations that characterize many individuals through no fault of their own.

Public policy and social justice debates might be more informed if what we know about intelligence, especially with respect to genetics, is part of the conversation. In the past, attempts to do this were met mostly with acrimony, as evidenced by the fierce criticisms of Arthur Jensen (Jensen, 1969; Snyderman & Rothman, 1988), Richard Herrnstein (Herrnstein, 1973), and Charles Murray (Herrnstein & Murray, 1994; Murray, 1995). After Jensen's 1969 article, both *IQ in the Meritocracy* and *The Bell Curve* raised this prospect in considerable detail. Advances in neuroscience research on intelligence now offer a different starting point for discussion. Given that approaches devoid of neuroscience input have failed for 50 years to minimize the root causes of poverty and the problems that go with it, is it not time to consider another perspective?

Here is the second most provocative sentence in this book: *The uncomfortable concept of "treating" neuro-poverty by enhancing intelligence based on neurobiology, in my view, affords an alternative, optimistic concept for positive change as neuroscience research advances.* This is in contrast to the view that programs that target only social/cultural influences on intelligence can diminish cognitive gaps and overcome biological/genetic influences. The weight of evidence suggests that a neuroscience approach might be even more effective as we learn more about the roots of intelligence. I am not arguing that neurobiology alone is the only approach, but it should not be ignored any longer in favor of SES-only approaches. What works best is an empirical question, although political context cannot be ignored. On the political level, the idea of treating neuro-poverty like it is a neurological disorder is supremely naïve. This might change in the long run if neuroscience research ever leads to ways to enhance intelligence, as I believe it will. For now, the inexact concept of epigenetics might help bridge both neuroscience and social science approaches. Nothing will advance epigenetic research faster than identifying specific genes related to intelligence so that the ways environmental (social/cultural) factors might influence those genes can be determined. There is common ground to discuss and that includes what we know about the neuroscience of intelligence from the weight of empirical evidence. It is time to bring "intelligence" back from a 50-year exile and into reasonable discussions about education and social policies without acrimony.

An insightful book explored this possibility. Authored by two behavioral genetics researchers, the starting point is an acknowledgment that all students enter the education system with different genetic propensities for learning reading, writing, and arithmetic (Asbury & Plomin, 2014). The authors propose policy ideas for tailoring the educational environment to help each student learn core material in a way that is likely best suited to that student's genetic endowment. This is a long way from the incorrect assumption that genes are always deterministic;

actually, genes are starting points. As the authors note, genetic research findings are uniquely excluded from discussions about education while at the same time genetic research has transformed aspects of medicine, public health, agriculture, energy, and the law. Individualized education is a longtime goal for educators and genetic research supports that goal. Asbury and Plomin conclude:

We aim to treat all children with equal respect and provide them with equal opportunities, but we do not believe that all our pupils are the same. Children come in all shapes and sizes, with all sorts of talents and personalities. It's time to use the lessons of behavioral genetics to create a school system that celebrates and encourages this wonderful diversity. (2014: 187)

This view is strikingly similar to Jensen's conclusion nearly 50 years ago (Jensen, 1969: 117): "Diversity rather than uniformity of approaches and aims would seem to be the key to making education rewarding for children of different patterns of ability. The reality of individual differences thus need not mean educational rewards for some children and frustration and defeat for others." Both views are common among neuroscientists who study intelligence and understand the probabilistic nature of genes. Nonetheless, failure to acknowledge the conclusive findings about the role of genetics for individual differences in intelligence and other cognitive abilities perpetuates the ineffective "one size fits all" approach to education reform. It is easy to see how ignoring what we know about intelligence has led, and will continue to lead, to frustration and failure for addressing any issue where intelligence matters (Gottfredson, 2005). Nonetheless, intelligence remains missing from public conversations.

In the United States, for example, considerable rancor pervades discussions about education reform even without any reference whatsoever to intelligence differences among students. The idea that every high school student should be held to a graduation standard of four-year college-ready, irrespective of mental ability, is naïve and grossly unfair to those students for whom this expectation is unrealistic. Remember, statistically half of the high school student population has an IQ score of 100 or lower, making college work considerably difficult even for highly motivated individuals; although there are exceptions (McGue et al., 2022). It is similarly naïve and unfair to evaluate teachers by student test score changes when many tests are largely de facto measures of general intelligence rather than of the amount of course material learned over a short time period. Perhaps the greatest disservice to students will come from purposefully increasing the difficulty of evaluation tests by requiring more complex thinking to get the right answers. The odds are that this change alone will *increase* performance gaps because the tests are now more *g*-loaded (e.g., on September 12, 2015, the *Los Angeles Times* reported a front-page story with the headline: "New Scores Show Wider Ethnic Gap").

In principle, there is nothing wrong with evaluation testing or having high expectations and standards. These examples, however, illustrate the consequences of ignoring what we know about intelligence from empirical studies when crafting well-intentioned policies for education, especially those policies that assume thinking skills can be taught to the same degree to all students, or that buying iPads for everyone in the education system will increase school achievement. As most teachers recognize, maximizing a student's cognitive strengths, whatever they may be, is a worthy goal. Everything we know from the research literature on intelligence supports this view, including why the *g*-factor is important, how the brain develops, and the major role genetics plays in explaining intelligence differences among individuals. A number of papers explore how research findings can inform education policies (Asbury and Fields, 2021; Asbury and Wai, 2019; Wai and Bailey, 2021; Wai and Worrell, 2021; Wai et al., 2019). In the future, the potential for enhancing intelligence based on neuroscience research just might make this goal more achievable for all students and result in greater school and life achievement. As the twenty-first century progresses, we all need to be aware of neuroscience research findings on intelligence and what they could mean for our lives.

6.7 Final Thoughts

I have focused this book on progress in neuroscience research on intelligence, especially based on genetic and neuroimaging methods. Many questions have yet to be answered by a solid weight of evidence. Some of the major outstanding issues include: more understanding of the mechanisms of how the brain develops in early childhood; how brain development relates to adult intelligence; whether the *g*-factor and other intelligence factors have specific sets of brain structural and/or functional networks that explain individual differences and whether they are network sex differences; what epigenetic factors influence intelligence. There are also bigger questions that will require new methods and technologies to work down the temporal and spatial resolution scales to circuits, neurons, and synapses to create an advanced molecular neurobiology of intelligence based on how genes function. Perhaps the most important questions to answer involve whether intelligence research findings can be used to inform education issues and public policy, especially regarding individuals who may lack the mental abilities to succeed in modern life. Also, it is not too early to discuss issues around any eventual enhancement of intelligence implied by neuroscience research.

Writing forces thinking. I have thought about the research I have reviewed as this book materialized and what I have learned from writing it. I believe my explicit bias toward biological explanations of intelligence, developed over my 40 plus years of conducting research, is supported by many of the newest findings from psychometrics, quantitative genetics, molecular genetics, and neuroimaging. Not all studies are consistent with this view but, as I see it, the weight of evidence continues to favor neuroscience approaches for understanding what intelligence is, where it comes from, and how it can be changed. That is the focus of this book and I will leave it to others to present alternative evaluations about the weight of evidence from other perspectives on these issues. I am open to compelling arguments about where I may be incorrect in my evaluation and I am prepared to change my mind if new data shift the weight of evidence. I also believe that neuroscience perspectives on intelligence offer the best hope to resolve pressing issues about education and public policy that have not yet been resolved or ameliorated after 50 years of attempts based on "Blank Slate" assumptions about individual differences in intelligence and where they come from. Neuroscience has the potential to change the status quo in ways that other approaches have yet to accomplish. You may not agree, but if you are now thinking about intelligence differently than when you started reading this book, my primary goal is met.

Speaking of you, reading also forces thinking. Even if you are convinced by my arguments, I challenge you to think critically about the studies I have presented throughout this book as representative of neuroscience progress and about what I think they mean. My challenge to you is to find weak links and loopholes in my presentation, and when you do, design a new research study to fix or falsify them.

I have a not-so-secret wish that I suspect many of you share. I would like to be transported 40 or 50 years into the future to see what has transpired. Perhaps you will be there working on brain research and nearing retirement. What have you learned? Are there specific intelligence genes? How many? How do they work? Can genetic engineering, drugs, or some experiences enhance intelligence dramatically? How does brain development during childhood or teenage years affect intelligence? Is there a realistic virtual human brain that can simulate all manner of cognition, especially intelligence? Are simulations the same for men and women? How smart is the most intelligent machine? Can we see intelligence in the structure and function of networks, circuits, neurons, and synapses? How has intelligence research been used to address problems in education and other social areas? Is there a new neuroscience-based definition of intelligence? Are chronometric testing, brain imaging, or polygenic scores now new standards for assessing intelligence? What do brain fingerprints predict and how will they be used? What new neuroscience research tools and methods are available to study intelligence?

It would please me to know these things, even if I learn that my bets written in 2015 and 2022 were badly misplaced. I was born at the midpoint of the twentieth century. As a college student from a modest background, I had no concept of the future I am now living, let alone developments in brain research. Now I can only imagine the answers and the new questions that will come by the midpoint of the twenty-first century. If you are thinking about whether to have a career studying intelligence and the brain, here is a statement that will always be true: Get started – science is a never-ending story – whenever you begin will be the most exciting time to work on the puzzles that define the neuroscience of intelligence.

Chapter 6 Summary

- Chronometrics refers to a method of measuring information processing in the brain while performing standard cognitive tasks. The measurements are made in units of time (milliseconds) and therefore provide a quantitative assessment of intelligence on a ratio scale.
- Memory is a key component of intelligence and neuroscience studies of memory can identify brain circuits that help explain individual differences.
- New neuroscience techniques such as optogenetics and chemogenetics allow animal studies of neurons and circuits that may be important for intelligence research in humans.
- A neuroscience understanding of actual brain circuits may lead to profound advances for building truly intelligent machines based on how the brain works.
- Brain fingerprints made from neuroimaging are stable and unique to individuals, and they predict intelligence.
- Neuroimaging studies of consciousness and creativity are providing some insights about intelligence.

- SES is confounded with intelligence, possibility on a neural level. The implications of this are provocative for public policy.
- To the extent that different patterns of cognitive strengths and weaknesses are rooted more in neurobiology and genetics than in childhood experience, it is incorrect to blame lack of economic or educational success entirely on poor motivation, poor education, or other social factors.
- Neuroscience progress offers exciting opportunities for intelligence researchers. It is a great time to enter the field.

Review Questions

- What is the sophistication of measurement gap?
- Would you take an IQ pill?
- Should the information in this book influence education and social policies?
- Has this book changed your mind about any aspect of intelligence?

Further Reading

Clocking the Mind: Mental Chronometry and Individual Differences (Jensen, 2006). This is a technical manifesto that lays out the promise of chronometrics and the challenges of implementing it.

"Creativity and Intelligence: Brain Networks that Link and Differentiate the Expression of Genius" (Jung & Haier, 2013). A summary of neuroimaging studies of intelligence and creativity that proposes how genius may emerge from specific brain networks.

"DREADDs (Designer Receptors Exclusively Activated by Designer Drugs): Chemogenetic Tools with Therapeutic Utility" (Urban & Roth, 2015). This is a highly technical neuroscience explanation of the topic.

G is for Genes: The Impact of Genetics on Education and Achievement (Asbury & Plomin, 2014). This is a highly readable, non-technical summary of genetic research on mental abilities. Specific policy recommendations for education reform are discussed.

"Cognitive Ability Has Powerful, Widespread and Robust Effects on Social Stratification: Evidence from the 1979 and 1997 US National Longitudinal Surveys of Youth" (Marks, 2022). This is a comprehensive statistical analysis that demonstrates how SES is confounded with intelligence. It is an important study, especially for students, because it includes a detailed discussion of implications for social policy.

Glossary

Allele. One of the alternative forms of a gene that is located at a specific location on a specific chromosome.

Autism. A complex neurodevelopment disorder with a range of various cognitive and behavioral symptoms often referred to as a spectrum of disorders.

Base pairs. Building blocks formed by pairing Adenine (A), Guanine (G), Cytosine (C), and Thymine (T) that link the two strands of DNA like rungs on a ladder. There are an estimated three billion base pairs in human DNA.

Behavioral genetics. The field of study that examines the role of genetic influences on behaviors and traits

Behaviorism. A psychology theory popular in the 1950s and 1960s that assumes a person is essentially passive, responding to environmental stimuli and that overt behavior is the only thing that can be studied.

Bell curve. Another name in statistics for any normal distribution of scores or traits. Also, the title of a provocative 1994 book about intelligence and society.

Blank Slate. A philosophical/psychological view that all traits in a person are formed mostly by nurture. Also, the title of a 2002 book that argues against this view based on modern science.

Bochumer Matrizen-Test (BOMAT). A standardized intelligence test based on solving abstract reasoning problems often used as an estimate of the g-factor.

Brain-derived neurotropic factor (BDNF). A protein implicated in learning and several aspects of neuron health and development.

Brodmann areas (BAs). A system for using numbers to define brain areas by anatomical location, originally based on autopsy studies of neuron structures (see Figure 3.6).

CAT scan. Computerized tomography is a procedure that uses X-rays to image body tissues and structures. These images provide no information about the functioning of tissues.

Chemogenetics. This is a technique used to experimentally turn neurons on and off using specially designed chemicals (see DREADD).

Chromosome. Thread-like structure that carries DNA in genes. Humans have 23 pairs of chromosomes.

Chronometrics. A method of measuring information-processing speed in the brain while performing standard cognitive tasks. The measurements are made in units of time and therefore may provide a quantitative assessment of intelligence on a ratio scale.

Continuity hypothesis. The idea that the genes related to high intelligence are the same as those related to low intelligence.

Correlation. A way to describe how strongly two things are related to each other (see Figure 1.2).

CRISPR/Cas9 (Clustered Regularly Interspaced Short Palindromic Repeats/Cas genes). This is a method for editing the genome using bacteria.

Cross-sectional study. A research study design that uses different subjects at different time points to establish a trend (see longitudinal study).

Cross-validation. A key step of replicating a finding to be sure the finding also applies to an independent sample.

Crystalized intelligence. The ability to learn facts and absorb information based on knowledge and experience.

Deep brain stimulation (DBS). A neurosurgical procedure involving the implantation of a medical device called a neuro-stimulator that delivers controlled mild electrical shocks to targeted brain areas.

Default network. The network of brain areas that is active while a person is not focused on any particular mental activity.

Diffusion tensor imaging (DTI). An MRI technique that uses water diffusion patterns to image white matter fibers.

Discontinuity hypothesis. The idea that the genes related to high intelligence are different to those related to low intelligence.

Dizygotic (DZ) twins. Fraternal twins; they have 50 percent of their genes in common.

DNA. Deoxyribonucleic acid is the hereditary material in genes.

Doogie mice. A strain of "smart" mice genetically engineered to solve mazes faster than controls.

Dopamine. A neurotransmitter that helps control the reward and pleasure centers and also helps regulate cognition, movement, and emotional responses.

Double helix. The structure formed by double-stranded molecules of DNA.

DREADD (Designer receptors exclusively activated by designer drugs). A system for activating brain receptors using synthetic molecules.

Edge. Refers to the association of two brain areas in graph analysis of brain connectivity.

EEG (electroencephalogram). A technique that measures the electrical activity of the brain by using electrodes attached to your scalp.

Elementary cognitive tasks (ECTs). These are tasks that require basic mental processes like attention.

Epigenetics. The field of study that investigates how genetic expression may be influenced by external factors.

Evoked potential (EP). A special application of EEG that records brain electrical activity that is induced by a specific stimulus such as a light flash. The EP is derived by averaging the EEG from the same stimulus repeated many times.

Factor analysis. A statistical method that describes patterns of relationships among many variables based on correlations.

False positive. A test result that erroneously indicates something that turns out to be untrue.

Fluid intelligence (Gf). Refers to inductive and deductive reasoning for novel problem solving. Fluid intelligence is closely associated with the *g*-factor.

Fluorescent proteins. Light-emitting chemicals that can be used to visualize the internal workings of neurons.

Flurodeoxyglucose (FDG). A radioactive substance used in PET to label metabolic activity.

Flynn effect. Refers to the gradual generational increase in raw IQ scores. The causes are not established and whether the effect is on the *g*-factor is not settled.

Fractional anisotropy (FA). A measure of water diffusion derived from MRI and used to image white matter fibers and assess their integrity.

Frontal Dis-inhibition Model (F-DIM). Based on neuroimaging studies of creativity, this framework suggests a system of brain areas that may be associated with creativity (see Figure 6.7).

Frontotemporal dementia (FTD). A rare degenerating illness similar to Alzheimer's disease characterized by progressive neuron loss especially in frontal lobes.

Full scale IQ. The total score based on summing all the subtest scores from a standardized intelligence test.

Functional MRI (fMRI). A neuroimaging procedure using MRI that measures regional brain activity by detecting aspects of blood flow.

***g*-factor (*g*).** Denotes the general feature of intelligence common to all tests of mental abilities and best estimated from a battery of tests.

Gene. A hereditary unit consisting of a DNA sequence at a specific location on a chromosome.

Gene expression. The process by which genetic instructions start or stop creating proteins.

Generalist genes hypothesis. The idea that the same genes affect most cognitive abilities rather than each cognitive ability being influenced by a different set of genes.

Genome. The entire set of DNA base pairs. All the genetic material of an organism.

Genome Wide Association Study (GWAS). This is a research approach used to identify genomic variants that are statistically associated with a risk for a disease or a particular trait. The method involves surveying the genomes of many people, looking for genomic variants (SNPs) that occur more frequently in those with a specific disease or trait compared to those without the disease or trait.

Genomic informatics. The field of managing and understanding vast data sets of genetic information from individual base pairs to genomes.

Genomics. The study of gene structure and function.

Graph analysis. A mathematical tool that is used to model brain connectivity and infer networks of brain areas that are structurally or functionally associated.

Heritability. A statistical estimate of how much variation in a trait or behavior in a population is due to genetic influences.

Hub. The term used in graph analysis to denote a brain area with many connections to other areas.

Intelligence. The ability to think and learn. The opposite of stupidity.

IQ. A measure of intelligence derived from a psychometric test but defined differently depending on the test. IQ points are not measures of a quantity such as distance or weight. IQ scores have meaning only relative to other people and are best understood in terms of percentiles.

Locus. The position of a gene or genes on a specific region of a chromosome.

Longitudinal study. A study where each subject is followed over time to investigate any changes (see contrast with cross-sectional study).

Magnetic resonance imaging (MRI). A technique based on pulsing radio wave energy through powerful magnetic fields to create detailed images of body tissue as water molecules react to the energy changes.

Magnetic resonance spectroscopy (MRS). A specialized MRI technique used to measure biochemicals in the brain.

Magneto-encephalogram (MEG). A technique for measuring localized brain activity based on detecting fluctuating magnetic fields that result from aggregate neuronal firings.

Meritocracy. A system based on ability.

Meta-analysis. A statistical method that examines data from many studies on a topic to determine trends in the data and help establish whether there is an overall finding.

Methylation. A chemical process that can change DNA. Of special interest in epigenetic investigations.

Microarray. A tool used to determine whether the DNA from a particular individual contains a mutation in genes (SNPs). Thousands of SNPs can be studied simultaneously.

Mnemonic methods. Techniques and strategies used to improve and augment memorization.

Molecular genetics. A field of study that investigates how genes function in terms of chemistry and physics.

Monozygotic (MZ) twins. Identical twins; they have 100 percent of their genes in common.

Mozart effect. The claim that listening to classical music increases intelligence.

Neuro-*g*. The concept that at least part of the general factor of intelligence has a specific basis in the brain (genetic or not).

Neuro-poverty. The concept that one of the many causes of poverty may be related to the genetic basis of intelligence.

Neuro-SES (social economic status). The concept that part of the overlap between intelligence and SES may be due to genetic influences.

Nonshared environment. Unique experiences that contribute to the environmental influences on heritability.

Optogenetics. Methods for controlling brain function with light.

Parieto-Frontal Integration Theory (PFIT). A framework proposed in 2007 that identifies specific areas distributed across the brain that are relevant for general intelligence.

Performance IQ. Nonverbal intelligence score derived from an IQ test.

Pleiotropy. Occurs when one gene influences variability in two or more seemingly unrelated traits.

Polygenic score (PGS). A computed aggregation of the number of SNPs (usually determined in GWASs) related to a trait or condition used to predict its probability.

Polygenicity. Occurs when many genes contribute to variance in a single trait.

Positive manifold. Term used to describe the robust finding that tests of all mental abilities are related to each other in the same direction: as scores on one test increase, scores on the others tend to also increase.

Positron emission tomography (PET). A technique to image body tissue functioning based on detecting the accumulation of low-level radioactive labels.

Proteomics. The study of proteins and how they work.

Psychometrics. A set of methods that uses various kinds of paper and pencil tests and statistical methods to study intelligence and personality.

Quantitative genetics. The study of genetic influences on variance of continuous traits (like intelligence or height) among individuals in a population.

Quantitative trait locus (QTL). Refers to a region of DNA related to a trait such as intelligence, as determined by statistical techniques.

Raven's Advanced Progressive Matrices (RAPM). A difficult test of nonverbal abstract reasoning widely used to estimate the general factor of intelligence.

Region of interest (ROI). A brain area defined for neuroimaging analysis.

Regression equation. A general statistical method with many varieties for estimating the relationship among variables. Often used to predict one variable from a set of other variables weighted to maximize the accuracy of prediction.

Restriction of range. A statistical problem referring to a lack of sufficient variance on a variable (like intelligence) for determining whether that variance is related to another variable.

SAT (Scholastic Assessment Test). A standardized test often used for college admission in the United States.

Savant. A person of unusual ability or of having profoundly detailed knowledge in a specialized or narrow topic.

Shared environment. Common experiences that contribute to the environmental influences on heritability.

Single nucleotide polymorphisms (SNPs). A change or variation in a base pair substituting one base for another. SNPs may be associated with traits or diseases and can be hints for identifying relevant genes.

Social economic status (SES). A measure that combines education and income in various ways to estimate social class for use in studies about how these variables may influence behavior or traits.

Standard deviation. A statistical measure that describes variation around the mean of a distribution of scores.

STEM. Abbreviation for the fields of science, technology, engineering, and math.

Structural MRI. An MRI technique that visualizes the makeup of tissue but contains no functional information.

Synesthesia. A rare neurological condition where sensory perception is mixed up. For example, hearing sounds may produce visual colors.

Termites. The slang term for the participants in Louis Terman's longitudinal study of high-IQ individuals.

Test of Nonverbal Intelligence (TONI). A nonverbal test of intelligence designed for children.

Transcranial alternating current stimulation (tACS). A noninvasive technique for applying weak alternating electrical current through the skull to stimulate brain areas.

Transcranial direct current stimulation (tDCS). A noninvasive technique for applying weak constant electrical current through the skull to stimulate brain areas.

Transcranial magnetic stimulation (TMS). A procedure that uses magnetic fields placed over the scalp to stimulate or suppress brain activity.

Val66Met. A gene associated with BDNF.

Voxel. The smallest unit in a neuroimage. A three-dimensional pixel.

Voxel-based morphometry (VBM). A technique for measuring brain characteristics at the level of individual voxels.

Wechsler Adult Intelligence Scale (WAIS). A widely used standardized battery of mental tests that estimates intelligence relative to other people with an IQ score.

Wechsler Intelligence Scale for Children. A version of the WAIS especially designed and normed for children.

References

Abdellaoui, A., Dolan, C. V., Verweij, K. J. H., & Nivard, M. G. (2022). Gene–environment correlations across geographic regions affect genome-wide association studies. *Nature Genetics.* doi:10.1038/s41588-022-01158-0

Ackerman, P. L., Beier, M. E., & Boyle, M. O. (2005). Working memory and intelligence: The same or different constructs? *Psychological Bulletin, 131*(1), 30–60. doi:10.1037/0033-2909.131.1.30

Adamczyk, A. K., & Zawadzki, P. (2020). The memory-modifying potential of optogenetics and the need for neuroethics. *NanoEthics, 14*(3), 207–225. doi:10.1007/s11569-020-00377-1

Alavash, M., Lim, S. J., Thiel, C., Sehm, B., Deserno, L., & Obleser, J. (2018). Dopaminergic modulation of hemodynamic signal variability and the functional connectome during cognitive performance. *Neuroimage, 172*, 341–356. doi:10.1016/j.neuroimage.2018.01.048

Alkire, M. T., & Haier, R. J. (2001). Correlating in vivo anaesthetic effects with ex vivo receptor density data supports a GABAergic mechanism of action for propofol, but not for isoflurane. *British Journal of Anaesthesia, 86*(5), 618–626. Retrieved from www.ncbi.nlm.nih.gov/pubmed/11575335

Alkire, M. T., Haier, R. J., & Fallon, J. H. (2000). Toward a unified theory of narcosis: Brain imaging evidence for a thalamocortical switch as the neurophysiologic basis of anesthetic-induced unconsciousness. *Consciousness and Cognition, 9*(3), 370–386.

Alkire, M. T., Haier, R. J., Barker, S. J., Shah, N. K., Wu, J. C., & Kao, Y. J. (1995). Cerebral metabolism during propofol anesthesia in humans studied with positron emission tomography. *Anesthesiology, 82*(2), 393–403; discussion 327A. Retrieved from www.ncbi.nlm.nih.gov/pubmed/7856898

Alkire, M. T., Pomfrett, C. J. D., Haier, R. J., Gianzero, M. V., Chan, C. M., Jacobsen, B. P., & Fallon, J. H. (1999). Functional brain imaging during anesthesia in humans – Effects of halothane on global and regional cerebral glucose metabolism. *Anesthesiology, 90*(3), 701–709.

Allegrini, A. G., Selzam, S., Rimfeld, K., von Stumm, S., Pingault, J. B., & Plomin, R. (2019). Genomic prediction of cognitive traits in childhood and adolescence. *Molecular Psychiatry, 24*(6), 819–827. doi:10.1038/s41380-019-0394-4

Anderson, D. J. (2012). Optogenetics, sex, and violence in the brain: Implications for psychiatry. *Biological Psychiatry, 71*(12), 1081–1089. doi:10.1016/j.biopsych.2011.11.012

Anderson, J. W., Johnstone, B. M., & Remley, D. T. (1999). Breast-feeding and cognitive development: A meta-analysis. *The American Journal of Clinical Nutrition, 70*(4), 525–535. Retrieved from www.ncbi.nlm.nih .gov/pubmed/10500022

Anderson, K. M., & Holmes, A. J. (2021). Predicting individual differences in cognitive ability from brain imaging and genetics. In A. K. Barbey, S. Karama, & R. J. Haier (Eds.), *Cambridge Handbook of Intelligence and Cognitive Neuroscience* (pp. 327–348). New York: Cambridge University Press.

Andreasen, N. C., Flaum, M., Swayze, V., Oleary, D. S., Alliger, R., Cohen, G., … Yuh, W. T. C. (1993). Intelligence and brain structure in normal individuals. *American Journal of Psychiatry, 150*(1), 130–134.

Antal, A., Luber, B., Brem, A. K., Bikson, M., Brunoni, A. R., Cohen Kadosh, R., … Paulus, W. (2022). Non-invasive brain stimulation and neuroenhancement. *Clinical Neurophysiological Practise, 7*, 146–165. doi:10.1016/j.cnp.2022.05.002

Arden, R. (2003). An arthurian romance. In H. Nyborg (Ed.), *The Scientific Study of General Intelligence* (pp. 533–553). Amsterdam: Pergamon.

Arden, R., Chavez, R. S., Grazioplene, R., & Jung, R. E. (2010). Neuroimaging creativity: A psychometric view. *Behavioral Brain Research, 214*(2), 143–156. doi:10.1016/j.bbr.2010.05.015

Arden, R., Luciano, M., Deary, I. J., Reynolds, C. A., Pedersen, N. L., Plassman, B. L., … Visscher, P. M. (2015). The association between intelligence and lifespan is mostly genetic. *International Journal of Epidemiology.* doi:10.1093/ije/dyv112

Ardlie, K. G., DeLuca, D. S., Segre, A. V., Sullivan, T. J., Young, T. R., Gelfand, E. T., … Consortium, G. (2015). The Genotype-Tissue Expression (GTEx) pilot analysis: Multitissue gene regulation in humans. *Science, 348*(6235), 648–660. doi:10.1126/science.1262110

Arevalo, A., Abusamra, V., & Lepski, G. (2022). Editorial: How to improve neuroscience education for the public and for a multi-professional audience in different parts of the globe. *Frontiers in Human Neuroscience*, 16, 973893.

Aristizabal, M. J., Anreiter, I., Halldorsdottir, T., Odgers, C. L., McDade, T. W., Goldenberg, A., … O'Donnell, K. J. (2020). Biological embedding of experience: A primer on epigenetics. *Proceedings of the National Academy of Sciences, 117*(38), 23261–23269. doi:10.1073/ pnas.1820838116

Asbury, K., & Fields, D. (2021). Implications of biological research on intelligence for education and public policy. In A. Barbey, S. Karama, & R. J. Haier (Eds.), *The Cambridge Handbook of Intelligence and Cognitive Neuroscience (pp. 399–415).* New York: Cambridge University Press.

Asbury, K., & Plomin, R. (2014). *G Is for Genes : The Impact of Genetics on Education and Achievement*. Chichester, West Sussex: Wiley-Blackwell.

Asbury, K., & Wai, J. (2019). Viewing education policy through a genetic lens. *Journal of School Choice*, 1–15. doi:10.1080/15582159.2019.1705008

Asbury, K., Wachs, T. D., & Plomin, R. (2005). Environmental moderators of genetic influence on verbal and nonverbal abilities in early childhood. *Intelligence, 33*(6), 643–661.

Ashburner, J., & Friston, K. (1997). Multimodal image coregistration and partitioning – a unified framework. *Neuroimage, 6*(3), 209–217.

Ashburner, J., & Friston, K. J. (2000). Voxel-based morphometry – the methods. *Neuroimage, 11*(6 Pt 1), 805–821.

Aston-Jones, G., & Deisseroth, K. (2013). Recent advances in optogenetics and pharmacogenetics. *Brain Research, 1511*, 1–5. doi:10.1016/j .brainres.2013.01.026

Atherton, M., Zhuang, J. C., Bart, W. M., Hu, X. P., & He, S. (2000). A functional magnetic resonance imaging study of chess expertise. *Journal of Cognitive Neuroscience*, 105–105.

Au, J., Buschkuehl, M., Duncan, G. J., & Jaeggi, S. M. (2016). There is no convincing evidence that working memory training is NOT effective: A reply to Melby-Lervag and Hulme (2015). *Psychonomic Bulletin and Review, 23*(1), 331–337. doi:10.3758/s13423-015-0967-4

Au, J., Gibson, B. C., Bunarjo, K., Buschkuehl, M., & Jaeggi, S. M. (2020). Quantifying the difference between active and passive control groups in cognitive interventions using two meta-analytical approaches. *Journal of Cognitive Enhancement, 4*(2), 192–210. doi:10.1007/s41465-020-00164-6

Au, J., Sheehan, E., Tsai, N., Duncan, G. J., Buschkuehl, M., & Jaeggi, S. M. (2015). Improving fluid intelligence with training on working memory: A meta-analysis. *Psychonomic Bulletin and Review, 22*(2), 366–377. doi:10.3758/s13423-014-0699-x

Bagot, K. S., & Kaminer, Y. (2014). Efficacy of stimulants for cognitive enhancement in non-attention deficit hyperactivity disorder youth: A systematic review. *Addiction, 109*(4), 547–557. Retrieved from www .ncbi.nlm.nih.gov/pubmed/24749160

Bailey, M. J., Sun, S., & Timpe, B. (2021). Prep School for Poor Kids: The long-run impacts of head start on human capital and economic self-sufficiency. *American Economic Review, 111*(12), 3963–4001. doi:10.1257/aer.20181801

Barbey, A. (2018). Network neuroscience theory of human intelligence. *Trends in Cognitive Sciences, 22*(1), 8–20.

Barbey, A. (2021). Human intelligence and network neuroscience. In A. Barbey, S. Karama, & R. J. Haier (Eds.), *Cambridge Handbook of Intelligence and Cognitive Neuroscience* (pp. 102–122). New York: Cambridge University Press.

Barbey, A. K., Colom, R., Paul, E. J., & Grafman, J. (2014). Architecture of fluid intelligence and working memory revealed by lesion mapping. *Brain Structure and Function, 219*(2), 485–494.

Barbey, A. K., Colom, R., Paul, E., Forbes, C., Krueger, F., Goldman, D., & Grafman, J. (2014). Preservation of general intelligence following traumatic brain injury: Contributions of the Met66 brain-derived neurotrophic factor. *Plos One, 9*(2).

Barbey, A. K., Colom, R., Solomon, J., Krueger, F., Forbes, C., & Grafman, J. (2012). An integrative architecture for general intelligence and executive function revealed by lesion mapping. *Brain, 135*(Pt 4), 1154–1164. doi:10.1093/brain/aws021

Barbey, A., Karama, S., & Haier, R. J. (2021). *Cambridge Handbook of Intelligence and Cognitive Neuroscience.* New York: Cambridge University Press.

Barnett, W. S., & Hustedt, J. T. (2005). Head start's lasting benefits. *Infants and Young Children, 18*(1), 16–24.

Bashwiner, D. M. (2018). The neuroscience of musical creativity. In J. R. E. & V. O. (Eds.), *The Cambridge Handbook of the Neuroscience of Creativity* (pp. 495–516). Cambridge: Cambridge University Press.

Bashwiner, D. M., Wertz, C. J., Flores, R. A., & Jung, R. E. (2016). Musical creativity "Revealed" in brain structure: Interplay between motor, default mode, and limbic networks. *Scientific Reports, 6*, 20482. doi:10.1038/srep20482

Bassett, D. S., Yang, M., Wymbs, N. F., & Grafton, S. T. (2015). Learning-induced autonomy of sensorimotor systems. *Natural Neuroscience, 18*(5), 744–751. doi:10.1038/nn.3993

Basso, A., De Renzi, E., Faglioni, P., Scotti, G., & Spinnler, H. (1973). Neuropsychological evidence for the existence of cerebral areas critical to the performance of intelligence tasks. *Brain, 96*(4), 715–728.

Basten, U., & Fiebach, C. J. (2021). Functional brain imaging of intelligence. In A. Barbey, S. Karama, & R. J. Haier (Eds.), *The Cambridge Handbook of Intelligence and Cognitive Neuroscience (pp. 235–260).* New York: Cambridge University Press.

Basten, U., Hilger, K., & Fiebach, C. J. (2015). Where smart brains are different: A quantitative meta-analysis of functional and structural brain imaging studies on intelligence. *Intelligence, 51*(0), 10–27. doi:10.1016/j .intell.2015.04.009

Basten, U., Stelzel, C., & Fiebach, C. J. (2013). Intelligence is differentially related to neural effort in the task-positive and the task-negative brain network. *Intelligence, 41*(5), 517–528.

Bates, T. C., & Gignac, G. E. (2022). Effort impacts IQ test scores in a minor way: A multi-study investigation with healthy adult volunteers. *Intelligence, 92*, 101652. doi:10.1016/j.intell.2022.101652

Bates, T. C., Lewis, G. J., & Weiss, A. (2013). Childhood socioeconomic status amplifies genetic effects on adult intelligence. *Psychological Science, 24*(10), 2111–2116.

Batty, G. D., Deary, I. J., & Gottfredson, L. S. (2007). Premorbid (early life) IQ and later mortality risk: Systematic review. *Annals of Epidemiology, 17*(4), 278–288.

Beaty, R. E. (2015). The neuroscience of musical improvisation. *Neuroscience and Biobehavioral Reviews, 51*, 108–117. doi:10.1016/j.neubiorev.2015.01.004

Beaujean, A. A., Firmin, M. W., Knoop, A. J., Michonski, J. D., Berry, T. P., & Lowrie, R. E. (2006). Validation of the Frey and Detterman (2004) IQ prediction equations using the Reynolds Intellectual Assessment Scales. *Personality and Individual Differences, 41*(2), 353–357. doi:10.1016/j.paid.2006.01.014

Bejjanki, V. R., Zhang, R., Li, R., Pouget, A., Green, C. S., Lu, Z. L., & Bavelier, D. (2014). Action video game play facilitates the development of better perceptual templates. *Proceedings of the National Academy Science USA, 111*(47), 16961–16966. doi:10.1073/pnas.1417056111

Bengtsson, S. L., Csikszentmihalyi, M., & Ullen, F. (2007). Cortical regions involved in the generation of musical structures during improvisation in pianists. *Journal of Cognition Neuroscience, 19*(5), 830–842. doi:10.1162/jocn.2007.19.5.830

Benyamin, B., Pourcain, B., Davis, O. S., Davies, G., Hansell, N. K., Brion, M. J., … Visscher, P. M. (2014). Childhood intelligence is heritable, highly polygenic and associated with FNBP1L. *Molecular Psychiatry, 19*(2), 253–258. doi:10.1038/mp.2012.184

Berkowitz, A. L., & Ansari, D. (2010). Expertise-related deactivation of the right temporoparietal junction during musical improvisation. *Neuroimage, 49*(1), 712–719. doi:10.1016/j.neuroimage.2009.08.042

Bernstein, B. O., Lubinski, D., & Benbow, C. P. (2019). Psychological constellations assessed at age 13 predict distinct forms of eminence 35 years later. *Psychological Science, 30*(3), 444–454. doi:10.1177/0956797618822524

Bhaduri, A., Sandoval-Espinosa, C., Otero-Garcia, M., Oh, I., Yin, R., Eze, U. C., … Kriegstein, A. R. (2021). An atlas of cortical arealization identifies dynamic molecular signatures. *Nature, 598*(7879), 200–204. doi:10.1038/s41586-021-03910-8

Biazoli, C. E., Jr., Salum, G. A., Pan, P. M., Zugman, A., Amaro, E., Jr., Rohde, L. A., … Sato, J. R. (2017). Commentary: Functional connectome fingerprint: Identifying individuals using patterns of brain connectivity. *Frontiers in Human Neuroscience, 11*, 47.

Bishop, S. J., Fossella, J., Croucher, C. J., & Duncan, J. (2008). COMT val158met genotype affects recruitment of neural mechanisms supporting fluid intelligence. *Cerebral Cortex, 18*(9), 2132–2140. doi:10.1093/cercor/bhm240

Bogg, T., & Lasecki, L. (2015). Reliable gains? Evidence for substantially underpowered designs in studies of working memory training transfer to fluid intelligence. *Frontiers in Psychology*, 5.

Bohlken, M. M., Brouwer, R. M., Mandl, R. C., van Haren, N. E., Brans, R. G., van Baal, G. C., … Hulshoff Pol, H. E. (2014). Genes contributing to subcortical volumes and intellectual ability implicate the thalamus. *Human Brain Mapping, 35*(6), 2632–2642. doi:10.1002/hbm.22356

Boivin, M. J., Giordani, B., Berent, S., Amato, D. A., Lehtinen, S., Koeppe, R. A., … Kuhl, D. E. (1992). Verbal fluency and positron emission tomographic mapping of regional cerebral glucose-metabolism. *Cortex, 28*(2), 231–239.

Bouchard, T. J., Jr. (1998). Genetic and environmental influences on adult intelligence and special mental abilities. *Human Biology, 70*(2), 257–279.

Bouchard, T. J. (2009). Genetic influence on human intelligence (Spearman's g): How much? *Annals of Human Biology, 36*(5), 527–544.

Bouchard, T. J., Jr., & McGue, M. (1981). Familial studies of intelligence: A review. *Science, 212*(4498), 1055–1059. Retrieved from www.ncbi.nlm.nih.gov/pubmed/7195071

Bowren, M., Adolphs, R., Bruss, J., Manzel, K., Corbetta, M., Tranel, D., & Boes, A. D. (2020). Multivariate lesion-behavior mapping of general cognitive ability and its psychometric constituents. *The Journal of Neuroscience, 40*(46), 8924. doi:10.1523/JNEUROSCI.1415-20.2020

Brans, R. G. H., Kahn, R. S., Schnack, H. G., van Baal, G. C. M., Posthuma, D., van Haren, N. E. M., … Pol, H. E. H. (2010). Brain plasticity and intellectual ability are influenced by shared genes. *Journal of Neuroscience, 30*(16), 5519–5524.

Brem, A. K., Almquist, J. N., Mansfield, K., Plessow, F., Sella, F., Santarnecchi, E., … Honeywell SHARP Team authors. (2018). Modulating fluid intelligence performance through combined cognitive training and brain stimulation. *Neuropsychologia, 118*(Pt A), 107–114. doi:10.1016/j.neuropsychologia.2018.04.008

Brodmann, K. (1909). *Vergleichende Lokalisationslehre der Grosshirnrinde in ihren Prinzipien dargestellt auf Grund des Zellenbaues*. Leipzig: Barth.

Brody, N. (2003). Construct validation of the Sternberg Triarchic Abilities Test Comment and reanalysis. *Intelligence, 31*(4), 319–329. doi:10.1016/S0160-2896(01)00087-3

Bruzzone, S. E. P., Lumaca, M., Brattico, E., Vuust, P., Kringelbach, M. L., & Bonetti, L. (2022). Dissociated brain functional connectivity of fast versus slow frequencies underlying individual differences in fluid intelligence: A DTI and MEG study. *Scientific Reports, 12*(1), 4746. doi:10.1038/s41598-022-08521-5

Burgaleta, M., & Colom, R. (2008). Short-term storage and mental speed account for the relationship between working memory and fluid intelligence. *Psicothema, 20*(4), 780–785.

Burgaleta, M., Johnson, W., Waber, D. P., Colom, R., & Karama, S. (2014). Cognitive ability changes and dynamics of cortical thickness development in healthy children and adolescents. *Neuroimage, 84*, 810–819. doi:10.1016/j.neuroimage.2013.09.038

Burgaleta, M., Macdonald, P. A., Martinez, K., Roman, F. J., Alvarez-Linera, J., Gonzalez, A. R., … Colom, R. (2013). Subcortical regional morphology correlates with fluid and spatial intelligence. *Human Brain Mapping.* doi:10.1002/hbm.22305

Burgoyne, A. P., & Engle, R. W. (2020). Mitochondrial functioning and its relation to higher-order cognitive processes. *Journal of Intelligence, 8*(2).

Burt, C. (1943). Ability and income. *British Journal of Educational Psychology, 13*, 83–98.

Burt, C. (1955). The evidence for the concept of intelligence. *British Journal of Educational Psychology, 25*, 158–177.

Burt, C. (1966). Genetic determination of differences in intelligence – A study of monozygotic twins reared together and apart. *British Journal of Psychology, 57*, 137-&.

Cabeza, R., & Nyberg, L. (2000). Imaging cognition II: An empirical review of 275 PET and fMRI studies. *Journal of Cognitive Neuroscience, 12*(1), 1–47.

Cajal, R. S. (1924). *Pensamientos Escogidos (Chosen Thoughts)*. Madrid: Cuadernos Literarios.

Campbell, F. A., Pungello, E., Miller-Johnson, S., Burchinal, M., & Ramey, C. T. (2001). The development of cognitive and academic abilities: Growth curves from an early childhood educational experiment. *Developmental Psychology, 37*(2), 231–242.

Cardoso-Leite, P., & Bavelier, D. (2014). Video game play, attention, and learning: How to shape the development of attention and influence learning? *Current Opinion in Neurology, 27*(2), 185–191. doi:10.1097/WCO.0000000000000077

Carl, N. (2018). How stifling debate around race, genes and IQ can do harm. *Evolutionary Psychological Science, 4*(4), 399–407. doi:10.1007/s40806-018-0152-x

Cattell, R. B. (1971). *Abilities : Their structure, growth, and action*. Boston: Houghton Mifflin.

Cattell, R. B. (1987). *Intelligence: Its structure, growth, and action*. North-Holland: Sole distributors for the U.S.A. and Canada, Elsevier Science Pub. Co.

Ceci, S. J. (1991). How much does schooling influence general intelligence and its cognitive components – A reassessment of the evidence. *Developmental Psychology, 27*(5), 703–722.

Ceci, S. J., & Williams, W. M. (1997). Schooling, intelligence, and income. *American Psychologist, 52*(10), 1051–1058.

Chabris, C. F. (1999). Prelude or requiem for the ʃ Mozart effect/'? *Nature, 400*(6747), 826–827.

Chabris, C. F., Hebert, B. M., Benjamin, D. J., Beauchamp, J., Cesarini, D., van der Loos, M., … Laibson, D. (2012). Most reported genetic associations with general intelligence are probably false positives. *Psychological Science, 23*(11), 1314–1323.

Champagne, F. A., & Curley, J. P. (2009). Epigenetic mechanisms mediating the long-term effects of maternal care on development. *Neuroscience Biobehavioral Review, 33*(4), 593–600. doi:10.1016/j.neubiorev.2007.10.009

Cheesman, R., Hunjan, A., Coleman, J. R. I., Ahmadzadeh, Y., Plomin, R., McAdams, T. A., … Breen, G. (2020). Comparison of adopted and nonadopted individuals reveals gene–environment interplay for education in the UK Biobank. *Psychological Science, 31*(5), 582–591. doi:10.1177/0956797620904450

Chen, C.-Y., Tian, R., Ge, T., Lam, M., Sanchez-Andrade, G., Singh, T., … Runz, H. (2022). The impact of rare protein coding genetic variation on adult cognitive function. medRxiv, 2022.2006.2024.22276728. doi:10.1101/2022.06.24.22276728

Chiang, M. C., Barysheva, M., McMahon, K. L., de Zubicaray, G. I., Johnson, K., Montgomery, G. W., … Thompson, P. M. (2012). Gene network effects on brain microstructure and intellectual performance identified in 472 twins. *Journal of Neuroscience, 32*(25), 8732–8745. doi:10.1523/Jneurosci.5993-11.2012

Chiang, M. C., Barysheva, M., Shattuck, D. W., Lee, A. D., Madsen, S. K., Avedissian, C., … Thompson, P. M. (2009). Genetics of brain fiber architecture and intellectual performance. *Journal of Neuroscience, 29*(7), 2212–2224. doi:10.1523/Jneurosci.4184-08.2009

Chiang, M. C., Barysheva, M., Toga, A. W., Medland, S. E., Hansell, N. K., James, M. R., … Thompson, P. M. (2011a). BDNF gene effects on brain circuitry replicated in 455 twins. *Neuroimage, 55*(2), 448–454. doi:10.1016/j.neuroimage.2010.12.053

Chiang, M. C., McMahon, K. L., de Zubicaray, G. I., Martin, N. G., Hickie, I., Toga, A. W., … Thompson, P. M. (2011b). Genetics of white matter development: A DTI study of 705 twins and their siblings aged 12 to 29. *Neuroimage, 54*(3), 2308–2317. doi:10.1016/j.neuroimage.2010.10.015

Choi, Y. Y., Shamosh, N. A., Cho, S. H., DeYoung, C. G., Lee, M. J., Lee, J. M., … Lee, K. H. (2008). Multiple bases of human intelligence revealed by cortical thickness and neural activation. *Journal of Neuroscience, 28*(41), 10323–10329. doi:10.1523/JNEUROSCI.3259-08.2008

Chollet, F. (2019). On the measure of intelligence. arXiv preprint arXiv: 1911.01547.

Chooi, W. T., & Thompson, L. A. (2012). Working memory training does not improve intelligence in healthy young adults. *Intelligence, 40*(6), 531–542.

Chugani, H. T., Phelps, M. E., & Mazziotta, J. C. (1987). Positron emission tomography study of human brain functional development. *Annals of Neurology, 22*(4), 487–497.

Clark, V. P., Coffman, B. A., Mayer, A. R., Weisend, M. P., Lane, T. D., Calhoun, V. D., … Wassermann, E. M. (2012). TDCS guided using fMRI significantly accelerates learning to identify concealed objects. *Neuroimage, 59*(1), 117–128. doi:10.1016/j.neuroimage.2010.11.036

Coffman, B. A., Clark, V. P., & Parasuraman, R. (2014). Battery powered thought: Enhancement of attention, learning, and memory in healthy adults using transcranial direct current stimulation. *Neuroimage*, 85(Pt 3), 895–908. doi:10.1016/j.neuroimage.2013.07.083

Cofnas, N. (2020). Research on group differences in intelligence: A defense of free inquiry. *Philosophical Psychology, 33*(1), 125–147. doi:10.1080/09515089.2019.1697803

Cohen, J. R., & D'Esposito, M. (2021). An integrated, dynamic functional connectome undies intelligence. In A. Barbey, S. Karama, & R. Haier (Eds.), *Cambridge Handbook of Intelligence and Cognitive Neuroscience* (pp. 261–281). New York: Cambridge University Press.

Cole, M. W., Yarkoni, T., Repovs, G., Anticevic, A., & Braver, T. S. (2012). Global connectivity of prefrontal cortex predicts cognitive control and intelligence. *Journal of Neuroscience, 32*(26), 8988–8999. doi:10.1523/JNEUROSCI.0536-12.2012

Colom, R., Abad, F. J., Quiroga, M. A., Shih, P. C., & Flores-Mendoza, C. (2008). Working memory and intelligence are highly related constructs, but why? *Intelligence, 36*(6), 584–606.

Colom, R., & Flores-Mendoza, C. E. (2007). Intelligence predicts scholastic achievement irrespective of SES factors: Evidence from Brazil. *Intelligence, 35*(3), 243–251.

Colom, R., Jung, R. E., & Haier, R. J. (2006a). Distributed brain sites for the g-factor of intelligence. *Neuroimage, 31*(3), 1359–1365.

Colom, R., Jung, R. E., & Haier, R. J. (2006b). Finding the g-factor in brain structure using the method of correlated vectors. *Intelligence, 34*(6), 561.

Colom, R., Jung, R. E., & Haier, R. J. (2007). General intelligence and memory span: Evidence for a common neuroanatomic framework. *Cognitive Neuropsychology, 24*(8), 867–878.

Colom, R., Karama, S., Jung, R. E., & Haier, R. J. (2010). Human intelligence and brain networks. *Dialogues in Clinical Neuroscience, 12*(4), 489–501. Retrieved from www.ncbi.nlm.nih.gov/pubmed/21319494

Colom, R., Rebollo, I., Palacios, A., Juan-Espinosa, M., & Kyllonen, P. C. (2004). Working memory is (almost) perfectly predicted by g. *Intelligence, 32*(3), 277–296.

Colom, R., Roman, F. J., Abad, F. J., Shih, P. C., Privado, J., Froufe, M., … Jaeggi, S. M. (2013). Adaptive n-back training does not improve fluid intelligence at the construct level: Gains on individual tests suggest that training may enhance visuospatial processing. *Intelligence, 41*(5), 712–727.

Conway, A. R. A., Kane, M. J., & Engle, R. W. (2003). Working memory capacity and its relation to general intelligence. *Trends in Cognitive Sciences, 7*(12), 547–552. doi:10.1016/J.Tics.2003.10.005

Covington, H. E., 3rd, Lobo, M. K., Maze, I., Vialou, V., Hyman, J. M., Zaman, S., … Nestler, E. J. (2010). Antidepressant effect of optogenetic stimulation of the medial prefrontal cortex. *Journal of Neuroscience, 30*(48), 16082–16090. doi:10.1523/JNEUROSCI.1731-10.2010

Cowan, N. (2014). Working memory underpins cognitive development, learning, and education. *Educational Psychology Review, 26*(2), 197–223. doi:10.1007/s10648-013-9246-y

Coyle, T. R. (2015). Relations among general intelligence (g), aptitude tests, and GPA: Linear effects dominate. *Intelligence, 53*, 16–22. doi:10.1016/j.intell.2015.08.005

Coyle, T. R. (2021). Defining and measuring intelligence: The psychometrics and neuroscience of g. In A. Barbey, S. Karama, & R. J. Haier (Eds.), *Cambridge Handbook of Intelligence and Cognitive Neuroscience* (pp. 3–25). New York: Cambridge University Press.

Crick, F. (1994). *The Astonishing Hypothesis : The Scientific Search for the Soul*. New York, Scribner : Maxwell Macmillan International.

Curlik, D. M., 2nd, & Shors, T. J. (2013). Training your brain: Do mental and physical (MAP) training enhance cognition through the process of neurogenesis in the hippocampus? *Neuropharmacology, 64*, 506–514. doi:10.1016/j.neuropharm.2012.07.027

Curlik, D. M., 2nd, Maeng, L. Y., Agarwal, P. R., & Shors, T. J. (2013). Physical skill training increases the number of surviving new cells in

the adult hippocampus. *Plos One, 8*(2), e55850. doi:10.1371/journal
.pone.0055850

Davies, G., Armstrong, N., Bis, J. C., Bressler, J., Chouraki, V., Giddaluru,
S., ... Deary, I. J. (2015). Genetic contributions to variation in general
cognitive function: A meta-analysis of genome-wide association studies
in the CHARGE consortium (N=53949). *Molecular Psychiatry, 20*(2),
183–192. doi:10.1038/mp.2014.188

Davies, G., Tenesa, A., Payton, A., Yang, J., Harris, S. E., Liewald, D.,
... Deary, I. J. (2011). Genome-wide association studies establish
that human intelligence is highly heritable and polygenic. *Molecular
Psychiatry, 16*(10), 996–1005.

Davis, J. M., Searles, V. B., Anderson, N., Keeney, J., Raznahan, A.,
Horwood, L. J., ... Sikela, J. M. (2015). DUF1220 copy number is
linearly associated with increased cognitive function as measured by
total IQ and mathematical aptitude scores. *Human Genetics, 134*(1),
67–75. doi:10.1007/s00439-014-1489-2

Dawkins, R. (2016). *The Selfish Gene : 40th Anniversary Edition.* New
York: Oxford University Press.

Deary, I. J. (2000). *Looking Down on Human Intelligence: From
Psychometrics to the Brain.* Oxford; New York: Oxford University Press.

Deary, I. J., Cox, S. R., & Hill, W. D. (2022). Genetic variation, brain,
and intelligence differences. *Molecular Psychiatry, 27*(1), 335–353.
doi:10.1038/s41380-021-01027-y

Deary, I. J., Penke, L., & Johnson, W. (2010). The neuroscience of human
intelligence differences. *Nature Reviews Neuroscience, 11*(3), 201–211.
doi:10.1038/Nrn2793

Deary, I. J., Whiteman, M. C., Starr, J. M., Whalley, L. J., & Fox, H. C.
(2004). The impact of childhood intelligence on later life: Following up
the Scottish Mental Surveys of 1932 and 1947. *Journal of Personality and
Social Psychology, 86*(1), 130–147.

Del Río, D., Cuesta, P., Bajo, R., García-Pacios, J., López-
Higes, R., del-Pozo, F., & Maestú, F. (2012). Efficiency at rest:
Magnetoencephalographic resting-state connectivity and individual
differences in verbal working memory. *International Journal of
Psychophysiology, 86*(2), 160–167.

Demetriou, A., Golino, H., Spanoudis, G., Makris, N., & Greiff, S.
(2021). The future of intelligence: The central meaning-making unit of
intelligence in the mind, the brain, and artificial intelligence. *Intelligence,
87*, 101562. doi:10.1016/j.intell.2021.101562

Der, G., Batty, G. D., & Deary, I. J. (2006). Effect of breast feeding on
intelligence in children: Prospective study, sibling pairs analysis, and
meta-analysis. *BMJ, 333*(7575), 945. doi:10.1136/bmj.38978.699583.55

Desrivieres, S., Lourdusamy, A., Tao, C., Toro, R., Jia, T., Loth, E., … Consortium, I. (2015). Single nucleotide polymorphism in the neuroplastin locus associates with cortical thickness and intellectual ability in adolescents. *Molecular Psychiatry, 20*(2), 263–274. doi:10.1038/mp.2013.197

Detterman, D. K. (1998). Kings of men: Introduction to a special issue. *Intelligence, 26*(3), 175–180. doi:10.1016/S0160-2896(99)80001-4

Detterman, D. K. (2014). Introduction to the intelligence special issue on the development of expertise: Is ability necessary? *Intelligence, 45*, 1–5.

Detterman, D. K. (2016). Education and intelligence: Pity the poor teacher because student characteristics are more significant than teachers or schools. *Spanish Journal of Psychology*, 19. doi:10.1017/sjp.2016.88

Dietrich, A., & Kanso, R. (2010). A review of EEG, ERP, and neuroimaging studies of creativity and insight. *Psychological Bulletin, 136*(5), 822–848. doi:10.1037/a0019749

Dietz, P., Soyka, M., & Franke, A. G. (2016). Pharmacological neuroenhancement in the field of economics-poll results from an online survey. *Frontiers in Psychology*, 7.

Dietz, P., Striegel, H., Franke, A. G., Lieb, K., Simon, P., & Ulrich, R. (2013). Randomized response estimates for the 12-month prevalence of cognitive-enhancing drug use in University Students. *Pharmacotherapy, 33*(1), 44–50.

Donnay, G. F., Rankin, S. K., Lopez-Gonzalez, M., Jiradejvong, P., & Limb, C. J. (2014). Neural substrates of interactive musical improvisation: An FMRI study of 'trading fours' in jazz. *Plos One, 9*(2), e88665. doi:10.1371/journal.pone.0088665

Douw, L., Nissen, I. A., Fitzsimmons, S. M. D. D., Santos, F. A. N., Hillebrand, A., van Straaten, E. C. W., … Goriounova, N. A. (2021). Cellular substrates of functional network integration and memory in temporal lobe epilepsy. *Cerebral Cortex, 32*(11), 2424–2436. doi:10.1093/cercor/bhab349

Drakulich, S., & Karama, S. (2021). Structural brain imaging of intelligence. In A. Barbey, S. Karama, & R. Haier (Eds.), *Cambridge Handbook of Intelligence and Cognitive Neuroscience* (pp. 210–234). New York: Cambridge University Press.

Drakulich, S., Sitartchouk, A., Olafson, E., Sarhani, R., Thiffault, A.-C., Chakravarty, M., … Karama, S. (2022). General cognitive ability and pericortical contrast. *Intelligence, 91*, 101633. doi:10.1016/j.intell.2022.101633

Dresler, M., Sandberg, A., Bublitz, C., Ohla, K., Trenado, C., Mroczko-Wąsowicz, A., … Repantis, D. (2019). Hacking the brain: Dimensions of cognitive enhancement. *ACS Chemical Neuroscience, 10*(3), 1137–1148. doi:10.1021/acschemneuro.8b00571

Dreszer, J., Grochowski, M., Lewandowska, M., Nikadon, J., Gorgol, J., Balaj, B., ... Piotrowski, T. (2020). Spatiotemporal complexity patterns of resting-state bioelectrical activity explain fluid intelligence: Sex matters. *Human Brain Mapping, 41*(17), 4846–4865. doi:10.1002/hbm.25162

Drury, S. S., Theall, K., Gleason, M. M., Smyke, A. T., De Vivo, I., Wong, J. Y., ... Nelson, C. A. (2012). Telomere length and early severe social deprivation: linking early adversity and cellular aging. *Molecular Psychiatry, 17*(7), 719–727. doi:10.1038/mp.2011.53

Dubois, J., Galdi, P., Paul, L. K., & Adolphs, R. (2018). A distributed brain network predicts general intelligence from resting-state human neuroimaging data. *Philosophical Transactions on Royal Society London B Biological Sciences, 373*(1756). doi:10.1098/rstb.2017.0284

Duncan, G. J., & Sojourner, A. J. (2013). Can intensive early childhood intervention programs eliminate income-based cognitive and achievement gaps? *Journal of Human Resources, 48*(4), 945–968.

Duncan, J. (2010). The multiple-demand (MD) system of the primate brain: Mental programs for intelligent behaviour. *Trends in Cognition Science, 14*(4), 172–179. doi:10.1016/j.tics.2010.01.004

Duncan, J., Burgess, P., & Emslie, H. (1995). Fluid intelligence after frontal lobe lesions. *Neuropsychologia, 33*(3), 261–268.

Duncan, J., Seitz, R. J., Kolodny, J., Bor, D., Herzog, H., Ahmed, A., ... Emslie, H. (2000). A neural basis for general intelligence. *Science, 289*(5478), 457–460.

Durkin, K., Lipsey, M. W., Farran, D. C., & Wiesen, S. E. (2022). Effects of a statewide pre-kindergarten program on children's achievement and behavior through sixth grade. *Developmental Psychology.* doi:10.1037/dev0001301

Editorial. (2017). Intelligence research should not be held back by its past. *Nature, 545*, 385–386.

Ericsson, K. A. (2014). Why expert performance is special and cannot be extrapolated from studies of performance in the general population: A response to criticisms. *Intelligence, 45*, 81–103.

Ericsson, K. A., & Towne, T. J. (2010). *Expertise. Wiley Interdisciplinary Reviews-Cognitive Science, 1*(3), 404–416

Esposito, G., Kirkby, B. S., Van Horn, J. D., Ellmore, T. M., & Berman, K. F. (1999). Context-dependent, neural system-specific neurophysiological concomitants of ageing: Mapping PET correlates during cognitive activation. *Brain, 122*, 963–979.

Estrada, E., Ferrer, E., Román, F. J., Karama, S., & Colom, R. (2019). Time-lagged associations between cognitive and cortical development from childhood to early adulthood. *Developmental Psychology, 55*(6), 1338–1352.

Euler, M. J., & McKinney, T. L. (2021). Evaluating the weight of the evidence: Cognitive neuroscience theories of intelligence. In A. Barbey, S. Karama, & R. J. Haier (Eds.), *The Cambridge Handbook of Intelligence and Cognitive Neuroscience (pp. 85–101)*. New York: Cambridge University Press.

Euler, M. J., McKinney, T. L., Schryver, H. M., & Okabe, H. (2017). ERP correlates of the decision time-IQ relationship: The role of complexity in task- and brain-IQ effects. *Intelligence, 65*, 1–10. doi:10.1016/j.intell.2017.08.003

Euler, M. J., Weisend, M. P., Jung, R. E., Thoma, R. J., & Yeo, R. A. (2015). Reliable activation to novel stimuli predicts higher fluid intelligence. *Neuroimage, 114*, 311–319. doi:10.1016/j.neuroimage.2015.03.078

Ezkurdia, I., Juan, D., Rodriguez, J. M., Frankish, A., Diekhans, M., Harrow, J., … Tress, M. L. (2014). Multiple evidence strands suggest that there may be as few as 19,000 human protein-coding genes. *Human Molecular Genetics, 23*(22), 5866–5878.

Falk, D., Lepore, F. E., & Noe, A. (2013). The cerebral cortex of Albert Einstein: A description and preliminary analysis of unpublished photographs. *Brain, 136*(Pt 4), 1304–1327. doi:10.1093/brain/aws295

Fangmeier, T., Knauff, M., Ruff, C. C., & Sloutsky, V. M. (2006). fMRI evidence for a three-stage model of deductive reasoning. *Journal of Cognitive Neuroscience, 18*(3), 320–334.

Farah, M. J., Betancourt, L., Shera, D. M., Savage, J. H., Giannetta, J. M., Brodsky, N. L., … Hurt, H. (2008). Environmental stimulation, parental nurturance and cognitive development in humans. *Developmental Science, 11*(5), 793–801. Retrieved from www.ncbi.nlm.nih.gov/pubmed/18810850

Farah, M. J., Shera, D. M., Savage, J. H., Betancourt, L., Giannetta, J. M., Brodsky, N. L., … Hurt, H. (2006). Childhood poverty: Specific associations with neurocognitive development. *Brain Research, 1110*(1), 166–174. doi:10.1016/j.brainres.2006.06.072

Farah, M. J., Smith, M. E., Ilieva, I., & Hamilton, R. H. (2014). Cognitive enhancement. *Wiley Interdisciplinary Reviews-Cognitive Science, 5*(1), 95–103.

Feilong, M., Guntupalli, J. S., & Haxby, J. V. (2021). The neural basis of intelligence in fine-grained cortical topographies. *Elife, 10*, e64058. doi:10.7554/eLife.64058

Ferrer, E., O'Hare, E. D., & Bunge, S. A. (2009). Fluid reasoning and the developing brain. *Frontiers in Neuroscience, 3*(1), 46–51. doi:10.3389/neuro.01.003.2009

Ferrero, M., Vadillo, M. A., & León, S. P. (2021). A valid evaluation of the theory of multiple intelligences is not yet possible: Problems of methodological quality for intervention studies. *Intelligence, 88*, 101566. doi:10.1016/j.intell.2021.101566

Fink, A., Grabner, R. H., Gebauer, D., Reishofer, G., Koschutnig, K., & Ebner, F. (2010). Enhancing creativity by means of cognitive stimulation: Evidence from an fMRI study. *Neuroimage, 52*(4), 1687–1695. doi:10.1016/j.neuroimage.2010.05.072

Finn, E. S., & Rosenberg, M. D. (2021). Beyond fingerprinting: Choosing predictive connectomes over reliable connectomes. *Neuroimage, 239*, 118254. doi:10.1016/j.neuroimage.2021.118254

Finn, E. S., Shen, X., Scheinost, D., Rosenberg, M. D., Huang, J., Chun, M. M., ... Constable, R. T. (2015). Functional connectome fingerprinting: Identifying individuals using patterns of brain connectivity. *Natural Neuroscience, 18*(11), 1664–1671. doi:10.1038/nn.4135

Firkowska, A., Ostrowska, A., Sokolowska, M., Stein, Z., Susser, M., & Wald, I. (1978). Cognitive-development and social-policy. *Science, 200*(4348), 1357–1362.

Flashman, L. A., Andreasen, N. C., Flaum, M., & Swayze, V. W. (1997). Intelligence and regional brain volumes in normal controls. *Intelligence, 25*(3), 149–160.

Flynn, J. R. (2013). The "Flynn Effect" and Flynn's paradox. *Intelligence, 41*(6), 851–857. doi:10.1016/J.Intell.2013.06.014

Fox, K. C. R., Girn, M., Parro, C. C., & Christoff, K. (2018). Functional neuroimaging of psychedelic experience: An overview of psychological and neural effects and their relevance to research on creativity, daydreaming, and dreaming. In R.E. Jung & O. Vartanian (Eds.), *The Cambridge Handbook of the Neuroscience of Creativity* (pp. 92–113): Cambridge University Press.

Fraenz, C., Schlüter, C., Friedrich, P., Jung, R. E., Güntürkün, O., & Genç, E. (2021). Interindividual differences in matrix reasoning are linked to functional connectivity between brain regions nominated by Parieto-Frontal Integration Theory. *Intelligence, 87*, 101545. doi:10.1016/j.intell.2021.101545

Frangou, S., Chitins, X., & Williams, S. C. R. (2004). Mapping IQ and gray matter density in healthy young people. *Neuroimage, 23*(3), 800–805.

Franke, A. G., Bagusat, C., Rust, S., Engel, A., & Lieb, K. (2014). Substances used and prevalence rates of pharmacological cognitive enhancement among healthy subjects. *European Archives of Psychiatry and Clinical Neuroscience, 264*, S83-S90. doi:10.1007/s00406-014-0537-1

Fregnac, Y., & Laurent, G. (2014). Neuroscience: Where is the brain in the human brain project? *Nature, 513*(7516), 27–29. doi:10.1038/513027a

Frey, M. C., & Detterman, D. K. (2004). Scholastic assessment or g? The relationship between the scholastic assessment test and general cognitive ability (vol 15, p. 373, 2004). *Psychological Science, 15*(9), 641–641.

Frischkorn, G. T., Schubert, A. L., & Hagemann, D. (2019). Processing speed, working memory, and executive functions: Independent or inter-related predictors of general intelligence. *Intelligence, 75*, 95–110. doi:10.1016/j.intell.2019.05.003

Galaburda, A. M. (1999). Albert Einstein's brain. *Lancet, 354*(9192), 1821; author reply 1822. Retrieved from www.ncbi.nlm.nih.gov/pubmed/10577668

Galton, F. (1869). *Hereditary Genius: An Inquiry into Its Laws and Consequences*. London: Macmillan.

Galton, F. (2006). *Hereditary Genius: An Inquiry into Its Laws and Consequences*. Amherst, NY: Prometheus Books.

Galton, F., & Prinzmetal, M. (1884). *Hereditary Genius : An Inquiry into Its Laws and Consequences* (New and revised edition, with an American preface. ed.). New York: D. Appleton and Company …

Gardner, H. (1987). The theory of multiple intelligences. *Annals of Dyslexia, 37*, 19–35.

Gardner, H., & Moran, S. (2006). The science of multiple intelligences theory: A response to Lynn Waterhouse. *Educational Psychologist, 41*(4), 227–232. doi:10.1207/s15326985ep4104_2

Geake, J. (2008). Neuromythologies in education. *Educational Research, 50*(2), 123–133.

Geake, J. (2011). Position statement on motivations, methodologies, and practical implications of educational neuroscience research: fMRI studies of the neural correlates of creative intelligence. *Educational Philosophy and Theory, 43*(1), 43–47.

Geake, J. G., & Hansen, P. C. (2005). Neural correlates of intelligence as revealed by fMRI of fluid analogies. *Neuroimage, 26*(2), 555–564.

Geary, D. C. (2018). Efficiency of mitochondrial functioning as the fundamental biological mechanism of general intelligence (g). *Psychological Review, 125*(6), 1028–1050.

Geary, D. C. (2019). Mitochondria as the Linchpin of general intelligence and the link between g, health, and aging. *Journal of Intelligence, 7*(4).

Genc, E., & Fraenz, C. (2021). Diffusion-weighted imaging of intelligence. In A. Barbey, S. Karama, & R. Haier (Eds.), *The Cambridge Handbook of Intelligence and Cognitive Neuroscience* (pp. 191–209). New York: Cambridge University Press.

Genc, E., Fraenz, C., Schluter, C., Friedrich, P., Hossiep, R., Voelkle, M. C., … Jung, R. E. (2018). Diffusion markers of dendritic density and arborization in gray matter predict differences in intelligence. *Nature Communication, 9*(1), 1905. doi:10.1038/s41467-018-04268-8

Genc, E., Schluter, C., Fraenz, C., Arning, L., Metzen, D., Nguyen, H. P., … Ocklenburg, S. (2021). Polygenic scores for cognitive abilities and their association with different aspects of general intelligence: A deep phenotyping approach. *Molecular Neurobiology, 58*(8), 4145–4156.

George, M. S., Nahas, Z., Molloy, M., Speer, A. M., Oliver, N. C., Li, X. B., … Ballenger, J. C. (2000). A controlled trial of daily left prefrontal cortex TMS for treating depression. *Biological Psychiatry, 48*(10), 962–970. doi:10.1016/s0006-3223(00)01048-9

Ghatan, P. H., Hsieh, J. C., Wirsenmeurling, A., Wredling, R., Eriksson, L., Stoneelander, S., … Ingvar, M. (1995). Brain activation-induced by the perceptual Maze Test – A Pet Study of Cognitive Performance. *Neuroimage, 2*(2), 112–124.

Gignac, G. E. (2015). Raven's is not a pure measure of general intelligence: Implications for g factor theory and the brief measurement of g. *Intelligence, 52*, 71–79. doi:10.1016/j.intell.2015.07.006

Girn, M., Mills, C., & Christoff, K. (2019). Linking brain network reconfiguration and intelligence: Are we there yet? *Trends in Neuroscience and Education, 15*, 62–70. doi:10.1016/j.tine.2019.04.001

Glascher, J., Rudrauf, D., Colom, R., Paul, L. K., Tranel, D., Damasio, H., & Adolphs, R. (2010). Distributed neural system for general intelligence revealed by lesion mapping. *Proceedings of the National Academy of Sciences of the United States of America, 107*(10), 4705–4709. doi:10.1073/Pnas.0910397107

Glascher, J., Tranel, D., Paul, L. K., Rudrauf, D., Rorden, C., Hornaday, A., … Adolphs, R. (2009). Lesion mapping of cognitive abilities linked to intelligence. *Neuron, 61*(5), 681–691. doi:10.1016/j.neuron.2009.01.026

Gobet, F., & Sala, G. (2022). Cognitive training: A field in search of a phenomenon. *Perspectives on Psychological Science*, 17456916221091830. doi:10.1177/17456916221091830

Goel, V., & Dolan, R. J. (2004). Differential involvement of left prefrontal cortex in inductive and deductive reasoning. *Cognition, 93*(3), B109–B121.

Goel, V., Gold, B., Kapur, S., & Houle, S. (1997). The seats of reason? An imaging study of deductive and inductive reasoning. *Neuroreport, 8*(5), 1305–1310.

Goel, V., Gold, B., Kapur, S., & Houle, S. (1998). Neuroanatomical correlates of human reasoning. *Journal of Cognitive Neuroscience, 10*(3), 293–302.

Gonen-Yaacovi, G., de Souza, L. C., Levy, R., Urbanski, M., Josse, G., & Volle, E. (2013). Rostral and caudal prefrontal contribution to creativity: A meta-analysis of functional imaging data. *Frontiers in Human Neuroscience, 7*, 465. doi:10.3389/fnhum.2013.00465

Gong, Q.-Y., Sluming, V., Mayes, A., Keller, S., Barrick, T., Cezayirli, E., & Roberts, N. (2005). Voxel-based morphometry and stereology provide convergent evidence of the importance of medial prefrontal cortex for fluid intelligence in healthy adults. *Neuroimage, 25*(4), 1175.

Gonzalez-Lima, F., & Barrett, D. W. (2014). Augmentation of cognitive brain functions with transcranial lasers. *Frontiers in Systematic Neuroscience, 8*, 36. doi:10.3389/fnsys.2014.00036

Gordon, R. A. (1997). Everyday life as an intelligence test: Effects of intelligence and intelligence context. *Intelligence, 24*(1), 203–320. doi:10.1016/S0160-2896(97)90017-9

Goriounova, N. A., & Mansvelder, H. D. (2019). Genes, Cells and Brain Areas of Intelligence. *Frontiers in Human Neuroscience, 13*, 44. doi:10.3389/fnhum.2019.00044

Goriounova, N. A., Heyer, D. B., Wilbers, R., Verhoog, M. B., Giugliano, M., Verbist, C., … Mansvelder, H. D. (2018). Large and fast human pyramidal neurons associate with intelligence. *Elife, 7*.

Gottfredson, L. S. (1997a). Mainstream science on intelligence: An editorial with 52 signatories, history, and bibliography (Reprinted from The Wall Street Journal, 1994). *Intelligence, 24*(1), 13–23.

Gottfredson, L. S. (1997b). Why g matters: The complexity of everyday life. *Intelligence, 24*(1), 79–132.

Gottfredson, L. S. (2002). Where and why g matters: Not a mystery. *Human Performance, 15*(1/2).

Gottfredson, L. S. (2003a). Dissecting practical intelligence theory: Its claims and evidence. *Intelligence, 31*(4), 343–397.

Gottfredson, L. S. (2003b). g, jobs and life. In H. Nyborg (Ed.), *The Scientific Study of General Intelligence* (pp. 293–342). New York: Elsevier Science.

Gottfredson, L. S. (2005). Suppressing intelligence research: Hurting those we intend to help. In R.H. Wright and N.A. Cummings (Ed.), *Destructive Trends in Mental Health: The Well-intentioned Path to Harm* (pp. 155–186). New York: Routledge.

Gozli, D. G., Bavelier, D., & Pratt, J. (2014). The effect of action video game playing on sensorimotor learning: Evidence from a movement tracking task. *Human Movement Science, 38C*, 152–162. doi:10.1016/j.humov.2014.09.004

Grabner, R. H. (2014). The role of intelligence for performance in the prototypical expertise domain of chess. *Intelligence, 45*, 26–33.

Grabner, R. H., Stern, E., & Neubauer, A. C. (2007). Individual differences in chess expertise: A psychometric investigation. *Acta Psychologica, 124*(3), 398–420.

Graham, S., Jiang, J., Manning, V., Nejad, A. B., Zhisheng, K., Salleh, S. R., … McKenna, P. J. (2010). IQ-related fMRI differences during cognitive set shifting. *Cerebral Cortex, 20*(3), 641–649. doi:10.1093/cercor/bhp130

Gray, J. R., Chabris, C. F., & Braver, T. S. (2003). Neural mechanisms of general fluid intelligence. *Nature Neuroscience, 6*(3), 316–322.

Greely, H., Sahakian, B., Harris, J., Kessler, R. C., Gazzaniga, M., Campbell, P., & Farah, M. J. (2008). Towards responsible use of cognitive-enhancing drugs by the healthy. *Nature, 456*(7223), 702–705. doi:10.1038/456702a

Green, A. E., Kraemer, D. J., Fugelsang, J. A., Gray, J. R., & Dunbar, K. N. (2012). Neural correlates of creativity in analogical reasoning. *Journal of Experimental Psychology: Learning, Memory, and Cognition, 38*(2), 264–272. doi:10.1037/a0025764

Gregory, H. (2015). *McNamara's Folly: The Use of Low IQ Troops in the Vietnam War*: West Conshohocken: Infinity.

Gullich, A., Macnamara, B. N., & Hambrick, D. Z. (2022). What makes a champion? Early multidisciplinary practice, not early specialization, predicts world-class performance. *Perspectives on Psychological Science, 17*(1), 6–29. doi:10.1177/1745691620974772

Gur, R. C., Butler, E. R., Moore, T. M., Rosen, A. F. G., Ruparel, K., Satterthwaite, T. D., … Gur, R. E. (2021). Structural and functional brain parameters related to cognitive performance across development: Replication and extension of the parieto-frontal integration theory in a single sample. *Cerebral Cortex, 31*(3), 1444–1463. doi:10.1093/cercor/bhaa282

Gur, R. C., Ragland, J. D., Resnick, S. M., Skolnick, B. E., Jaggi, J., Muenz, L., & Gur, R. E. (1994). Lateralized increases in cerebral blood-flow during performance of verbal and spatial tasks – Relationship with performance-level. *Brain and Cognition, 24*(2), 244–258.

Haasz, J., Westlye, E. T., Fjaer, S., Espeseth, T., Lundervold, A., & Lundervold, A. J. (2013). General fluid-type intelligence is related to indices of white matter structure in middle-aged and old adults. *Neuroimage, 83*, 372–383. doi:10.1016/j.neuroimage.2013.06.040

Hackman, D. A., Farah, M. J., & Meaney, M. J. (2010). Socioeconomic status and the brain: Mechanistic insights from human and animal research. *Nature Review Neuroscience, 11*(9), 651–659. doi:10.1038/nrn2897

Haggarty, P., Hoad, G., Harris, S. E., Starr, J. M., Fox, H. C., Deary, I. J., & Whalley, L. J. (2010). Human intelligence and polymorphisms in the DNA methyltransferase genes involved in epigenetic marking. *Plos One, 5*(6), e11329. doi:10.1371/journal.pone.0011329

Haier, R. J. (1990). The end of intelligence research. *Intelligence, 14*(4), 371–374.

Haier, R. J. (2009a). Neuro-intelligence, neuro-metrics and the next phase of brain imaging studies. *Intelligence, 37*(2), 121–123.

Haier, R. J. (2009b). What does a smart brain look like? *Scientific American Mind, November/December*, 26–33.

Haier, R. J. (Producer). (2013). The intelligent brain. [lecture course] Retrieved from www.thegreatcourses.com/courses/the-intelligent-brain

Haier, R. J. (2014). Increased intelligence is a myth (so far). *Frontiers in Systematic Neuroscience, 8*, 34. doi:10.3389/fnsys.2014.00034

Haier, R. J. (2021). Are we thinking big enough about the road ahead? Overview of the special issue on the future of intelligence research. *Intelligence*, 89, 101603. doi:10.1016/j.intell.2021.101603

Haier, R. J., & Benbow, C. P. (1995). Sex differences and lateralization in temporal lobe glucose metabolism during mathematical reasoning. *Developmental Neuropsychology, 11*(4), 405–414.

Haier, R. J., & Colom, R. (2023). *The Science of Human Intelligence*. Oxford: Cambridge University Press.

Haier, R. J., & Jung, R. E. (2007). Beautiful minds (i.e., brains) and the neural basis of intelligence. *Behavioral and Brain Sciences, 30*(02), 174–178.

Haier, R. J., & Jung, R. E. (2008). Brain imaging studies of intelligence and creativity – What is the picture for education?. *Roeper Review, 30*(3), 171–180.

Haier, R. J., Jung, R. E., Yeo, R. A., Head, K., & Alkire, M. T. (2004). Structural brain variation and general intelligence. *Neuroimage, 23*(1), 425–433.

Haier, R. J., Jung, R. E., Yeo, R. A., Head, K., & Alkire, M. T. (2005). The neuroanatomy of general intelligence: Sex matters. *Neuroimage, 25*(1), 320–327.

Haier, R. J., Robinson, D. L., Braden, W., & Williams, D. (1983). Electrical potentials of the cerebral cortex and psychometric intelligence. *Personality & Individual Differences*, 4(6), 591–599.

Haier, R. J., Siegel, B. V., Jr., Crinella, F. M., & Buchsbaum, M. S. (1993). Biological and psychometric intelligence: Testing an animal model in humans with positron emission tomography. In Douglas K. Detterman (Ed.), *Individual Differences and Cognition* (pp. 317–331). Norwood: Ablex Publishing Corp.

Haier, R. J., Siegel, B. V., Jr., MacLachlan, A., Soderling, E., Lottenberg, S., & Buchsbaum, M. S. (1992a). Regional glucose metabolic changes after learning a complex visuospatial/motor task: A positron emission tomographic study. *Brain Research, 570*(1–2), 134–143.

Haier, R. J., Siegel, B. V., Nuechterlein, K. H., Hazlett, E., Wu, J. C., Paek, J., … Buchsbaum, M. S. (1988). Cortical glucose metabolic-rate correlates of abstract reasoning and attention studied with positron emission tomography. *Intelligence, 12*(2), 199–217.

Haier, R. J., Siegel, B., Tang, C., Abel, L., & Buchsbaum, M. S. (1992b). Intelligence and changes in regional cerebral glucose metabolic-rate following learning. *Intelligence, 16*(3–4), 415–426.

Haier, R. J., White, N. S., & Alkire, M. T. (2003). Individual differences in general intelligence correlate with brain function during nonreasoning tasks. *Intelligence, 31*(5), 429–441. (2007).

Halpern, Diane F., Camilla P. Benbow, David C. Geary, Ruben C. Gur, Janet Shibley Hyde, and Morton Ann Gernsbacher. The science of sex differences in science and mathematics. *Psychological Science in the Public Interest, 8*(1), 1–51. doi:10.1111/j.1529-1006.2007.00032.x

Halstead, W. C. (1947). *Brain and Intelligence; A Quantitative Study of the Frontal Lobes*. Chicago: University of Chicago Press.

Hambrick, D. Z., Macnamara, B. N., & Oswald, F. L. (2020). Is the deliberate practice view defensible? A review of evidence and discussion of issues. *Frontiers in Psychology, 11*, 1134. doi:10.3389/fpsyg.2020.01134

Hampshire, A., Thompson, R., Duncan, J., & Owen, A. M. (2011). Lateral prefrontal cortex subregions make dissociable contributions during fluid reasoning. *Cerebral Cortex, 21*(1), 1–10. doi:10.1093/cercor/bhq085

Hanscombe, K. B., Trzaskowski, M., Haworth, C. M. A., Davis, O. S. P., Dale, P. S., & Plomin, R. (2012). Socioeconomic Status (SES) and Children's Intelligence (IQ): In a UK-Representative sample SES moderates the environmental, not genetic, effect on IQ. *Plos One, 7*(2).

Hansen, J. Y., Shafiei, G., Markello, R. D., Smart, K., Cox, S. M. L., Wu, Y., … Misic, B. (2021). Mapping neurotransmitter systems to the structural and functional organization of the human neocortex. bioRxiv, 2021.2010.2028.466336. doi:10.1101/2021.10.28.466336

Harrison, T. L., Shipstead, Z., Hicks, K. L., Hambrick, D. Z., Redick, T. S., & Engle, R. W. (2013). Working memory training may increase working memory capacity but not fluid intelligence. *Psychological Science, 24*(12), 2409–2419. doi:10.1177/0956797613492984

Hartshorne, J. K. & L. T. Germine (2015). When does cognitive functioning peak? The asynchronous rise and fall of different cognitive abilities across the life span. *Psychological Science, 26*(4), 433–443.

Hawkins, J. (2021). *A Thousand Brains : A New Theory of Intelligence* (First edition. ed.). New York: Basic Books.

Hawkins, J., & Blakeslee, S. (2004). *On intelligence* (1st ed.). New York: Times Books.

Haworth, C. M. A., Wright, M. J., Luciano, M., Martin, N. G., de Geus, E. J. C., van Beijsterveldt, C. E. M., … Plomin, R. (2010). The heritability of general cognitive ability increases linearly from childhood to young adulthood. *Molecular Psychiatry, 15*(11), 1112–1120.

Hayes, T. R., Petrov, A. A., & Sederberg, P. B. (2015). Do we really become smarter when our fluid-intelligence test scores improve? *Intelligence, 48*, 1–14.

Hebling Vieira, B., Dubois, J., Calhoun, V. D., & Garrido Salmon, C. E. (2021). A deep learning based approach identifies regions more relevant than resting-state networks to the prediction of general intelligence from resting-state fMRI. *Human Brain Mapping, 42*(18), 5873–5887. doi:10.1002/hbm.25656

Heishman, S. J., Kleykamp, B. A., & Singleton, E. G. (2010). Meta-analysis of the acute effects of nicotine and smoking on human performance. *Psychopharmacology (Berl), 210*(4), 453–469. doi:10.1007/s00213-010-1848-1

Herrnstein, R. J. (1973). *I.Q. in the Meritocracy* (1st ed.). Boston: Little.

Herrnstein, R. J., & Murray, C. A. (1994). *The Bell Curve : Intelligence and Class Structure in American life*. New York: Free Press.

Hescham, S., Liu, H., Jahanshahi, A., & Temel, Y. (2020). Deep brain stimulation and cognition: Translational aspects. *Neurobiology of Learning and Memory, 174*, 107283. doi:10.1016/j.nlm.2020.107283

Heyer, D. B., Wilbers, R., Galakhova, A. A., Hartsema, E., Braak, S., Hunt, S., … Goriounova, N. A. (2021). Verbal and general IQ associate with supragranular layer Thickness and cell properties of the left temporal cortex. *Cerebral Cortex, 32*(11), 2343–2357. doi:10.1093/cercor/bhab330

Heyward, F. D., & Sweatt, J. D. (2015). DNA methylation in memory formation: Emerging insights. *Neuroscientist.* doi:10.1177/1073858415579635

Hilger, K., Ekman, M., Fiebach, C. J., & Basten, U. (2017a). Efficient hubs in the intelligent brain: Nodal efficiency of hub regions in the salience network is associated with general intelligence. *Intelligence, 60*, 10–25. doi:10.1016/j.intell.2016.11.001

Hilger, K., Ekman, M., Fiebach, C. J., & Basten, U. (2017b). Intelligence is associated with the modular structure of intrinsic brain networks. *Scientific Reports, 7*(1), 16088. doi:10.1038/s41598-017-15795-7

Hilger, K., Fukushima, M., Sporns, O., & Fiebach, C. J. (2020). Temporal stability of functional brain modules associated with human intelligence. *Human Brain Mapping, 41*(2), 362–372. doi:10.1002/hbm.24807

Hilger, K., Spinath, F. M., Troche, S., & Schubert, A. L. (2022). The biological basis of intelligence: Benchmark findings. *Intelligence*, 93. doi:ARTN 101665

Hill, W. D., Davies, G., van de Lagemaat, L. N., Christoforou, A., Marioni, R. E., Fernandes, C. P., … Deary, I. J. (2014). Human cognitive ability is influenced by genetic variation in components of postsynaptic signalling complexes assembled by NMDA receptors and MAGUK proteins. *Translational Psychiatry, 4*, e341. doi:10.1038/tp.2013.114

Hines, T. (1998). Further on Einstein's brain. *Experiment in Neurology, 150*(2), 343–344. doi:10.1006/exnr.1997.6759

Hopkins, W. D., Russe, J. L., & Schaeffer, J. (2014). Chimpanzee Intelligence Is Heritable. *Current Biology, 24*(14), 1649–1652.

Horvath, J. C., Forte, J. D., & Carter, O. (2015a). Evidence that transcranial direct current stimulation (tDCS) generates little-to-no reliable neurophysiologic effect beyond MEP amplitude modulation in healthy human subjects: A systematic review. *Neuropsychologia, 66,* 213–236. doi:10.1016/j.neuropsychologia.2014.11.021

Horvath, J. C., Forte, J. D., & Carter, O. (2015b). Quantitative review finds no evidence of cognitive effects in healthy populations from single-session transcranial direct current stimulation (tDCS). *Brain Stimulation.* doi:10.1016/j.brs.2015.01.400

Howard-Jones, P. A. (2014). Neuroscience and education: Myths and messages. *Natral Review in Neuroscience, 15*(12), 817–824. doi:10.1038/nrn3817

Hulshoff-Pol, H. E., Schnack, H. G., Posthuma, D., Mandl, R. C. W., Baare, W. F., van Oel, C., … Kahn, R. S. (2006). Genetic contributions to human brain morphology and intelligence. *Journal of Neuroscience, 26*(40), 10235–10242.

Hunt, E. (2011). *Human Intelligence.* Cambridge; New York: Cambridge University Press.

Husain, M., & Mehta, M. A. (2011). Cognitive enhancement by drugs in health and disease. *Trends in Cognition Science, 15*(1), 28–36. doi:10.1016/j.tics.2010.11.002

Huttenlocher, P. R. (1975). Snyaptic and dendritic development and mental defect. *UCLA Forum Medical Science, 18,* 123–140.

Ilieva, I. P., & Farah, M. J. (2013). Enhancement stimulants: Perceived motivational and cognitive advantages. *Frontiers in Neuroscience, 7.*

Jacobs, B., Schall, M., & Scheibel, A. B. (1993). A quantitative dendritic analysis of Wernicke's area in humans. II. Gender, hemispheric, and environmental factors. *Journal of Comparative Neurology, 327*(1), 97–111. doi:10.1002/cne.903270108

Jaeggi, S. M., Buschkuehl, M., Jonides, J., & Perrig, W. J. (2008). Improving fluid intelligence with training on working memory. *Proceedings of the National Academy Science USA, 105*(19), 6829–6833. doi:10.1073/pnas.0801268105

Jaeggi, S. M., Buschkuehl, M., Jonides, J., & Shah, P. (2011). Short- and long-term benefits of cognitive training. *Proceedings of the National Academy Science USA, 108*(25), 10081–10086. doi:10.1073/pnas.1103228108

Jaeggi, S. M., Buschkuehl, M., Shah, P., & Jonides, J. (2013). The role of individual differences in cognitive training and transfer. *Memory and Cognition.* doi:10.3758/s13421-013-0364-z

Jaeggi, S. M., Buschkuehl, M., Shah, P., & Jonides, J. (2014). The role of individual differences in cognitive training and transfer. *Memory and Cognition , 42*(3), 464–480. doi:10.3758/s13421-013-0364-z

Jaeggi, S. M., Studer-Luethi, B., Buschkuehl, M., Su, Y. F., Jonides, J., & Perrig, W. J. (2010). The relationship between n-back performance and matrix reasoning – implications for training and transfer. *Intelligence, 38*(6), 625–635.

Jaenisch, R., & Bird, A. (2003). Epigenetic regulation of gene expression: How the genome integrates intrinsic and environmental signals. *Nature Genetics*, 33(Suppl), 245–254. doi:10.1038/ng1089

Jauk, E. (2018). Intelligence and creativity from the neuroscience perspective. In R. E. Jung and O. Vartanian (Eds.), *The Cambridge Handbook of the Neuroscience of Creativity* (pp. 421–436). Cambridge: Cambridge University Press.

Jauk, E., Neubauer, A. C., Dunst, B., Fink, A., & Benedek, M. (2015). Gray matter correlates of creative potential: A latent variable voxel-based morphometry study. *Neuroimage, 111*, 312–320. doi:10.1016/j.neuroimage.2015.02.002

Jensen, A. R. (1980). *Bias in Mental Testing*. New York: The Free Press.

Jensen, A. R. (1969). How much can we boost IQ and scholastic achievement. *Harvard Educational Review, 39*(1), 1–123.

Jensen, A. R. (1974). Kinship correlations reported by Burt,*C Sir. Behavior Genetics, 4*(1), 1–28.

Jensen, A. R. (1981). *Straight Talk about Mental Tests*. New York: Free Press.

Jensen, A. R. (1998a). Jensen on "Jensenism." *Intelligence, 26*(3), 181–208. doi:10.1016/S0160-2896(99)80002-6

Jensen, A. R. (1998b). *The g Factor: The Science of Mental Ability*. Westport: Praeger.

Jensen, A. R. (2006). *Clocking the Mind: Mental Chronometry and Individual Differences*. New York: Elsevier.

Jensen, A. R., & Miele, F. (2002). *Intelligence, Race, and Genetics : Conversations with Arthur R. Jensen*. Boulder, CO: Westview.

Jiang, L., Cui, H., Zhang, C., Cao, X., Gu, N., Zhu, Y., … Li, C. (2020). Repetitive transcranial magnetic stimulation for improving cognitive function in patients with mild cognitive impairment: A systematic review. *Frontiers in Aging Neuroscience, 12*, 593000. doi:10.3389/fnagi.2020.593000

Jiang, R., Calhoun, V. D., Fan, L., Zuo, N., Jung, R., Qi, S., … Sui, J. (2020). Gender differences in connectome-based predictions of individualized intelligence quotient and sub-domain scores. *Cerebral Cortex, 30*(3), 888–900. doi:10.1093/cercor/bhz134

Johnson, M. R., Shkura, K., Langley, S. R., Delahaye-Duriez, A., Srivastava, P., Hill, W. D., … Petretto, E. (2016). Systems genetics identifies a convergent gene network for cognition and neurodevelopmental disease. *Nature Neuroscience, 19*(2), 223–232. doi:10.1038/nn.4205

Johnson, W. (2012). How much can we boost IQ? An updated look at Jensen's (1969) question and answer.

Johnson, W., & Bouchard, T. J. (2005). The structure of human intelligence: It is verbal, perceptual, and image rotation (VPR), not fluid and crystallized. *Intelligence, 33*(4), 393–416.

Johnson, W., Bouchard, T. J., Krueger, R. F., McGue, M., & Gottesman, I. I. (2004). Just one g: Consistent results from three test batteries. *Intelligence, 32*(1), 95–107.

Johnson, W., Jung, R. E., Colom, R., & Haier, R. J. (2008). Cognitive abilities independent of IQ correlate with regional brain structure. *Intelligence, 36*(1), 18–28.

Johnson, W., te Nijenhuis, J., & Bouchard, T. J. (2008). Still just 1 g: Consistent results from five test batteries. *Intelligence, 36*(1), 81–95.

Jung, R. E. (2014). Evolution, creativity, intelligence, and madness: "Here Be Dragons." *Frontiers in Psychology, 5.*

Jung, R. E., & Haier, R. J. (2007). The Parieto-Frontal Integration Theory (P-FIT) of intelligence: Converging neuroimaging evidence. *Behavioral and Brain Sciences, 30*(02), 135–154.

Jung, R. E., & Haier, R. J. (2013). Creativity and intelligence: Brain networks that link and differentiate the expression of genius. In O. Vartanian, A. S. Bristol, & J. C. Kaufman (Eds.), *Neuroscience of creativity* (pp. 233–254). Cambridge, MA: The MIT Press.

Jung, R. E., & Vartanian, O. (Eds.) (2018). *The Cambridge Handbook of the Neuroscience of Creativity*. Cambridge, New York: Cambridge University Press.

Jung, R. E., Brooks, W. M., Yeo, R. A., Chiulli, S. J., Weers, D. C., & Sibbitt, W. L. (1999a). Biochemical markers of intelligence: A proton MR spectroscopy study of normal human brain. *Proceedings of the Royal Society of London Series B-Biological Sciences, 266*(1426), 1375–1379.

Jung, R. E., Haier, R. J., Yeo, R. A., Rowland, L. M., Petropoulos, H., Levine, A. S., … Brooks, W. M. (2005). Sex differences in N-acetylaspartate correlates of general intelligence: An H-1-MRS study of normal human brain. *Neuroimage, 26*(3), 965–972.

Jung, R. E., Yeo, R. A., Chiulli, S. J., Sibbitt, W. L., Weers, D. C., Hart, B. L., & Brooks, W. M. (1999b). Biochemical markers of cognition: A proton MR spectroscopy study of normal human brain. *Neuroreport, 10*(16), 3327–3331.

Kaminski, J. A., Schlagenhauf, F., Rapp, M., Awasthi, S., Ruggeri, B., Deserno, L., … the, IMAGEN consortium. (2018). Epigenetic variance in dopamine D2 receptor: A marker of IQ malleability? *Translational Psychiatry, 8*(1), 169. doi:10.1038/s41398-018-0222-7

Kanai, R., & Rees, G. (2011). The structural basis of inter-individual differences in human behaviour and cognition. *Nature Review in Neuroscience, 12*(4), 231–242. doi:10.1038/nrn3000

Kane, M. J., & Engle, R. W. (2002). The role of prefrontal cortex in working-memory capacity, executive attention, and general fluid intelligence: An individual-differences perspective. *Psychonomic Bulletin & Review, 9*(4), 637–671. doi:10.3758/Bf03196323

Kane, M. J., Hambrick, D. Z., & Conway, A. R. A. (2005). Working memory capacity and fluid intelligence are strongly related constructs: Comment on Ackerman, Beier, and Boyle (2005). *Psychological Bulletin, 131*(1), 66–71. doi:10.1037/0033-2909.131.1.66

Karalija, N., Köhncke, Y., Düzel, S., Bertram, L., Papenberg, G., Demuth, I., ... Brandmaier, A. M. (2021). A common polymorphism in the dopamine transporter gene predicts working memory performance and in vivo dopamine integrity in aging. *Neuroimage, 245*, 118707. DOI: 10.1016/j.neuroimage.2021.118707

Karama, S., Ad-Dab'bagh, Y., Haier, R. J., Deary, I. J., Lyttelton, O. C., Lepage, C., & Evans, A. C. (2009b). Positive association between cognitive ability and cortical thickness in a representative US sample of healthy 6 to 18 year-olds (vol 37, p. 145, 2009). *Intelligence, 37*(4), 431–442.

Karama, S., Ad-Dab'bagh, Y., Haier, R. J., Deary, I. J., Lyttelton, O. C., Lepage, C., ... Grp, B. D. C. (2009a). Positive association between cognitive ability and cortical thickness in a representative US sample of healthy 6 to 18 year-olds. *Intelligence, 37*(2), 145–155.

Karama, S., Bastin, M. E., Murray, C., Royle, N. A., Penke, L., Muñoz Maniega, S., ... Deary, I. J. (2014). Childhood cognitive ability accounts for associations between cognitive ability and brain cortical thickness in old age. *Molecular Psychiatry, 19*(5), 555–559. doi:10.1038/mp.2013.64

Karama, S., Colom, R., Johnson, W., Deary, I. J., Haier, R., Waber, D. P., ... Grp, B. D. C. (2011). Cortical thickness correlates of specific cognitive performance accounted for by the general factor of intelligence in healthy children aged 6 to 18. *Neuroimage, 55*(4), 1443–1453. doi:10.1016/J.Neuroimage.2011.01.016

Kendler, K. S., Turkheimer, E., Ohlsson, H., Sundquist, J., & Sundquist, K. (2015). Family environment and the malleability of cognitive ability: A Swedish national home-reared and adopted-away cosibling control study. *Proceedings of the National Academy Science USA, 112*(15), 4612–4617. doi:10.1073/pnas.1417106112

Keyes, D. (1966). *Flowers for Algernon* (1st ed.). New York: Harcourt.

Khundrakpam, B. S., Poline, J., & Evans, A. (2021). Research consortia and large-scale data repositories for studying intelligence. In A. Barbey, S.

Karama, & R. J. Haier (Eds.), *The Cambridge Handbook of Intelligence and Cognitive Neuroscience* (pp. 70–82). New York: Cambridge University Press.

Kievit, R. A., & Simpson-Kent, I. L. (2021). It's about time: Towards a longitudinal cognitive neuroscience of intelligence. In A. Barbey, S. Karama, & R. J. Haier (Eds.), *The Cambridge Handbook of Intelligence and Cognitive Neuroscience* (pp. 123–146). New York: Cambridge University Press.

Kievit, R. A., Romeijn, J. W., Waldorp, L. J., Wicherts, J. M., Scholte, H. S., & Borsboom, D. (2011). Mind the gap: A psychometric approach to the reduction problem. *Psychological Inquiry, 22*(2), 67–87. doi:10.1080/1047 840x.2011.550181

Kievit, R. A., van Rooijen, H., Wicherts, J. M., Waldorp, L. J., Kan, K. J., Scholte, H. S., & Borsboom, D. (2012). Intelligence and the brain: A model-based approach. *Cognition Neuroscience, 3*(2), 89–97. doi:10.1080 /17588928.2011.628383

Kim, D.-J., Davis, E. P., Sandman, C. A., Sporns, O., O'Donnell, B. F., Buss, C., & Hetrick, W. P. (2016). Children's intellectual ability is associated with structural network integrity. *Neuroimage, 124*(Part A), 550–556. doi:10.1016/j.neuroimage.2015.09.012

Kim, T. D., Hong, G., Kim, J., & Yoon, S. (2019). Cognitive enhancement in neurological and psychiatric disorders using transcranial magnetic stimulation (TMS): A review of modalities, potential mechanisms and future implications. *Experimental Neurobiology, 28*(1), 1–16. doi:10.5607/ en.2019.28.1.1

Knafo, S., & Venero, C. (2015). *Cognitive Enhancement : Pharmacologic, Environmental, and Genetic factors*. Amsterdam; Boston: Elsevier/AP, Academic Press is an imprint of Elsevier.

Knowles, E. E. M., Mathias, S. R., McKay, D. R., Sprooten, E., Blangero, J., Almasy, L., & Glahn, D. C. (2014). Genome-wide analyses of working-memory ability: A review. *Current Behavioral Neuroscience Reports, 1*(4), 224–233. doi:10.1007/s40473-014-0028-8

Koenig, K. A., Frey, M. C., & Detterman, D. K. (2008). ACT and general cognitive ability. *Intelligence, 36*(2), 153–160.

Koenis, M. M., Brouwer, R. M., van den Heuvel, M. P., Mandl, R. C., van Soelen, I. L., Kahn, R. S., … Hulshoff Pol, H. E. (2015). Development of the brain's structural network efficiency in early adolescence: A longitudinal DTI twin study. *Human Brain Mapping.* doi:10.1002/hbm.22988

Kohannim, O., Hibar, D. P., Stein, J. L., Jahanshad, N., Hua, X., Rajagopalan, P., … Alzheimers Disease Neuroimaging, I. (2012a). Discovery and replication of gene influences on brain structure using LASSO regression. *Frontiers in Neuroscience, 6*, 115.

Kohannim, O., Jahanshad, N., Braskie, M. N., Stein, J. L., Chiang, M.-C., Reese, A. H., ... Thompson, P. M. (2012b). Predicting white matter integrity from multiple common genetic variants. *Neuropsychopharmacology : Official Publication of the American College of Neuropsychopharmacology, 37*(9), 2012–2019.

Kokkinakis, A. V., Cowling, P. I., Drachen, A., & Wade, A. R. (2017). Exploring the relationship between video game expertise and fluid intelligence. *Plos One, 12*(11), e0186621. doi:10.1371/journal .pone.0186621

Kolata, S., Light, K., Wass, C. D., Colas-Zelin, D., Roy, D., & Matzel, L. D. (2010). A dopaminergic gene cluster in the prefrontal cortex predicts performance indicative of general intelligence in genetically heterogeneous mice. *Plos One, 5*(11), e14036. doi:10.1371/journal .pone.0014036

Kovas, Y., & Plomin, R. (2006). Generalist genes: Implications for the cognitive sciences. *Trends in Cognitive Sciences, 10*(5), 198–203.

Krause, B., & Cohen Kadosh, R. (2014). Not all brains are created equal: The relevance of individual differences in responsiveness to transcranial electrical stimulation. *Frontiers in Systematic Neuroscience, 8,* 25. doi:10.3389/fnsys.2014.00025

Kuhl, P. K. (2000). A new view of language acquisition. *Proceedings of the National Academy of Sciences of the United States of America, 97*(22), 11850–11857.

Kuhl, P. K. (2004). Early language acquisition: Cracking the speech code. *Nature Reviews Neuroscience, 5*(11), 831–843.

Kyllonen, P. C., & Christal, R. E. (1990). Reasoning ability is (Little More Than) working-memory capacity. *Intelligence, 14*(4), 389–433.

Langer, N., Pedroni, A., Gianotti, L. R., Hanggi, J., Knoch, D., & Jancke, L. (2012). Functional brain network efficiency predicts intelligence. *Hum Brain Mappings, 33*(6), 1393–1406. doi:10.1002/hbm.21297

Lashley, K. S. (1964). *Brain Mechanisms and Intelligence*. New York: Hafner.

Lee, J. J. (2010). Review of intelligence and how to get it: Why schools and cultures count. *Personality and Individual Differences, 48,* 247–255.

Lee, J. J., & Willoughby, E. A. (2021). Predicting cognitive-ability differences from genetic and brain-imaging data. In A. K. Barbey, S. Karama, & R. J. Haier (Eds.), *Cambridge Handbook of Intelligence and Cognitive Neuroscience* (pp. 349–363). New York: Cambridge University Press.

Lee, J. J., McGue, M., Iacono, W. G., Michael, A. M., & Chabris, C. F. (2019). The causal influence of brain size on human intelligence: Evidence from within-family phenotypic associations and GWAS modeling. *Intelligence, 75,* 48–58. doi:10.1016/j.intell.2019.01.011

Lee, J. J., Wedow, R., Okbay, A., Kong, E., Maghzian, O., Zacher, M., ... Consortiu, S. S. G. A. (2018). Gene discovery and polygenic prediction from a genome-wide association study of educational attainment in 1.1 million individuals. *Nature Genetics, 50*(8), 1112-+. doi:10.1038/s41588-018-0147-3

Lee, J. Y., Jun, H., Soma, S., Nakazono, T., Shiraiwa, K., Dasgupta, A., ... Igarashi, K. M. (2021). Dopamine facilitates associative memory encoding in the entorhinal cortex. *Nature, 598*(7880), 321–326. doi:10.1038/s41586-021-03948-8

Lee, K. H., Choi, Y. Y., Gray, J. R., Cho, S. H., Chae, J. H., Lee, S., & Kim, K. (2006). Neural correlates of superior intelligence: Stronger recruitment of posterior parietal cortex. *Neuroimage, 29*(2), 578–586.

Lemos, G. C., Almeida, L. S., & Colom, R. (2011). Intelligence of adolescents is related to their parents' educational level but not to family income. *Personality and Individual Differences, 50*(7), 1062–1067.

Lerner, B. (1980). The war on testing – Detroit Edison in perspective. *Personnel Psychology, 33*(1), 11–16.

Lett, T. A., Vogel, B. O., Ripke, S., Wackerhagen, C., Erk, S., Awasthi, S., ... consortium, I. (2019). Cortical surfaces mediate the relationship between polygenic scores for intelligence and general intelligence. *Cerebral Cortex, 30*(4), 2708–2719. doi:10.1093/cercor/bhz270

Li, H., Namburi, P., Olson, J. M., Borio, M., Lemieux, M. E., Beyeler, A., ... Tye, K. M. (2022). Neurotensin orchestrates valence assignment in the amygdala. *Nature.* doi:10.1038/s41586-022-04964-y

Li, Y., Liu, Y., Li, J., Qin, W., Li, K., Yu, C., & Jiang, T. (2009). Brain anatomical network and intelligence. *PLoS Computational Biology, 5*(5), e1000395. doi:10.1371/journal.pcbi.1000395

Limb, C. J., & Braun, A. R. (2008). Neural substrates of spontaneous musical performance: An FMRI study of jazz improvisation. *Plos One, 3*(2), e1679. doi:10.1371/journal.pone.0001679

Lipp, I., Benedek, M., Fink, A., Koschutnig, K., Reishofer, G., Bergner, S., ... Neubauer, A. (2012). Investigating neural efficiency in the visuo-spatial domain: An FMRI study. *Plos One, 7*(12), e51316. doi:10.1371/journal.pone.0051316

Liu, S., Chow, H. M., Xu, Y., Erkkinen, M. G., Swett, K. E., Eagle, M. W., ... Braun, A. R. (2012). Neural correlates of lyrical improvisation: An FMRI study of freestyle rap. *Scientific Reports, 2*, 834. doi:10.1038/srep00834

Loehlin, J. C. (1989). Partitioning environmental and genetic contributions to behavioral development. *American Psychology, 44*(10), 1285–1292. Retrieved from www.ncbi.nlm.nih.gov/pubmed/2679255

Loehlin, J. C., & Nichols, R. C. (1976). *Heredity, Environment, & Personality : A Study of 850 Sets of Twins.* Austin: University of Texas Press.

Loo, C. K., & Mitchell, P. B. (2005). A review of the efficacy of transcranial magnetic stimulation (TMS) treatment for depression, and current and future strategies to optimize efficacy. *Journal of Affected Disorders, 88*(3), 255–267. doi:10.1016/j.jad.2005.08.001

Luber, B., & Lisanby, S. H. (2014). Enhancement of human cognitive performance using transcranial magnetic stimulation (TMS). *Neuroimage*, 85(Pt 3), 961–970. doi:10.1016/j.neuroimage.2013.06.007

Lubinski, D. (2009). Cognitive epidemiology: With emphasis on untangling cognitive ability and socioeconomic status. *Intelligence, 37*(6), 625–633.

Lubinski, D., Benbow, C. P., & Kell, H. J. (2014). Life paths and accomplishments of mathematically precocious males and females four decades later. *Psychological Science, 25*(12), 2217–2232. doi:10.1177/0956797614551371

Lubinski, D., Benbow, C. P., Webb, R. M., & Bleske-Rechek, A. (2006). Tracking exceptional human capital over two decades. *Psychological Science, 17*(3), 194–199.

Lubinski, D., Schmidt, D. B., & Benbow, C. P. (1996). A 20-year stability analysis of the study of values for intellectually gifted individuals from adolescence to adulthood. *Journal of Applied Psychology, 81*(4), 443–451.

Luciano, M., Wright, M. J., Smith, G. A., Geffen, G. M., Geffen, L. B., & Martin, N. G. (2001). Genetic covariance among measures of information processing speed, working memory, and IQ. *Behavior Genetics, 31*(6), 581–592.

Luders, E., Harr, K. L., Thompson, P. M., Rex, D. E., Woods, R. P., DeLuca, H., … Toga, A. W. (2006). Gender effects on cortical thickness and the influence of scaling. *Human Brain Mapping, 27*(4), 314–324.

Luders, E., Narr, K. L., Bilder, R. M., Thompson, P. M., Szeszko, P. R., Hamilton, L., & Toga, A. W. (2007). Positive correlations between corpus callosum thickness and intelligence. *Neuroimage, 37*(4), 1457–1464. doi:10.1016/j.neuroimage.2007.06.028

Luders, E., Narr, K. L., Thompson, P. M., Rex, D. E., Jancke, L., Steinmetz, H., & Toga, A. W. (2004). Gender differences in cortical complexity. Nat*ural* Neurosci*ence*, advanced *online publication*. Retrieved from http://dx.doi.org/10.1038/nn1277

Luo, Q., Perry, C., Peng, D. L., Jin, Z., Xu, D., Ding, G. S., & Xu, S. Y. (2003). The neural substrate of analogical reasoning: An fMRI study. *Cognitive Brain Research, 17*(3), 527–534.

Lynn, R. (2009). What has caused the Flynn effect? Secular increases in the Development Quotients of infants. *Intelligence, 37*(1), 16–24.

Mackey, A. P., Finn, A. S., Leonard, J. A., Jacoby-Senghor, D. S., West, M. R., Gabrieli, C. F., & Gabrieli, J. D. (2015). Neuroanatomical correlates of the income-achievement gap. *Psychological Science, 26*(6), 925–933. doi:10.1177/0956797615572233

Mackey, A. P., Hill, S. S., Stone, S. I., & Bunge, S. A. (2011). Differential effects of reasoning and speed training in children. *Developmental Science, 14*(3), 582–590.

Mackintosh, N. J. (1995). *Cyril Burt : Fraud or Framed?* Oxford; New York: Oxford University Press.

Mackintosh, N. J. (2011). *IQ and Human Intelligence* (2nd ed.). Oxford; New York: Oxford University Press.

Macnamara, B. N., & Maitra, M. (2019). The role of deliberate practice in expert performance: Revisiting Ericsson, Krampe & Tesch-Romer (1993). *Royal Society on Open Science, 6*(8), 190327. doi:10.1098/rsos.190327

Maguire, E. A., Valentine, E. R., Wilding, J. M., & Kapur, N. (2003). Routes to remembering: The brains behind superior memory. *Nature Neuroscience, 6*(1), 90–95. doi:10.1038/nn988

Maher, B. (2008). Poll results: Look who's doping. *Nature, 452*(7188), 674–675. doi:10.1038/452674a

Makel, M. C., Kell, H. J., Lubinski, D., Putallaz, M., & Benbow, C. P. (2016). When lightning strikes twice: Profoundly gifted, profoundly accomplished. *Psychological Science, 27*(7), 1004–1018. doi:10.1177/0956797616644735

Makowski, C., Meer, D. v. d., Dong, W., Wang, H., Wu, Y., Zou, J., ... Chen, C.-H. (2022). Discovery of genomic loci of the human cerebral cortex using genetically informed brain atlases. *Science, 375*(6580), 522–528. doi:10.1126/science.abe8457

Malanchini, M., Rimfeld, K., Gidziela, A., Cheesman, R., Allegrini, A. G., Shakeshaft, N., ... Plomin, R. (2021). Pathfinder: A gamified measure to integrate general cognitive ability into the biological, medical, and behavioural sciences. *Molecular Psychiatry.* doi:10.1038/s41380-021-01300-0

Maldjian, J. A., Davenport, E. M., & Whitlow, C. T. (2014). Graph theoretical analysis of resting-state MEG data: Identifying interhemispheric connectivity and the default mode. *Neuroimage, 96*, 88–94. doi:10.1016/j.neuroimage.2014.03.065

Mardis, E. R. (2008). Next-generation DNA sequencing methods. *Annual Review of Genomics and Human Genetics, 9*, 387–402. doi:10.1146/annurev.genom.9.081307.164359

Marioni, R. E., Davies, G., Hayward, C., Liewald, D., Kerr, S. M., Campbell, A., ... Deary, I. J. (2014). Molecular genetic contributions to socioeconomic status and intelligence. *Intelligence, 44*(100), 26–32. doi:10.1016/j.intell.2014.02.006

Marks, G. N. (2022). Cognitive ability has powerful, widespread and robust effects on social stratification: Evidence from the 1979 and 1997 US National Longitudinal Surveys of Youth. *Intelligence, 94*, 101686. doi:10.1016/j.intell.2022.101686

Martinez, K., & Colom, R. (2021). Imaging the intelligence of humans. In A. Barbey, S. Karama, & H. R. J. (Eds.), *The Cambridge Handbook of Intelligence and Cognitive Neuroscience* (pp. 44–69). New York: Cambridge University Press.

Martinez, K., Janssen, J., Pineda-Pardo, J. A., Carmona, S., Roman, F. J., Aleman-Gomez, Y., ... Colom, R. (2017). Individual differences in the dominance of interhemispheric connections predict cognitive ability beyond sex and brain size. *Neuroimage, 155*, 234–244.

Maslen, H., Faulmuller, N., & Savulescu, J. (2014). Pharmacological cognitive enhancement-how neuroscientific research could advance ethical debate. *Frontier in Systematic Neuroscience, 8*, 107. doi:10.3389/fnsys.2014.00107

Matzel, L. D., & Kolata, S. (2010). Selective attention, working memory, and animal intelligence. *Neuroscience Biobehavioral Review, 34*(1), 23–30. doi:10.1016/j.neubiorev.2009.07.002

Matzel, L. D., Crawford, D. W., & Sauce, B. (2020). Déjà vu All Over Again: A unitary biological mechanism for intelligence is (Probably) untenable. *Journal of Intelligence, 8*(24).

Matzel, L. D., Han, Y. R., Grossman, H., Karnik, M. S., Patel, D., Scott, N., ... Gandhi, C. C. (2003). Individual differences in the expression of a "General" learning ability in mice. *Journal of Neuroscience, 23*(16), 6423–6433. Retrieved from www.jneurosci.org/cgi/content/abstract/23/16/6423

Matzel, L. D., Sauce, B., & Wass, C. (2013). The architecture of intelligence: Converging evidence from studies of humans and animals. *Current Directions in Psychological Science, 22*(5), 342–348. doi:10.1177/0963721413491764

Mayseless, N., & Shamay-Tsoory, S. G. (2015). Enhancing verbal creativity: Modulating creativity by altering the balance between right and left inferior frontal gyrus with tDCS. *Neuroscience, 291*, 167–176. doi:10.1016/j.neuroscience.2015.01.061

McCabe, K. O., Lubinski, D., & Benbow, C. P. (2020). Who shines most among the brightest?: A 25-year longitudinal study of elite STEM graduate students. *Journal of Personality and Social Psychology, 119*(2), 390–416. doi:10.1037/pspp0000239

McDaniel, M. A. (2005). Big-brained people are smarter: A meta-analysis of the relationship between in vivo brain volume and intelligence. *Intelligence, 33*(4), 337–346.

McGue, M., Anderson, E. L., Willoughby, E., Giannelis, A., Iacono, W. G., & Lee, J. J. (2022). Not by g alone: The benefits of a college education among individuals with low levels of general cognitive ability. *Intelligence, 92*, 101642. doi:10.1016/j.intell.2022.101642

McGue, M., Bouchard, T. J., Iacono, W. G., & Lykken, D. T. (1993). Age effects on heritability of intelligence. In R. Plomin & G. E. McClearn (Eds.), *Nature, Nurture, and Psychology* (pp. 59–76). Washington, DC: American Psychological Association.

McKinley, R. A., Bridges, N., Walters, C. M., & Nelson, J. (2012). Modulating the brain at work using noninvasive transcranial stimulation. *Neuroimage, 59*(1), 129–137. doi:10.1016/j.neuroimage.2011.07.075

Melby-Lervag, M., & Hulme, C. (2013). Is working memory training effective? A meta-analytic review. *Developmental Psychology, 49*(2), 270–291. doi:10.1037/a0028228

Melby-Lervåg, M., Redick, T. S., & Hulme, C. (2016). Working memory training does not improve performance on measures of intelligence or other measures of "Far Transfer": Evidence from a meta-analytic review. *Perspectives on Psychological Science, 11*(4), 512–534. doi:10.1177/1745691616635612

Miller, B. L., Boone, K., Cummings, J. L., Read, S. L., & Mishkin, F. (2000). Functional correlates of musical and visual ability in frontotemporal dementia. *British Journal of Psychiatry, 176*, 458–463. Retrieved from www.ncbi.nlm.nih.gov/pubmed/10912222

Miller, B. L., Cummings, J., Mishkin, F., Boone, K., Prince, F., Ponton, M., & Cotman, C. (1998). Emergence of artistic talent in frontotemporal dementia. *Neurology, 51*(4), 978–982. Retrieved from www.ncbi.nlm.nih.gov/pubmed/9781516

Miller, E. B., Farkas, G., & Duncan, G. J. (2016). Does Head Start differentially benefit children with risks targeted by the program's service model? *Early Childhood Research Quarterly, 34*, 1–12. doi:10.1016/j.ecresq.2015.08.001

Mitchell, B. L., Hansell, N. K., McAloney, K., Martin, N. G., Wright, M. J., Renteria, M. E., & Grasby, K. L. (2022). Polygenic influences associated with adolescent cognitive skills. *Intelligence, 94*, 101680. doi:10.1016/j.intell.2022.101680

Mitchell, K. J. (2018). *Innate : How the Wiring of Our Brains Shapes Who We Are.* Princeton, NJ: Princeton University Press.

Moody, D. E. (2009). Can intelligence be increased by training on a task of working memory? *Intelligence, 37*(4), 327–328.

Moreau, D. (2022). How malleable are cognitive abilities? A critical perspective on popular brief interventions. *American Psychologist, 77*(3), 409–423. doi:10.1037/amp0000872

Moreau, D., Macnamara, B. N., & Hambrick, D. Z. (2019). Overstating the role of environmental factors in success: A cautionary note. *Current Directions in Psychological Science, 28*(1), 28–33. doi:10.1177/0963721418797300

Mountjoy, E., Schmidt, E. M., Carmona, M., Schwartzentruber, J., Peat, G., Miranda, A., … Ghoussaini, M. (2021). An open approach to systematically prioritize causal variants and genes at all published human GWAS trait-associated loci. *Nature Genetics.* doi:10.1038/s41588-021-00945-5

Muetzel, R. L., Mous, S. E., van der Ende, J., Blanken, L. M., van der Lugt, A., Jaddoe, V. W., … White, T. (2015). White matter integrity and cognitive performance in school-age children: A population-based neuroimaging study. *Neuroimage, 119*, 119–128. doi:10.1016/j.neuroimage.2015.06.014

Murray, C. (1995). The-bell-curve and its critics. *Commentary, 99*(5), 23–30.

Murray, C. A. (2013). *Coming Apart : The State of White America, 1960–2010* (First paperback edition. ed.). New York: Crown Forum.

Murray, C., Pattie, A., Starr, J. M., & Deary, I. J. (2012). Does cognitive ability predict mortality in the ninth decade? The Lothian Birth Cohort 1921. *Intelligence, 40*(5), 490–498.

Muzur, A., Pace-Schott, E. F., & Hobson, J. A. (2002). The prefrontal cortex in sleep. *Trends in Cognition Science, 6*(11), 475–481. Retrieved from www.ncbi.nlm.nih.gov/pubmed/12457899

Neisser, U., Boodoo, G., Bouchard, T. J., Boykin, A. W., Brody, N., Ceci, S. J., … Urbina, S. (1996). Intelligence: Knowns and unknowns. *American Psychologist, 51*(2), 77–101.

Neubauer, A. C. (2021). The future of intelligence research in the coming age of artificial intelligence – With a special consideration of the philosophical movements of trans- and posthumanism. *Intelligence, 87*, 101563. doi:10.1016/j.intell.2021.101563

Neubauer, A. C., & Fink, A. (2009). Intelligence and neural efficiency. *Neuroscience and Biobehavioral Reviews, 33*(7), 1004–1023.

Neville, H., Stevens, C., Pakulak, E., & Bell, T. A. (2013). Commentary: Neurocognitive consequences of socioeconomic disparities. *Developmental Science, 16*(5), 708–712. doi:10.1111/desc.12081

Newman, S. D., & Just, M. A. (2005). The neural bases of intelligence: A perspective based on functional neuroimaging. In Robert J. Sternberg & Jean E. Pretz (Eds.), *Cognition and Intelligence: Identifying the Mechanisms of the Mind* (pp. 88–103). New York: Cambridge University Press.

Nihongaki, Y., Kawano, F., Nakajima, T., & Sato, M. (2015). Photoactivatable CRISPR-Cas9 for optogenetic genome editing. *Nature Biotechnology.* doi:10.1038/nbt.3245

Nisbett, R. E. (2009). *Intelligence and How to Get It: Why Schools and Cultures Count* (1st ed.). New York: W.W. Norton & Co.

Nisbett, R. E., Aronson, J., Blair, C., Dickens, W., Flynn, J., Halpern, D. F., & Turkheimer, E. (2012). Intelligence new findings and theoretical developments. *American Psychologist, 67*(2), 130–159. doi:10.1037/a0026699

Noble, K. G., & Giebler, M. A. (2020). The neuroscience of socioeconomic inequality. *Current Opinion in Behavioral Sciences, 36*, 23–28. doi:10.1016/j.cobeha.2020.05.007

Noble, K. G., Houston, S. M., Brito, N. H., Bartsch, H., Kan, E., Kuperman, J. M., … Sowell, E. R. (2015). Family income, parental education and brain structure in children and adolescents. *Nature Neuroscience, 18*(5), 773–778. doi:10.1038/nn.3983

Noble, K. G., Wolmetz, M. E., Ochs, L. G., Farah, M. J., & McCandliss, B. D. (2006). Brain-behavior relationships in reading acquisition are modulated by socioeconomic factors. *Developmental Science, 9*(6), 642–654. doi:10.1111/j.1467-7687.2006.00542.x

Okbay, A., Wu, Y., Wang, N., Jayashankar, H., Bennett, M., Nehzati, S. M., … LifeLines Cohort, S. (2022). Polygenic prediction of educational attainment within and between families from genome-wide association analyses in 3 million individuals. *Nature Genetics, 54*(4), 437–449. doi:10.1038/s41588-022-01016-z

Ozawa, A., & Arakawa, H. (2021). Chemogenetics drives paradigm change in the investigation of behavioral circuits and neural mechanisms underlying drug action. *Behavioural Brain Research, 406*, 113234. doi:10.1016/j.bbr.2021.113234

Pages, R., Lukes, D. J., Bailey, D. H., & Duncan, G. J. (2020). Elusive longer-run impacts of head start: Replications within and across cohorts. *Educational Evaluation and Policy Analysis, 42*(4), 471–492. doi:10.3102/0162373720948884

Pages, R., Protzko, J., & Bailey, D. H. (2021). The breadth of impacts from the Abecedarian project early intervention on cognitive skills. *Journal of Research on Educational Effectiveness*, 1–20. doi:10.1080/19345747.2021.1969711

Pagnaer, T., Siermann, M., Borry, P., & Tsuiko, O. (2021). Polygenic risk scoring of human embryos: A qualitative study of media coverage. *BMC Medical Ethics, 22*(1), 125. doi:10.1186/s12910-021-00694-4

Pahor, A., & Jausovec, N. (2014). The effects of theta transcranial alternating current stimulation (tACS) on fluid intelligence. *International Journal of Psychophysiology, 93*(3), 322–331. doi:10.1016/j.ijpsycho.2014.06.015

Pahor, A., Seitz, A. R., & Jaeggi, S. M. (2022). Near transfer to an unrelated N-back task mediates the effect of N-back working memory

training on matrix reasoning. *Nature Human Behaviour.* doi:10.1038/
s41562-022-01384-w

Panizzon, M. S., Vuoksimaa, E., Spoon, K. M., Jacobson, K. C., Lyons, M.
J., Franz, C. E., ... Kremen, W. S. (2014). Genetic and environmental
influences of general cognitive ability: Is g a valid latent construct?
Intelligence, 43, 65–76. doi:10.1016/j.intell.2014.01.008

Parasuraman, R., & Jiang, Y. (2012). Individual differences in cognition,
affect, and performance: Behavioral, neuroimaging, and molecular
genetic approaches. *Neuroimage, 59*(1), 70–82.

Parks, R. W., Loewenstein, D. A., Dodrill, K. L., Barker, W. W., Yoshii, F.,
Chang, J. Y., ... Duara, R. (1988). Cerebral metabolic effects of a verbal
fluency test – A Pet Scan Study. *Journal of Clinical and Experimental
Neuropsychology, 10*(5), 565–575.

Pascoli, V., Turiault, M., & Luscher, C. (2012). Reversal of cocaine-evoked
synaptic potentiation resets drug-induced adaptive behaviour. *Nature,
481*(7379), 71–75. doi:10.1038/nature10709

Pedersen, N. L., Plomin, R., Nesselroade, J. R., & Mcclearn, G. E. (1992).
A quantitative genetic-analysis of cognitive-abilities during the 2nd-half
of the life-span. *Psychological Science, 3*(6), 346–353.

Penke, L., Maniega, S. M., Bastin, M. E., Hernandez, M. C. V., Murray, C.,
Royle, N. A., ... Deary, I. J. (2012). Brain white matter tract integrity
as a neural foundation for general intelligence. *Molecular Psychiatry,
17*(10), 1026–1030. doi:10.1038/Mp.2012.66

Perez, P., Chavret-Reculon, E., Ravassard, P., & Bouret, S. (2022). Using
inhibitory DREADDs to silence LC neurons in Monkeys. *Brain Sciences,
12*(2), 206. Retrieved from www.mdpi.com/2076-3425/12/2/206

Perfetti, B., Saggino, A., Ferretti, A., Caulo, M., Romani, G. L., & Onofrj,
M. (2009). Differential patterns of cortical activation as a function
of fluid reasoning complexity. *Hum Brain Mapping, 30*(2), 497–510.
doi:10.1002/hbm.20519

Perobelli, S., Alessandrini, F., Zoccatelli, G., Nicolis, E., Beltramello, A.,
Assael, B. M., & Cipolli, M. (2015). Diffuse alterations in grey and white
matter associated with cognitive impairment in Shwachman-Diamond
syndrome: Evidence from a multimodal approach. *Neuroimage Clinic,
7*, 721–731. doi:10.1016/j.nicl.2015.02.014

Pesenti, M., Zago, L., Crivello, F., Mellet, E., Samson, D., Duroux, B., ...
Tzourio-Mazoyer, N. (2001). Mental calculation in a prodigy is sustained
by right prefrontal and medial temporal areas. *Nature Neuroscience,
4*(1), 103–107.

Petrill, S. A., & Deater-Deckard, K. (2004). The heritability of general
cognitive ability: A within-family adoption design. *Intelligence, 32*(4),
403–409.

Pfleiderer, B., Ohrmann, P., Suslow, T., Wolgast, M., Gerlach, A. L., Heindel, W., & Michael, N. (2004). N-acetylaspartate levels of left frontal cortex are associated with verbal intelligence in women but not in men: A proton magnetic resonance spectroscopy study. *Neuroscience, 123*(4), 1053–1058.

Pietschnig, J., & Voracek, M. (2015). One century of global IQ gains: A formal meta-analysis of the Flynn Effect (1909–2013). *Perspectives on Psychological Science, 10*(3), 282–306.

Pietschnig, J., Voracek, M., & Formann, A. K. (2010). Mozart effect-Shmozart effect: A meta-analysis. *Intelligence, 38*(3), 314–323.

Pineda-Pardo, J. A., Bruna, R., Woolrich, M., Marcos, A., Nobre, A. C., Maestu, F., & Vidaurre, D. (2014). Guiding functional connectivity estimation by structural connectivity in MEG: An application to discrimination of conditions of mild cognitive impairment. *Neuroimage, 101*, 765–777. doi:10.1016/j.neuroimage.2014.08.002

Pinho, A. L., de Manzano, O., Fransson, P., Eriksson, H., & Ullen, F. (2014). Connecting to create: Expertise in musical improvisation is associated with increased functional connectivity between premotor and prefrontal areas. *Journal of Neuroscience, 34*(18), 6156–6163. doi:10.1523/JNEUROSCI.4769-13.2014

Pinker, S. (2002). *The Blank Slate : The Modern Denial of Human Nature.* New York: Viking.

Plis, S. M., Weisend, M. P., Damaraju, E., Eichele, T., Mayer, A., Clark, V. P., ... Calhoun, V. D. (2011). Effective connectivity analysis of fMRI and MEG data collected under identical paradigms. *Computational Biology Medicine, 41*(12), 1156–1165. doi:10.1016/j.compbiomed.2011.04.011

Plomin, R. (1999). Genetics and general cognitive ability. *Nature, 402*(6761 Suppl), C25–29. doi:10.1038/35011520

Plomin, R. (2018). *Blueprint : How DNA Makes us Who We Are.* Cambridge, MA: The MIT Press.

Plomin, R., & Deary, I. J. (2015). Genetics and intelligence differences: Five special findings. *Molecular Psychiatry, 20*(1), 98–108. doi:10.1038/mp.2014.105

Plomin, R., DeFries, J. C., Knopik, V. S., & Neiderhiser, J. M. (2016). Top 10 replicated findings from behavioral genetics. *Perspectives on Psychological Science : A Journal of the Association for Psychological Science, 11*(1), 3–23. doi:10.1177/1745691615617439

Plomin, R., & Kosslyn, S. M. (2001). Genes, brain and cognition. *Nature Neuroscience, 4*(12), 1153–1154.

Plomin, R., & Petrill, S. A. (1997). Genetics and intelligence: What's new? *Intelligence, 24*(1), 53–77.

Plomin, R., Shakeshaft, N. G., McMillan, A., & Trzaskowski, M. (2014a). Nature, nurture, and expertise. *Intelligence, 45*, 46–59.

Plomin, R., Shakeshaft, N. G., McMillan, A., & Trzaskowski, M. (2014b). Nature, nurture, and expertise: Response to Ericsson. *Intelligence, 45*, 115–117.

Plomin, R., & von Stumm, S. (2018). The new genetics of intelligence. *Nature Review Genetics, 19*(3), 148–159. doi:10.1038/nrg.2017.104

Pol, H. E. H., Posthuma, D., Baare, W. F. C., De Geus, E. J. C., Schnack, H. G., van Haren, N. E. M., … Boomsma, D. I. (2002). Twin-singleton differences in brain structure using structural equation modelling. *Brain, 125*, 384–390.

Polderman, T. J., Benyamin, B., de Leeuw, C. A., Sullivan, P. F., van Bochoven, A., Visscher, P. M., & Posthuma, D. (2015). Meta-analysis of the heritability of human traits based on fifty years of twin studies. *Nature Genetics, 47*(7), 702–709. doi:10.1038/ng.3285

Poldrack, R. A. (2015). Is "efficiency" a useful concept in cognitive neuroscience? *Developmental Cognition Neuroscience, 11*, 12–17. doi:10.1016/j.dcn.2014.06.001

Posthuma, D., & de Geus, E. J. C. (2006). Progress in the molecular-genetic study of intelligence. *Current Directions in Psychological Science, 15*(4), 151–155.

Posthuma, D., Baare, W. F. C., Pol, H. E. H., Kahn, R. S., Boomsma, D. I., & De Geus, E. J. C. (2003). Genetic correlations between brain volumes and the WAIS-III dimensions of verbal comprehension, working memory, perceptual organization, and processing speed. *Twin Research, 6*(2), 131–139.

Posthuma, D., De Geus, E. J., Baare, W. F., Hulshoff Pol, H. E., Kahn, R. S., & Boomsma, D. I. (2002). The association between brain volume and intelligence is of genetic origin. *Natural Neuroscience, 5*(2), 83–84.

Posthuma, D., De Geus, E., & Boomsma, D. (2003). Genetic contributions to anatomical, behavioral, and neurophysiological indices of cognition. In R. Plomin, J. DeFries, I. W. Craig, & P. McGuffin (Eds.), *Behavioral Genetics in the Postgenomic Era* (pp. 141–161). Washington, DC: American psychological Association.

Prabhakaran, V., Smith, J. A., Desmond, J. E., Glover, G. H., & Gabrieli, J. D. (1997). Neural substrates of fluid reasoning: An fMRI study of neocortical activation during performance of the Raven's Progressive Matrices Test. *Cognitive Psychology, 33*(1), 43–63.

Prat, C. S., Mason, R. A., & Just, M. A. (2012). An fMRI investigation of analogical mapping in metaphor comprehension: The influence of context and individual cognitive capacities on processing demands. *Journal of Experimental Psychology: Learning, Memory, and Cognition, 38*(2), 282–294. doi:10.1037/a0026037

Preusse, F., Van Der Meer, E., Deshpande, G., Krueger, F., & Wartenburger, I. (2011). Frontiers: Fluid intelligence allows flexible recruitment of the parieto-frontal network in analogical reasoning. *Frontier in Human Neuroscience*, 5.

Protzko, J., Aronson, J., & Blair, C. (2013). How to make a young child smarter: Evidence from the database of raising intelligence. *Perspectives on Psychological Science, 8*(1), 25–40.

Quiroga, M. A., Escorial, S., Román, F. J., Morillo, D., Jarabo, A., Privado, J., ... Colom, R. (2015). Can we reliably measure the general factor of intelligence (g) through commercial video games? Yes, we can! *Intelligence, 53*, 1–7. doi:10.1016/j.intell.2015.08.004

Quiroga, M. Á., & Colom, R. (2020). Intelligence and video games. In R. J. Sternberg (Ed.), *The Cambridge Handbook of Intelligence* (2 ed., pp. 626–656). Cambridge: Cambridge University Press.

Quiroga, M. A., Diaz, A., Roman, F. J., Privado, J., & Colom, R. (2019). Intelligence and video games: Beyond "brain-games." *Intelligence, 75*, 85–94. doi:10.1016/j.intell.2019.05.001

Ramey, C. T., & Ramey, S. L. (2004). Early learning and school readiness: Can early intervention make a difference? *Merrill-Palmer Quarterly-Journal of Developmental Psychology, 50*(4), 471–491.

Rankin, K. P., Liu, A. A., Howard, S., Slama, H., Hou, C. E., Shuster, K., & Miller, B. L. (2007). A case-controlled study of altered visual art production in Alzheimer's and FTLD. *Cognitive Behavioral Neurology, 20*(1), 48–61. doi:10.1097/WNN.0b013e31803141dd

Rauscher, F. H., Shaw, G. L., & Ky, K. N. (1993). Music and spatial task performance. *Nature, 365*(6447), 611. doi:10.1038/365611a0

Redick, T. S. (2015). Working memory training and interpreting interactions in intelligence interventions. *Intelligence, 50*(0), 14–20. doi:10.1016/j.intell.2015.01.014

Redick, T. S. (2019). The Hype Cycle of working memory training. *Current Directions in Psychological Sciences, 28*(5), 423–429. doi:10.1177/0963721419848668

Redick, T. S., Shipstead, Z., Harrison, T. L., Hicks, K. L., Fried, D. E., Hambrick, D. Z., ... Engle, R. W. (2013). No evidence of intelligence improvement after working memory training: A Randomized, Placebo-Controlled Study. *Journal of Experimental Psychology-General, 142*(2), 359–379. doi:10.1037/A0029082

Ree, M. J., & Carretta, T. R. (1996). Central role of g in military pilot selection. *International Journal of Aviation Psychology, 6*(2), 111–123.

Ree, M. J., & Carretta, T. R. (2022). Thirty years of research on general and specific abilities: Still not much more than g. *Intelligence, 91*, 101617. doi:10.1016/j.intell.2021.101617

Ree, M. J., & Earles, J. A. (1991). Predicting training success – Not much more than G. *Personnel Psychology, 44*(2), 321–332.

Reijneveld, J. C., Ponten, S. C., Berendse, H. W., & Stam, C. J. (2007). The application of graph theoretical analysis to complex networks in the brain. *Clinical Neurophysiology, 118*(11), 2317–2331. doi:10.1016/j.clinph.2007.08.010

Reverberi, C., Bonatti, L. L., Frackowiak, R. S., Paulesu, E., Cherubini, P., & Macaluso, E. (2012). Large scale brain activations predict reasoning profiles. *Neuroimage, 59*(2), 1752–1764. doi:10.1016/j.neuroimage.2011.08.027

Rhein, C., Muhle, C., Richter-Schmidinger, T., Alexopoulos, P., Doerfler, A., & Kornhuber, J. (2014). Neuroanatomical correlates of intelligence in healthy young adults: The role of basal ganglia volume. *Plos One, 9*(4), e93623. doi:10.1371/journal.pone.0093623

Rietveld, C. A., Esko, T., Davies, G., Pers, T. H., Turley, P., Benyamin, B., … Koellinger, P. D. (2014). Common genetic variants associated with cognitive performance identified using the proxy-phenotype method. *Proceedings of the National Academy of Sciences of the United States of America, 111*(38), 13790–13794.

Ritchie, S. (2022). Everything you need to know about breastfeeding and intelligence. *Substack*: https://stuarttritchie.substack.com/p/breastfeeding-iq?utm_source=%2Fprofile%2F1881468-stuart-ritchie&utm_medium=reader2.

Ritchie, S. J., & Tucker-Drob, E. M. (2018). How much does education improve intelligence? A Meta-Analysis. *Psychological Science, 29*(8), 1358–1369. doi:10.1177/0956797618774253

Ritchie, S. J., Booth, T., Valdés Hernández, M. d. C., Corley, J., Maniega, S. M., Gow, A. J., … Deary, I. J. (2015). Beyond a bigger brain: Multivariable structural brain imaging and intelligence. *Intelligence, 51*(0), 47–56. doi:10.1016/j.intell.2015.05.001

Ritchie, S. J., Cox, S. R., Shen, X. Y., Lombardo, M. V., Reus, L. M., Alloza, C., … Deary, I. J. (2018). Sex differences in the adult human brain: Evidence from 5216 UK Biobank participants. *Cerebral Cortex, 28*(8), 2959–2975. doi:10.1093/cercor/bhy109

Robertson, K. F., Smeets, S., Lubinski, D., & Benbow, C. P. (2010). Beyond the threshold hypothesis: Even among the gifted and top math/science graduate students, cognitive abilities, vocational interests, and lifestyle preferences matter for career choice, performance, and persistence. *Current Directions in Psychological Science, 19*(6), 346–351.

Román, F. J., Morillo, D., Estrada, E., Escorial, S., Karama, S., & Colom, R. (2018). Brain-intelligence relationships across childhood and adolescence: A latent-variable approach. *Intelligence, 68*, 21–29. doi:10.1016/j.intell.2018.02.006

Romeo, R. R., Leonard, J. A., Scherer, E., Robinson, S., Takada, M., Mackey, A. P., … Gabrieli, J. D. E. (2021). Replication and extension of family-based training program to improve cognitive abilities in young children. *Journal of Research on Educational Effectiveness, 14*(4), 792–811. doi:10.1080/19345747.2021.1931999

Ryman, S. G., Yeo, R. A., Witkiewitz, K., Vakhtin, A. A., van den Heuvel, M., de Reus, M., ... Jung, R. E. (2016). Fronto-Parietal gray matter and white matter efficiency differentially predict intelligence in males and females. *Human Brain Mapping, 37*(11), 4006–4016. doi:10.1002/hbm.23291

Sackett, P. R., Kuncel, N. R., Arneson, J. J., Cooper, S. R., & Waters, S. D. (2009). Does socioeconomic status explain the relationship between admissions tests and post-secondary academic performance? *Psychological Bulletin, 135*(1), 1–22. doi:10.1037/a0013978

Sahakian, B. J., & Kramer, A. F. (2015). Editorial overview: Cognitive enhancement. *Current Opinion in Behavioral Sciences, 4*, V-vii. doi:10.1016/j.cobeha.2015.06.006

Sahakian, B., & Morein-Zamir, S. (2007). Professor's little helper. *Nature, 450*(7173), 1157–1159. doi:10.1038/4501157a

Sander, J. D., & Joung, J. K. (2014). CRISPR-Cas systems for genome editing, regulation and targeting. *Nature Biotechnology, 32*(4), 347–355. doi:10.1038/nbt.2842

Santarnecchi, E., Brem, A.-K., Levenbaum, E., Thompson, T., Kadosh, R. C., & Pascual-Leone, A. (2015). Enhancing cognition using transcranial electrical stimulation. *Current Opinion in Behavioral Sciences, 4*, 171–178. doi:10.1016/j.cobeha.2015.06.003

Santarnecchi, E., Emmendorfer, A., & Pascual-Leone, A. (2017). Dissecting the parieto-frontal correlates of fluid intelligence: A comprehensive ALE meta-analysis study. *Intelligence, 63*, 9–28. doi:10.1016/j.intell.2017.04.008

Santarnecchi, E., Emmendorfer, A., Tadayon, S., Rossi, S., Rossi, A., Pascual-Leone, A., & Team, H. S. (2017). Network connectivity correlates of variability in fluid intelligence performance. *Intelligence, 65*, 35–47.

Santarnecchi, E., Galli, G., Polizzotto, N. R., Rossi, A., & Rossi, S. (2014). Efficiency of weak brain connections support general cognitive functioning. *Hum Brain Mapping, 35*(9), 4566–4582. doi:10.1002/hbm.22495

Santarnecchi, E., Muller, T., Rossi, S., Sarkar, A., Polizzotto, N. R., Rossi, A., & Cohen Kadosh, R. (2016). Individual differences and specificity of prefrontal gamma frequency-tACS on fluid intelligence capabilities. *Cortex, 75*, 33–43. doi:10.1016/j.cortex.2015.11.003

Santarnecchi, E., Polizzotto, N. R., Godone, M., Giovannelli, F., Feurra, M., Matzen, L., ... Rossi, S. (2013). Frequency-dependent enhancement of fluid intelligence induced by transcranial oscillatory potentials. *Current Biology, 23*(15), 1449–1453. doi:10.1016/j.cub.2013.06.022

Santarnecchi, E., Rossi, S., & Rossi, A. (2015a). The smarter, the stronger: Intelligence level correlates with brain resilience to systematic insults. *Cortex, 64*, 293–309. doi:10.1016/j.cortex.2014.11.005

Santarnecchi, E., Tatti, E., Rossi, S., Serino, V., & Rossi, A. (2015b). Intelligence-related differences in the asymmetry of spontaneous cerebral activity. *Human Brain Mapping.* doi:10.1002/hbm.22864

Sauce, B., & Matzel, L. D. (2013). The causes of variation in learning and behavior: Why individual differences matter. *Frontiers in Psychology*, 4.

Sauce, B., & Matzel, L. D. (2018). The paradox of intelligence: Heritability and malleability coexist in hidden gene-environment interplay. *Psychology Bulletin, 144*(1), 26–47. doi:10.1037/bul0000131

Sauce, B., Bendrath, S., Herzfeld, M., Siegel, D., Style, C., Rab, S., … Matzel, L. D. (2018). The impact of environmental interventions among mouse siblings on the heritability and malleability of general cognitive ability. *Philosophical Transactions of the Royal Society B-Biological Sciences, 373*(1756). doi:ARTN 20170289

Sauce, B., Liebherr, M., Judd, N., & Klingberg, T. (2022). The impact of digital media on children's intelligence while controlling for genetic differences in cognition and socioeconomic background. *Scientific Reports, 12*(1), 7720. doi:10.1038/s41598-022-11341-2

Sawyer, K. (2011). The cognitive neuroscience of creativity: A critical review. *Creativity Research Journal, 23*(2), 137–154.

Schaie, K. W. (1993). The Seattle Longitudinal Study: A thirty-five-year inquiry of adult intellectual development. *Zeitschrift fur Gerontologie, 26*(3), 129–137. Retrieved from www.ncbi.nlm.nih.gov/pubmed/8337905

Schmidt, F. (2016). The validity and utility of selection methods in personnel psychology: Practical and theoretical implications of 100 years of research findings.

Schmidt, F. L., & Hunter, J. (2004). General mental ability in the world of work: Occupational attainment and job performance. *Journal of Personality and Social Psychology, 86*(1), 162–173. doi:10.1037/0022-3514.86.1.162

Schmidt, F. L., & Hunter, J. E. (1998). The validity and utility of selection methods in personnel psychology: Practical and theoretical implications of 85 years of research findings. *Psychological Bulletin, 124*(2), 262–274.

Schmithorst, V. J., & Holland, S. K. (2006). Functional MRI evidence for disparate developmental processes underlying intelligence in boys and girls. *Neuroimage, 31*(3), 1366–1379.

Schmithorst, V. J., Wilke, M., Dardzinski, B. J., & Holland, S. K. (2005). Cognitive functions correlate with white matter architecture in a normal pediatric population: A diffusion tensor MRI study. *Human Brain Mapping, 26*(2), 139–147.

Schubert, A.-L., Hagemann, D., & Frischkorn, G. T. (2017). Is general intelligence little more than the speed of higher-order processing? *Journal of Experimental Psychology-General, 146*(10), 1498–1512.

Schubert, A.-L., Hagemann, D., Frischkorn, G. T., & Herpertz, S. C. (2018). Faster, but not smarter: An experimental analysis of the relationship between mental speed and mental abilities. *Intelligence, 71*, 66–75. doi:10.1016/j.intell.2018.10.005

Schubert, A.-L., & Frischkorn, G. T. (2020). Neurocognitive psychometrics of intelligence: How measurement advancements unveiled the role of mental speed in intelligence differences. *Current Directions in Psychological Science, 29*(2), 140–146. doi:10.1177/0963721419896365

Schubert, A.-L., Hagemann, D., Löffler, C., & Frischkorn, G. T. (2020). Disentangling the effects of processing speed on the association between age differences and fluid intelligence. *Journal of Intelligence, 8*(1), 1. Retrieved from www.mdpi.com/2079-3200/8/1/1

Schubert, A.-L., Nunez, M. D., Hagemann, D., & Vandekerckhove, J. (2019). Individual differences in cortical processing speed predict cognitive abilities: A model-based cognitive neuroscience account. *Computational Brain & Behavior, 2*(2), 64–84. doi:10.1007/s42113-018-0021-5

Schwaighofer, M., Fischer, F., & Buhner, M. (2015). Does working memory training transfer? A meta-analysis including training conditions as moderators. *Educational Psychologist, 50*(2), 138–166.

Sellers, K. K., Mellin, J. M., Lustenberger, C. M., Boyle, M. R., Lee, W. H., Peterchev, A. V., & Frohlich, F. (2015). Transcranial direct current stimulation (tDCS) of frontal cortex decreases performance on the WAIS-IV intelligence test. *Behavioral Brain Research, 290*, 32–44. doi:10.1016/j.bbr.2015.04.031

Shakeshaft, N. G., Trzaskowski, M., McMillan, A., Krapohl, E., Simpson, M. A., Reichenberg, A., … Plomin, R. (2015). Thinking positively: The genetics of high intelligence. *Intelligence, 48*, 123–132. doi:10.1016/j .intell.2014.11.005

Shamay-Tsoory, S. G., Adler, N., Aharon-Peretz, J., Perry, D., & Mayseless, N. (2011). The origins of originality: The neural bases of creative thinking and originality. *Neuropsychologia, 49*(2), 178–185. doi:10.1016/j.neuropsychologia.2010.11.020

Sharif, S., Guirguis, A., Fergus, S., & Schifano, F. (2021). The Use and Impact of Cognitive Enhancers among University Students: A Systematic Review. *Brain Sciences, 11*(3), 355. doi:10.3390/ brainsci11030355

Shaw, P., Greenstein, D., Lerch, J., Clasen, L., Lenroot, R., Gogtay, N., …
Giedd, J. (2006). Intellectual ability and cortical development in children
and adolescents. *Nature, 440*(7084), 676–679.

Shehzad, Z., Kelly, C., Reiss, P. T., Cameron Craddock, R., Emerson, J.
W., McMahon, K., … Milham, M. P. (2014). A multivariate distance-
based analytic framework for connectome-wide association studies.
Neuroimage, 93 Pt 1, 74–94. doi:10.1016/j.neuroimage.2014.02.024

Shi, M., Li, Y., Sun, J., Li, X., Han, Y., Liu, Z., & Qiu, J. (2022). Intelligence
correlates with the temporal variability of brain networks. *Neuroscience.*
doi:10.1016/j.neuroscience.2022.08.001

Shipstead, Z., Redick, T. S., & Engle, R. W. (2012). Is working memory training
effective? *Psychology Bulletin, 138*(4), 628–654. doi:10.1037/a0027473

Shonkoff, J. P., Phillips, D., & National Research Council (U.S.).
Committee on Integrating the Science of Early Childhood Development.
(2000). *From Neurons to Neighborhoods : The Science of Early
Childhood Development.* Washington, DC: National Academy Press.

Siebner, H. R., Funke, K., Aberra, A. S., Antal, A., Bestmann, S., Chen,
R., … Ugawa, Y. (2022). Transcranial magnetic stimulation of the brain:
What is stimulated? – A consensus and critical position paper. *Clinical
Neurophysiology, 140*, 59–97. doi:10.1016/j.clinph.2022.04.022

Sigala, N. (2015). Effects of memory training or task design? A
Commentary on "Neural evidence for the use of digit-image mnemonic
in a superior memorist: An fMRI study." *Frontier in Human
Neuroscience, 9*, 183. doi:10.3389/fnhum.2015.00183

Sigman, M., Pena, M., Goldin, A. P., & Ribeiro, S. (2014). Neuroscience
and education: Prime time to build the bridge. *Nature Neuroscience,
17*(4), 497–502. doi:10.1038/nn.3672

Silverman, P. H. (2004). Rethinking genetic determinism. *The Scientist,
18*(10), 32–33.

Simos, P. G., Rezaie, R., Papanicolaou, A. C., & Fletcher, J. M. (2014).
Does IQ affect the functional brain network involved in pseudoword
reading in students with reading disability? A magnetoencephalography
study. *Frontier in Human Neuroscience, 7*, 932. doi:10.3389/
fnhum.2013.00932

Smith, M. E., & Farah, M. J. (2011). Are prescription stimulants "smart
pills"? The epidemiology and cognitive neuroscience of prescription
stimulant use by normal healthy individuals. *Psychology Bulletin, 137*(5),
717–741. doi:10.1037/a0023825

Smith, S. M., Nichols, T. E., Vidaurre, D., Winkler, A. M., Behrens, T.
E. J., Glasser, M. F., … Miller, K. L. (2015). A positive-negative mode
of population covariation links brain connectivity, demographics and
behavior. *Nature Neuroscience, 18*(11), 1565–1567. doi:10.1038/nn.4125

Sniekers, S., Stringer, S., Watanabe, K., Jansen, P. R., Coleman, J. R. I., Krapohl, E., ... Posthuma, D. (2017). Genome-wide association meta-analysis of 78,308 individuals identifies new loci and genes influencing human intelligence. *Nature Genetics, 49*(7), 1107–1112. doi:10.1038/ng.3869

Snyderman, M., & Rothman, S. (1988). *The IQ Controversy, the Media and Public Policy*. New Brunswick, NJ, USA: Transaction Books.

Song, M., Liu, Y., Zhou, Y., Wang, K., Yu, C., & Jiang, T. (2009). Default network and intelligence difference. *Conference Proceedings of the IEEE Engineering in Medicine and Biological Society, 2009*, 2212–2215. doi:10.1109/IEMBS.2009.5334874

Song, M., Zhou, Y., Li, J., Liu, Y., Tian, L., Yu, C., & Jiang, T. (2008). Brain spontaneous functional connectivity and intelligence. *Neuroimage, 41*(3), 1168–1176. doi:10.1016/j.neuroimage.2008.02.036

Sonmez, A. I., Camsari, D. D., Nandakumar, A. L., Voort, J. L. V., Kung, S., Lewis, C. P., & Croarkin, P. E. (2019). Accelerated TMS for Depression: A systematic review and meta-analysis. *Psychiatry Research, 273*, 770–781. doi:10.1016/j.psychres.2018.12.041

Soreq, E., Violante, I. R., Daws, R. E., & Hampshire, A. (2021). Neuroimaging evidence for a network sampling theory of individual differences in human intelligence test performance. *Nature Communication, 12*(1), 2072. doi:10.1038/s41467-021-22199-9

Spearman, C. (1904). General intelligence objectively determined and measured. *American Journal of Psychology, 15*, 201–293.

Sripada, C., Angstadt, M., Rutherford, S., Taxali, A., & Shedden, K. (2020). Toward a "treadmill test" for cognition: Improved prediction of general cognitive ability from the task activated brain. *Hum Brain Mapp 41*(12): 3186–3197.

Stam, C. J., & Reijneveld, J. C. (2007). Graph theoretical analysis of complex networks in the brain. *Nonlinear Biomedical Physics, 1*(1), 3. doi:10.1186/1753-4631-1-3

Stammen, C., Fraenz, C., Grazioplene, R. G., Schlüter, C., Merhof, V., Johnson, W., ... Genç, E. (2022). Robust associations between white matter microstructure and general intelligence. bioRxiv, 2022.2005.2002.490274. doi:10.1101/2022.05.02.490274

Stanley, J., Keating, D. P., and Fox L. H. (1974). *Mathematical Talent: Discovery, Description, and Development*. Baltimore: The Johns Hopkins University Press.

Stein, J. L., Medland, S. E., Vasquez, A. A., Hibar, D. P., Senstad, R. E., Winkler, A. M., ... Enhancing Neuro Imaging Genetics through Meta-Analysis, C. (2012). Identification of common variants associated with human hippocampal and intracranial volumes. *Nature Genetics, 44*(5), 552–561. doi:10.1038/ng.2250

Sternberg, R. J. (2000). *Practical Intelligence in Everyday Life*. Cambridge, UK; New York: Cambridge University Press.

Sternberg, R. J. (2003). Our research program validating the triarchic theory of successful intelligence: Reply to Gottfredson. *Intelligence, 31*(4), 399–413.

Sternberg, R. J. (2008). Increasing fluid intelligence is possible after all. *Proceedings of the National Academy Science USA, 105*(19), 6791–6792. doi:10.1073/pnas.0803396105

Sternberg, R. J. (2014). Teaching about the nature of intelligence. *Intelligence, 42*, 176–179. doi:10.1016/j.intell.2013.08.010

Sternberg, R. J. (2018). *The Nature of Human Intelligence* (R. J. Sternberg Ed.). New York: Cambridge University Press.

Strenze, T. (2007). Intelligence and socioeconomic success: A meta-analytic review of longitudinal research. *Intelligence, 35*(5), 401–426. doi:10.1016/j.intell.2006.09.004

Suthana, N., & Fried, I. (2014). Deep brain stimulation for enhancement of learning and memory. *Neuroimage, 85 Pt 3*, 996–1002. doi:10.1016/j.neuroimage.2013.07.066

Tammet, D. (2007). Born on a blue day : Inside the extraordinary mind of an autistic savant : A memoir (1st Free Press pbk. ed.). New York: Free Press.

Tang, C. Y., Eaves, E. L., Ng, J. C., Carpenter, D. M., Mai, X., Schroeder, D. H., … Haier, R. J. (2010). Brain networks for working memory and factors of intelligence assessed in males and females with fMRI and DTI. *Intelligence, 38*(3), 293–303.

Tang, Y. P., Shimizu, E., Dube, G. R., Rampon, C., Kerchner, G. A., Zhuo, M., … Tsien, J. Z. (1999). Genetic enhancement of learning and memory in mice. *Nature, 401*(6748), 63–69.

te Nijenhuis, J., Jongeneel-Grimen, B., & Kirkegaard, E. O. W. (2014). Are Headstart gains on the g factor? A meta-analysis. *Intelligence, 46*, 209–215.

Tellier, L. C. A. M., Eccles, J., Treff, N. R., Lello, L., Fishel, S., & Hsu, S. (2021). Embryo screening for polygenic disease risk: Recent advances and ethical considerations. *Genes, 12*(8), 1105. Retrieved from www.mdpi.com/2073-4425/12/8/1105

Terman, L. M. (1925). *Genetic Studies of Genius*. Stanford, CA: Stanford University Press.

Terman, L. M. (1954). *Scientists and Nonscientists in a Group of 800 Gifted Men*. Washington, DC: American Psychological Association.

Thiele, J. A., Faskowitz, J., Sporns, O., & Hilger, K. (2022). Multitask brain network reconfiguration is inversely associated with human intelligence. *Cerebral Cortex.*, 32(19), 4172–4182. doi:10.1093/cercor/bhab473

Thoma, R. J., Yeo, R. A., Gangestad, S., Halgren, E., Davis, J., Paulson, K. M., & Lewine, J. D. (2006). Developmental instability and the neural dynamics of the speed-intelligence relationship. *Neuroimage, 32*(3), 1456–1464.

Thomas, P., Rammsayer, T., Schweizer, K., & Troche, S. (2015). Elucidating the functional relationship between working memory capacity and psychometric intelligence: A fixed-links modeling approach for experimental repeated-measures designs. *Advances in Cognitive Psychology, 11*(1), 3–13.

Thompson, P. M., Cannon, T. D., Narr, K. L., van Erp, T., Poutanen, V. P., Huttunen, M., ... Toga, A. W. (2001). Genetic influences on brain structure. *Nature Neuroscience, 4*(12), 1253–1258.

Thompson, R., Crinella, F. M., & Yu, J. (1990). *Brain Mechanisms in Problem Solving and Intelligence: A Survey of the Rat Brain*. New York: Plenum Press.

Thompson, T. W., Waskom, M. L., Garel, K. L., Cardenas-Iniguez, C., Reynolds, G. O., Winter, R., ... Gabrieli, J. D. (2013). Failure of working memory training to enhance cognition or intelligence. *Plos One, 8*(5), e63614. doi:10.1371/journal.pone.0063614

Thurstone, L. L. (1938). *Primary Mental Abilities*. Chicago, IL: University of Chicago Press.

Thurstone, L. L., & Thurstone, T. (1941). *Factorial Studies of Intelligence*. Chicago, IL: University of Chicago Press.

Tidwell, J. W., Dougherty, M. R., Chrabaszcz, J. R., Thomas, R. P., & Mendoza, J. L. (2013). What counts as evidence for working memory training? Problems with correlated gains and dichotomization. *Psychonomic Bulletin Review*. doi:10.3758/s13423-013-0560-7

Toga, A. W., & Thompson, P. M. (2005). Genetics of brain structure and intelligence. *Annual Review of Neuroscience, 28*, 1–23.

Tommasi, M., Pezzuti, L., Colom, R., Abad, F. J., Saggino, A., & Orsini, A. (2015). Increased educational level is related with higher IQ scores but lower g-variance: Evidence from the standardization of the WAIS-R for Italy. *Intelligence, 50*, 68–74.

Trahan, L. H., Stuebing, K. K., Fletcher, J. M., & Hiscock, M. (2014). The Flynn effect: A meta-analysis. *Psychological Bulletin, 140*(5), 1332–1360. doi:10.1037/a0037173

Troller-Renfree, S. V., Costanzo, M. A., Duncan, G. J., Magnuson, K., Gennetian, L. A., Yoshikawa, H., ... Noble, K. G. (2022). The impact of a poverty reduction intervention on infant brain activity. *Proceedings of the National Academy of Sciences*, *119*(5), e2115649119. doi:10.1073/pnas.2115649119

Trzaskowski, M., Davis, O. S. P., DeFries, J. C., Yang, J., Visscher, P. M., & Plomin, R. (2013). DNA evidence for strong genome-wide pleiotropy of cognitive and learning abilities. *Behavior Genetics, 43*(4), 267–273.

Trzaskowski, M., Harlaar, N., Arden, R., Krapohl, E., Rimfeld, K., McMillan, A., … Plomin, R. (2014). Genetic influence on family socioeconomic status and children's intelligence. *Intelligence, 42*(100), 83–88. doi:10.1016/j.intell.2013.11.002

Trzaskowski, M., Shakeshaft, N. G., & Plomin, R. (2013). Intelligence indexes generalist genes for cognitive abilities. *Intelligence, 41*(5), 560–565.

Tsukahara, J. S., & Engle, R. W. (2021). Fluid intelligence and the locus coeruleus-norepinephrine system. *Proceedings of the National Academy Science USA, 118*(46). doi:10.1073/pnas.2110630118

Turkheimer, E. (2000). Three laws of behavior genetics and what they mean. *Current Directions in Psychological Science, 9*(5), 160–164.

Turkheimer, E., Haley, A., Waldron, M., D'Onofrio, B., & Gottesman, II. (2003). Socioeconomic status modifies heritability of IQ in young children. *Psychological Science, 14*(6), 623–628.

Turley, P., Meyer, M. N., Wang, N., Cesarini, D., Hammonds, E., Martin, A. R., … Visscher, P. M. (2021). Problems with Using Polygenic Scores to Select Embryos. *New England Journal of Medicine, 385*(1), 78–86. doi:10.1056/NEJMsr2105065

Ukkola-Vuoti, L., Kanduri, C., Oikkonen, J., Buck, G., Blancher, C., Raijas, P., … Jarvela, I. (2013). Genome-wide copy number variation analysis in extended families and unrelated individuals characterized for musical aptitude and creativity in music. *Plos One, 8*(2), e56356. doi:10.1371/journal.pone.0056356

Unsworth, N., Redick, T. S., McMillan, B. D., Hambrick, D. Z., Kane, M. J., & Engle, R. W. (2015). Is playing video games related to cognitive abilities? *Psychological Science, 26*(6), 759–774. doi:10.1177/0956797615570367

Urban, D. J., & Roth, B. L. (2015). DREADDs (designer receptors exclusively activated by designer drugs): Chemogenetic tools with therapeutic utility. *Annual Review of Pharmacological Toxicology, 55*, 399–417. doi:10.1146/annurev-pharmtox-010814-124803

Utz, K. S., Dimova, V., Oppenlander, K., & Kerkhoff, G. (2010). Electrified minds: Transcranial direct current stimulation (tDCS) and galvanic vestibular stimulation (GVS) as methods of non-invasive brain stimulation in neuropsychology – a review of current data and future implications. *Neuropsychologia, 48*(10), 2789–2810. doi:10.1016/j .neuropsychologia.2010.06.002

Vakhtin, A. A., Ryman, S. G., Flores, R. A., & Jung, R. E. (2014). Functional brain networks contributing to the Parieto-Frontal Integration Theory of Intelligence. *Neuroimage, 103*, 349–354. doi:10.1016/j.neuroimage.2014.09.055

van den Heuvel, M. P., & Sporns, O. (2011). Rich-club organization of the human connectome. *Journal of Neuroscience, 31*(44), 15775–15786. doi:10.1523/JNEUROSCI.3539-11.2011

van den Heuvel, M. P., Kahn, R. S., Goni, J., & Sporns, O. (2012). High-cost, high-capacity backbone for global brain communication. *Proceedings of the National Academy Science USA, 109*(28), 11372–11377. doi:10.1073/pnas.1203593109

van den Heuvel, M. P., Stam, C. J., Kahn, R. S., & Pol, H. E. H. (2009). Efficiency of functional brain networks and intellectual performance. *Journal of Neuroscience, 29*(23), 7619–7624. doi:10.1523/jneurosci.1443-09.2009

van der Linden, D., Dunkel, C. S., & Madison, G. (2017). Sex differences in brain size and general intelligence (g). *Intelligence, 63*, 78–88. doi:10.1016/j.intell.2017.04.007

van der Maas, H. L. J., Snoek, L., & Stevenson, C. E. (2021). How much intelligence is there in artificial intelligence? A 2020 update. *Intelligence, 87*, 101548. DOI: 10.1016/j.intell.2021.101548

van der Sluis, S., Willemsen, G., de Geus, E. J. C., Boomsma, D. I., & Posthuma, D. (2008). Gene-environment interaction in adults' IQ scores: Measures of past and present environment. *Behavior Genetics, 38*(4), 348–360.

van Leeuwen, M., van den Berg, S. M., & Boomsma, D. I. (2008). A twin-family study of general IQ. *Learning and Individual Differences, 18*(1), 76–88. doi:10.1016/j.lindif.2007.04.006

Vardy, E., Robinson, J. E., Li, C., Olsen, R. H., DiBerto, J. F., Giguere, P. M., … Roth, B. L. (2015). A new DREADD facilitates the multiplexed chemogenetic interrogation of behavior. *Neuron, 86*(4), 936–946. doi:10.1016/j.neuron.2015.03.065

Vendetti, M. S., & Bunge, S. A. (2014). Evolutionary and developmental changes in the lateral frontoparietal network: A little goes a long way for higher-level cognition. *Neuron, 84*(5), 906–917. doi:10.1016/j.neuron.2014.09.035

Vernon, P. A. (1983). Speed of information processing and general intelligence. *Intelligence, 7*(1), 53–70.

Vieira, B. H., Pamplona, G. S. P., Fachinello, K., Silva, A. K., Foss, M. P., & Salmon, C. E. G. (2022). On the prediction of human intelligence from neuroimaging: A systematic review of methods and reporting. *Intelligence, 93*, 101654. doi:10.1016/j.intell.2022.101654

Villarreal, M. F., Cerquetti, D., Caruso, S., Schwarcz Lopez Aranguren, V., Gerschcovich, E. R., Frega, A. L., & Leiguarda, R. C. (2013). Neural correlates of musical creativity: Differences between high and low creative subjects. *Plos One, 8*(9), e75427. doi:10.1371/journal.pone.0075427

Visscher, P. M. (2022). Genetics of cognitive performance, education and learning: From research to policy? *npj Science of Learning, 7*(1), 8. doi:10.1038/s41539-022-00124-z

von Bastian, C. C., & Oberauer, K. (2013). Distinct transfer effects of training different facets of working memory capacity. *Journal of Memory and Language, 69*(1), 36–58.

von Bastian, C. C., & Oberauer, K. (2014). Effects and mechanisms of working memory training: A review. *Psychological Research-Psychologische Forschung, 78*(6), 803–820.

von Stumm, S., & Deary, I. J. (2013). Intellect and cognitive performance in the Lothian Birth Cohort 1936. *Psychology and Aging, 28*(3), 680–684. doi:10.1037/A0033924

von Stumm, S., & Plomin, R. (2021). Using DNA to predict intelligence. *Intelligence, 86*, 101530. doi:10.1016/j.intell.2021.101530

Vuoksimaa, E., Panizzon, M. S., Chen, C. H., Fiecas, M., Eyler, L. T., Fennema-Notestine, C., ... Kremen, W. S. (2015). The genetic association between neocortical volume and general cognitive ability is driven by global surface area rather than thickness. *Cerebral Cortex, 25*(8), 2127–2137. doi:10.1093/cercor/bhu018

Wagner, T., Robaa, D., Sippl, W., & Jung, M. (2014). Mind the methyl: Methyllysine binding proteins in epigenetic regulation. *ChemMedChem, 9*(3), 466–483. doi:10.1002/cmdc.201300422

Wai, J., & Bailey, D. H. (2021). How intelligence research can inform education and public policy. In A. Barbey, S. Karama, & R. J. Haier (Eds.), *The Cambridge Handbook of Intelligence and Cognitive Neuroscience*. New York: Cambridge University Press.

Wai, J., & Worrell, F. C. (2021). The future of intelligence research and gifted education. *Intelligence, 87*, 101546. doi:10.1016/j.intell.2021.101546

Wai, J., Brown, M., & Chabris, C. F. (2019). No one likes the SAT. It's still the fairest thing about admissions. In. *Washington Post editorial.*

Wai, J., Lubinski, D., & Benbow, C. P. (2005). Creativity and occupational accomplishments among intellectually precocious youths: An age 13 to age 33 longitudinal study. *Journal of Educational Psychology, 97*(3), 484–492.

Walfisch, A., Sermer, C., Cressman, A., & Koren, G. (2013). Breast milk and cognitive development – the role of confounders: A systematic review. *BMJ Open, 3*(8), e003259. doi:10.1136/bmjopen-2013-003259

Wang, C., Jaeggi, S. M., Yang, L., Zhang, T., He, X., Buschkuehl, M., & Zhang, Q. (2019). Narrowing the achievement gap in low-achieving children by targeted executive function training. *Journal of Applied Developmental Psychology, 63*, 87–95. doi:10.1016/j.appdev.2019.06.002

Wang, L., Wee, C. Y., Suk, H. I., Tang, X., & Shen, D. (2015). MRI-Based Intelligence Quotient (IQ) estimation with sparse learning. *Plos One, 10*(3), e0117295. doi:10.1371/journal.pone.0117295

Warne, R. T. (2019). An evaluation (and Vindication?) of Lewis Terman: What the father of gifted education can teach the 21st century. *Gifted Child Quarterly, 63*(1), 3–21.

Warne, R. T. (2020). *In the Know : Debunking 35 Myths about Human Intelligence*. Cambridge, United Kingdom; New York, NY: Cambridge University Press.

Waterhouse, L. (2006). Inadequate evidence for multiple intelligences, Mozart effect, and emotional intelligence theories. *Educational Psychologist, 41*(4), 247–255. doi:10.1207/s15326985ep4104_5

Watrin, L., Hulur, G., & Wilhelm, O. (2022). Training working memory for two years-No evidence of transfer to intelligence. *Journal of Experimental Psychology: Learning, Memory, and Cognition, 48*(5), 717–733. doi:10.1037/xlm0001135

Watson, J. B. (1930). *Behaviorism*. New York,: W.W. Norton & Company.

Wax, A. L. (2017). The poverty of the neuroscience of poverty: Policy payoff or false promise? *Jurimetrics, 57*(2), 239–287. Retrieved from www.jstor.org/stable/26322667

Weiss, D., Haier, R., & Keating, D. (1974). Personality characteristics of mathematically precocious boys. In Stanley, Keating, & Fox (Eds.), *Mathematical Talent: Discovery, Description, and Development* (pp. 126–139). Baltimore, MD: The Johns Hopkins University Press.

Wendelken, C., Ferrer, E., Whitaker, K. J., & Bunge, S. A. (2015). Fronto-parietal network reconfiguration supports the development of reasoning ability. *Cerebral Cortex*. doi:10.1093/cercor/bhv050

Whalley, L. J., & Deary, I. J. (2001). Longitudinal cohort study of childhood IQ and survival up to age 76. *Bmj, 322*(7290), 819. Retrieved from www.ncbi.nlm.nih.gov/pubmed/11290633

Wharton, C. M., Grafman, J., Flitman, S. S., Hansen, E. K., Brauner, J., Marks, A., & Honda, M. (2000). Toward neuroanatomical models of analogy: A positron emission tomography study of analogical mapping. *Cognitive Psychology, 40*(3), 173–197.

Widge, A. S., Zorowitz, S., Basu, I., Paulk, A. C., Cash, S. S., Eskandar, E. N., … Dougherty, D. D. (2019). Deep brain stimulation of the internal capsule enhances human cognitive control and prefrontal cortex function. *Nature Communications, 10*(1), 1536. doi:10.1038/s41467-019-09557-4

Wiemers, E. A., Redick, T. S., & Morrison, A. B. (2019). The influence of individual differences in cognitive ability on working memory training gains. *Journal of Cognition Enhancement, 3*(2), 174–185. doi:10.1007/s41465-018-0111-2

Wilke, M., Sohn, J. H., Byars, A. W., & Holland, S. K. (2003). Bright spots: Correlations of gray matter volume with IQ in a normal pediatric population. *Neuroimage, 20*(1), 202–215.

Willerman, L., Schultz, R., Rutledge, J. N., & Bigler, E. D. (1991). In vivo brain size and intelligence. *Intelligence, 15*(2), 223–228.

Willoughby, E. A., McGue, M., Iacono, W. G., & Lee, J. J. (2021). Genetic and environmental contributions to IQ in adoptive and biological families with 30-year-old offspring. *Intelligence, 88.* doi:10.1016/j .intell.2021.101579

Wilson, E. O. (1975). *Sociobiology: The New Synthesis.* Cambridge: Belknap Press of Harvard Univ. Press.

Witelson, S. F., Beresh, H., & Kigar, D. L. (2006). Intelligence and brain size in 100 postmortem brains: Sex, lateralization and age factors. *Brain, 129*(Pt 2), 386–398. doi:10.1093/brain/awh696

Witelson, S. F., Kigar, D. L., & Harvey, T. (1999a). Albert Einstein's brain – Reply. *Lancet, 354*(9192), 1822–1822.

Witelson, S. F., Kigar, D. L., & Harvey, T. (1999b). The exceptional brain of Albert Einstein. *Lancet, 353*(9170), 2149–2153.

Wolff, S. B., Grundemann, J., Tovote, P., Krabbe, S., Jacobson, G. A., Muller, C., … Luthi, A. (2014). Amygdala interneuron subtypes control fear learning through disinhibition. *Nature, 509*(7501), 453–458. doi:10.1038/nature13258

Wooldridge, A. (2021). *The Aristocracy of Talent: How Meritocracy Made the Modern World.* New York: Skyhorse Publishing.

Wu, X., Yang, W., Tong, D., Sun, J., Chen, Q., Wei, D., … Qiu, J. (2015). A meta-analysis of neuroimaging studies on divergent thinking using activation likelihood estimation. *Human Brain Mapping, 36*(7), 2703–2718. doi:10.1002/hbm.22801

Yang, J. J., Yoon, U., Yun, H. J., Im, K., Choi, Y. Y., Lee, K. H., … Lee, J. M. (2013). Prediction for human intelligence using morphometric characteristics of cortical surface: Partial least square analysis. *Neuroscience, 246*, 351–361. doi:10.1016/j.neuroscience.2013.04.051

Yin, L. J., Lou, Y. T., Fan, M. X., Wang, Z. X., & Hu, Y. (2015). Neural evidence for the use of digit-image mnemonic in a superior memorist: an fMRI study. *Frontier in Human Neuroscience, 9*, 109. doi:10.3389/ fnhum.2015.00109

Yu, C. C., Furukawa, M., Kobayashi, K., Shikishima, C., Cha, P. C., Sese, J., … Toda, T. (2012). Genome-wide DNA methylation and gene expression analyses of monozygotic twins discordant for intelligence levels. *Plos One, 7*(10).

Zhao, M., Kong, L., & Qu, H. (2014). A systems biology approach to identify intelligence quotient score-related genomic regions, and pathways relevant to potential therapeutic treatments. *Scientific Reports, 4*, 4176. doi:10.1038/srep04176

Zhao, T., Zhu, Y., Tang, H., Xie, R., Zhu, J., & Zhang, J. H. (2019). Consciousness: New concepts and neural networks. *Frontiers in Cellular Neuroscience, 13*. doi:10.3389/fncel.2019.00302

Zisman, C., & Ganzach, Y. (2022). The claim that personality is more important than intelligence in predicting important life outcomes has been greatly exaggerated. *Intelligence, 92*, 101631. doi:10.1016/j.intell.2022.101631

Index

academic achievement
  influence of intelligence level, 21–22
achievement tests, 17
active reading for children, effect on IQ,
  170
adenine (A), 64
adoption studies, 45
  Denmark Adoption Studies of
    schizophrenia, 50
  Sweden Adoption Study, 51
  twin studies of intelligence, 50–54
  UK Biobank Project, 51
aging and IQ score, 32–35
Alkire, Michael, 203
alleles of genes, 64
Alzheimer's disease, 63, 67, 70, 172
amino acids, 64
analogy tests, 17
animal studies, bridging animal and human
    research at the level of neurons,
    194–198
aptitude tests, 17
arcuate fasciculus, working memory and,
  115
Armed Forces Qualification Test (AFQT),
  67–68
artificial intelligence (AI), 11
  artificial general intelligence (AGI), 11,
    199–200
  human intelligence-based AI, 198–203
attention deficit hyperactivity disorder
    (ADHD), 172
autism, 3
autism research, 45

Baby's First Years (BFY) study, 216–218
Barbey, Aron, 114
base pairs (nucleotides), 64–65
Basten, Ulrike, 116
BDNF (brain-derived neurotrophic
    factor), 67–68, 145

behavioral genetics, 44–48
Behavioral Genetics Institute, China, 72
Behaviorist view of human potential, 42
The Bell Curve (Herrnstein and Murray),
    26–27, 215, 221
bell curve distribution of IQ scores, 14–15
Benbow, Camilla, 32, 85
bias in intelligence tests, 18–19
Biden, Joe, 184
Big Data analysis, 65
Binet, Alfred, 12
Binet-Simon intelligence test, 12–13
bioinformatics, 65
The Blank Slate (Pinker), 49
Blank Slate view of human potential, 42,
    219
Bochumer Matrizen-Test (BOMAT), 158,
    160–162
boosting IQ. see increasing intelligence
Bouchard, Thomas, 54
brain activity
  evidence for individual differences, 84–87
  multiple areas involved in intelligence,
    84–87
brain-altering technologies, 175–179
brain anatomy
  Einstein's brain, 102. see also Brodmann
    Areas
brain-based products, growth in the
    industry, 168
brain efficiency and intelligence
  brain activity in low-IQ groups, 83–84
  complexity of the concept, 123–129
  effects of learning, 79–83
  functional connectivity study, 125
  functional neuroimaging studies,
    123–124
  MEG studies, 126–128
  PET studies, 78–80
brain fingerprint study (Finn), 120, 136
brain imaging. see neuroimaging

BRAIN Initiative (Brain Research
through Advancing Innovative
Neurotechnologies), 184, 202
brain lesion patients, evidence for brain
networks, 114–115
brain mapping, 201
brain networks
connectivity analysis techniques, 107–122
default network, 108, 109
evidence from brain lesion patients,
114–115
evidence of reconfiguration between
tasks, 126
homotopic connectivity, 111
and intelligence, 107–122
PFIT studies, 108–122
rich club networks, 108
small world networks, 108
brain proteins and IQ, 69
brain resilience after traumatic brain
injury, 109
brain size and intelligence, 69
size of brain regions and intelligence, 92
whole brain size/volume, 91–92
breastfeeding, 169–170
Brin, Sergey, 30
Brodmann Areas (BAs), 92, 108, 119
Buchsbaum, Monte, 78
Burt, Sir Cyril, 73
twin studies, 50–54

Cajal, Santiago Ramon, xi
candidate gene studies, 62–63, 69
CAT scan imaging of the brain, 77
Chabris, Christopher, 63, 69, 155
CHARGE (Cohorts for Heart and
Aging Research in Genomic
Epidemiology), 70
chemogenetic technique, 197
CHIC (Childhood Intelligence
Consortium), 66, 68
China
commitment to molecular genetic
research, 68–69
DNA sequencing capacity, 72
The CHIPS and Science Act (2022), 184

chromosomes, 64, 65, 69
chronometric testing, 188–190, 225
classical music, claims for increasing
intelligence, 152–156
Clemons, Alonso, 3
Clinton, Bill, 184
cognitive enhancing (CE) drugs, ethical
issues, 173–174
cognitive malleability, 171
cognitive processing, intelligence measures
and speed of, 126
cognitive segregation, 24, 27
college admission, standardized tests for,
18, 130
*Coming apart : the state of white America,
1960–2010* (Murray), 27
compensatory education programs, 46–49,
179
complex traits, three laws of heritability, 59
computer games, claims for increasing
intelligence, 164–171
computers, Watson (IBM computer), 4, 11
connectivity
functional connectivity study, 125
homotopic, 111
structural, 128
connectivity analyses studies, PFIT,
120–121
connectivity analysis techniques, 107–122
connectivity differences, related to
Wechsler IQ scores, 122
connectivity patterns, sex differences, 136
consciousness and creativity, 203–212, 225
Continuity Hypothesis, 59–60
corpus callosum, relationship to
intelligence, 112
correlations
between mental ability tests, 5–9
the concept, 7–8
cortical thickness and IQ, 95
effects of restricted range of scores,
35–36
cortical thickness
and intelligence, 95, 118, 141, 146
social economic status and, 215
twin studies, 95, 141

creativity and consciousness, 203–212
Crick, Francis, 203
CRISPR/Cas9 method for genome editing,
    181, 197
crystallized intelligence, 9–10, 33
cytosine (C), 63–64

Database of Raising Intelligence (NYU),
    169
    meta-analyses of interventions, 169–170
de Geus, Eco J.C., 40
Deary, Ian, 32–34, 40
deep brain stimulation (DBS), 179
deGrasse Tyson, Neil, 1
Denmark Adoption Studies, 50
Detterman, Doug, 22
diffusion tensor imaging (DTI), 97–98, 112,
    128, 141
Discontinuity Hypothesis, 59–60
DNA
    analysis techniques, 61–63
    double helix structure, 61, 64
    sequencing, 64, 72
    sequencing, Chinese sequencing
        capacity, 72
    technologies and methods, 45
Doogie strain of mice, 61–62
Down's syndrome, 83
DREADD (designer receptors exclusively
    activated by designer drugs),
    197–198
drugs
    to boost intelligence, 171–174
    ethical issues for cognitive enhancement
        (CE), 173–174
    psycho stimulant drugs, 172
DUF1220 brain protein subtypes and IQ,
    69
Duncan, John, 88
Dutch twin study, 55–56

early education, effect on IQ, 170
ECT (electric convulsive therapy), 176,
    189–190
education policy, neuro-poverty and the
    achievement gap, 219–223

educational achievement, influential
    factors, 21–22
Einstein, Albert, 4, 10, 11
    brain anatomy, 102
elementary cognitive tasks (ECTs), 189
emotional intelligence, 23
enhancing intelligence. see increasing
    intelligence
ENIGMA group, 146
environment and intelligence
    quantitative genetics studies, 54–61
    shared and non-shared environmental
        factors, 56–57, 140
    three-component model, 56–57, 59
epigenetics, 41–44, 221
ethical issues, cognitive enhancement,
    173–174
eugenics, 32, 44
everyday life functioning, predictive
        validity of intelligence tests, 24–27
expertise
    discontinuity hypothesis, 59
    level of practice required for, 23

Facebook, 30
factor analysis
    alternative models of intelligence, 9–10
    the concept, 7
    mental ability tests, 5–9
fairness of intelligence tests, 18–19
FDG (flurodeoxyglucose) PET, 77, 107
Finn brain fingerprint study, 120, 136
Flowers for Algernon (Keyes), 150
fluid intelligence, 9, 10, 157–164
fluorescent protein studies, 196
Flynn Effect, 54
fMRI (functional MRI), 99
    brain efficiency and intelligence studies,
        123–124
fractional anisotropy (FA) studies,
        141–144, 146. see also diffusion
        tensor imaging
frontal disinhibition model (F-DIM) of
        creativity, 209–211
frontal temporal dementia (FTD),
        204–205, 209

Frontiers of Human Neuroscience, 168
full scale IQ (FSIQ), 15
functional literacy score
  and the challenges of daily life, 25, 26
future of intelligence research, 184–186
  bridging animal and human research at
    the level of neurons, 194–198
  challenges for the future, 223–225
  chemogenetic technique, 197
  chronometric testing, 188, 225
  cognitive neuroscience of memory and
    super memory, 190–193
  consciousness and creativity, 203–212
  machine intelligence based on human
    intelligence, 198–203
  neuro-poverty, 212, 220–221, 223
  neuro-social economic status, 58,
    212–223
  optogenetic techniques, 196–197
  public policy on neuro-poverty, 219–223

Galton, Francis, 27
Gamm, R., 193
Gattaca (1996), 174
gender differences. *see* sex differences
gene expression, 43
  regulation of, 64
  role of methylation, 64
  types of, 41–42
genes
  coding for proteins, 43
  definition of a gene, 43, 64
  forms of (alleles), 64
  generalist genes, 45, 63
  locus on a chromosome, 64–65
  molecular genetics research, 61–66
  pleiotropy, 45, 144
  polygenicity of intelligence, 45
  protein formation, 64
  repeat copies at a locus, 65
  structure of (base pairs), 64
genetic code, 64
genetic engineering
  CRISPR/Cas9 method for genome
    editing, 181, 197
  Doogie strain of mice, 61–62

genetics, basic concepts, 64–65
genetics and intelligence
  anti-genetic feeling, 44
  behavioral genetics, 44–48
  common genes for brain structure and
    intelligence, 139–145
  debate over, 46–49
  heritability of intelligence, 45, 59–60
  three-component model, 56–59
  twin studies, 50–54
genius, 14, 204, 209–210
genome, 64
genome-wide association studies (GWAS),
  65, 67, 71, 189
genomic informatics, 65
genomics, 64
genotype, 45
g-factor
  in alternative factor analysis models,
    9–10
  the concept, 10–12
  distinction from IQ, 10–12
  heritability, 60
  influence on daily life functioning,
    24–27
  reasons for myths about, 35–37
  relationships to specific mental abilities,
    5–9
  and savant abilities, 10–11
gifted children, 27–32, 34
glucose metabolic rate (GMR), 19, 78, 80,
  107, 123
Google, 30
Gordon, Robert, 24
Graduate Record Exam (GRE), 21
graph analysis, 110, 112, 127, 143, 148
graph analysis of neuroimaging data,
  108–109
Gray, Jeremy, 99
guanine (G), 64
guanylate kinase (MAGUK), 67
Gur, Ruben, 118

Haier, Richard, 75, 105, 209–210
Halstead, Ward C., 105
Havana Syndrome, 179

Hawkins, Jeff, 199
Head Start education program (USA), 46,
    48, 49
heritability, the concept of, 55–56
heritability of intelligence, 44–45, 58
    Continuity Hypothesis, 59–60
    Discontinuity Hypothesis, 59–60
    effect of age at testing, 55
    social economic status and, 214
    three laws of complex traits, 59
Herrnstein, Richard, 26, 47, 213, 221
Hilger, Kirsten, 122, 125
Holden, Constance, 89
homotopic analysis of neuroimaging data,
    111
Human Brain Project (HBP), 202
Human Connectome Project, 120, 202
Human Connectome Project, Young
    Adult Human Connectome
    Project, 136
human genome, 64
Human Genome Project, 43, 184
Hunt, Earl, 24

IMAGEN consortium, 146
increasing intelligence, 20, 22, 69, 151–152,
    167
    active reading for children, 170
    assessing claims for, 150–152, 168–169
    brain-altering technologies, 175–179
    childhood nutrition studies, 169–170
    claims for classical music, 152–156
    compensatory education programs,
        46–49, 179
    computer games, 164–171
    Database of Raising Intelligence
        (NYU), meta-analyses of
        interventions, 169–170
    deep brain stimulation (DBS), 179
    drugs to boost intelligence, 171–174
    early failures, 46–49
    effect of early education, 170
    effect of preschool attendance, 170
    ethical issues, 173–174
    fundamental problem of measurement,
        151, 152

    future possibilities, 181–182
    Head Start education program (USA),
        46–49, 179
    IQ pill, 171–174
    light from low power "cold" lasers,
        179–180
    memory training, 157–163
    missing weight of evidence for, 151, 152,
        180–182
    Mozart Effect, 152–156
    need for independent replication of
        studies, 154–169
    role of neurotransmitters, 171
    transcranial alternating current
        stimulation (tACS), 178–179
    transcranial direct current stimulation
        (tDCS), 176–179
    transcranial magnetic stimulation
        (TMS), 175–179
independent component analysis, 111
independent replication of studies, 154–169
intelligence
    brain networks and, 107–122
    confounding with neuro-social economic
        status, 58, 212–223
    cortical thickness and, 95, 118, 141, 146
    defining, 4–5, 135–136
    as a general mental ability, 5
    g-factor, 10–12
    g-factor relationships, 5–9
    influence on longevity, 33
    involvement of multiple areas of the
        brain, 84–87
    Parieto-Frontal Integration Theory
        (PFIT), 99–102
    relationship to reasoning, 137–139
    savant abilities, 2–4, 9–10
Intelligence (Special Issue, 1998), 48
intelligence genes
    DNA analysis techniques, 62–63
    evidence from neuroimaging and
        molecular genetics, 145–147
    and genes for brain structure, 139–145
    molecular genetics research, 61–66
    problems with early candidate gene
        studies, 62–63, 69

intelligence measurement
  development of IQ testing, 12–15
  key problem for, 20
  reasons for myths about, 35–37
  relative score problem, 20
intelligence research
  defining intelligence, 4–5
  negative connotations, 46–49
  three laws governing, 186
  ultimate purpose, xii
intelligence tests/testing
  achievement tests, 17
  alternatives to IQ tests, 15–18
  analogy tests, 17
  aptitude tests, 17
  Binet-Simon intelligence test, 12–13
  chronometric testing, 188, 225
  correlations between, 5–9
  disadvantages of interval scales, 187
  fairness of, 18–19
  influence of age at testing, 55
  limitations of psychometric tests,
    152–172, 186, 187
  meaningfulness of, 18–19
  mental age concept, 12–14
  Moray House Test, 33
  myths about, 18–19
  need for a ratio scale of measurement,
    152–172, 186–187
  original purpose of, 13
  predictive validity
    for everyday life functioning, 24–27
    for job performance, 22–23
    for learning ability, 21–22
    in longitudinal studies, 27–35
  predictive value of, 19
  question of bias, 18–19
  Raven's Advanced Progressive Matrices
    (RAPM) test, 16–17, 79–80
  reasons for myths about, 35–37
  SAT (Scholastic Assessment Test), 17
  Stanford Binet test, 13, 27
  Time limited tests, 16, 161–162
  Wechsler Adult Intelligence Scale
    (WAIS), 13–15
  Wechsler Intelligence Scale for Children
    (WISC), 15

The Intelligent Brain (Haier, Great
    Courses), 11, 176
International Society for Intelligence
    Research (ISIR), 89, 99
IQ (intelligence quotient)
  cortical thickness and, 95
  development of IQ testing, 12–15
  distinction from g-factor, 10–12
IQ (intelligence quotient) score
  average test score differences between
    groups, 37
  calculation of deviation scores, 14–15
  difference in group average scores,
    47–49
  Flynn Effect, 54
  generation of norms, 15
  genius range, 14
  normal distribution, 15
  original calculation for children, 12–13
  as relative measure, 13
  relative scores issue, 20
  stability over time, 32–33
IQ in The Meritocracy (Herrnstein), 26, 47,
    213, 221

Jaeggi, Suzanne M., 157–164
James Webb telescope, xv
Jensen, Arthur, 1, 44, 73, 221, 222
  chronometric testing, 189–190
  genetic basis of intelligence, 47–49
  report on Burt's twin studies, 52
  review of compensatory education
    programs, 46–49
job performance
  predictive ability of intelligence tests,
    22–23
  requirements for expertise, 23
Jobs, Steve, 184
Jung, Rex, 99, 131, 204, 209–210

Kennedy, John F., 184

Lady Gaga, 30
laser light, low-power "cold" laser brain
    stimulation, 179–180
Lashley, Karl, 195
Law School Admission Test (LSAT), 21

learning ability, predictive validity of
  intelligence tests, 21–22
learning and brain efficiency, 79–83
Lerner, Barbara, 1
life events, relative risk related to IQ, 24
Limitless (2011), 150, 174
longevity, influence of intelligence, 32–33
longitudinal studies, predictive validity of
  intelligence testing, 27–35
low-power "cold" laser brain stimulation,
  179–180
Lubinski, David, 32, 213

machine intelligence, based on human
  intelligence, 198–203
mathematical reasoning, sex differences in
  brain activity, 85–86
The Matrix (1999), 150
meaningfulness of intelligence tests, 18–19
Medical College Admission Test (MCAT),
  21
MEG (magneto-encephalogram), 77
  brain efficiency studies, 126–128
memory, cognitive neuroscience of
  memory and super memory,
  190–193
memory training
  claims to increase intelligence, 156–163
  mnemonic methods, 191–193
  super-memory cases, 191–194
mental abilities, structure of, 5–9
mental ability tests
  correlations between, 5–9
  factor analysis, 5–9
  positive manifold, 7, 9, 126
mental age, the concept of, 12–14
mental calculators, 193
methylation, role in gene expression
  regulation, 64
mice, Doogie strain, 61–62
Miller, Zell, 154
Minnesota Multiphasic Personality
  Inventory (MMPI), 134
mnemonic methods of memory training,
  191–193
molecular genetic studies, 66–71
  benefits of a consortium approach, 66–71

brain proteins and IQ, 69
combined with neuroimaging studies,
  145–147
construction of IQ-related neural
  pathways, 68–69
DUF1220 brain protein subtypes and
  IQ, 69
factors in recovery after traumatic brain
  injury, 67–68
GWAS search for intelligence genes, 67
neurobiology of intelligence, 66–67
research commitment in China, 68–69
SNPs
  associated with general cognitive
    ability, 68–70
  associated with variation in
    intelligence in children, 68, 70
  and intelligence genes, 65
molecular genetics, 45–46
  basic genetics concepts, 64–65
  hunt for intelligence genes, 61–66
Montreal Neurological Institute (MNI)
  coordinates, 93
Moody, David E., 161–162
Moray House Test, 33
Moreau, David, 171
Mozart Effect, 152–156
MR spectroscopy (MRS), 97–98
MRI (magnetic resonance imaging)
  basic structural MRI findings, 91–92
  consistency issues, 92–96
  diffusion tensor imaging (DTI), 97–98
  functional MRI (fMRI), 98–99
  imaging white matter tracts, 97–98
  improved MRI analyses, 92–96
  MR spectroscopy (MRS), 97–98
  NODDI technique, 129
  principles and techniques, 89–91
  size of brain regions and intelligence, 92
  voxel-based morphometry (VBM),
    92–96
  voxels, 90
  whole brain size/volume and
    intelligence, 91–92
multiple demand theory, 120
multiple regression equations, 132–135
Murray, Charles, 26, 221

NAA (*N*-acetylaspartate), correlation with IQ, 97
nature–nurture debate
  all (or mostly) environment scenario, 42
  all (or mostly) gene scenario, 42
  Behaviorist view, 42
  Blank Slate view, 40, 42
  epigenetic view, 42–43
  middle position, 42
  types of gene expression, 41–42
n-back test, 158–160, 163
network neuroscience, 110
Neubauer, Aljoscha, 99
neural pruning, 83–84
neuro-*g*, 116
neuroimaging
  brain networks and intelligence, 107–122
  CAT scans, 77
  combined with molecular genetics, 145–147
  common genes for brain structure and intelligence, 139–145
  defining intelligence, 136
  early applications in intelligence research, 75–76
  findings from recent studies, 106
  functional brain efficiency, 123–130
  graph analysis of neuroimaging data, 108–109
  homotopic analysis of neuroimaging data, 111
  Parieto-Frontal Integration Theory (PFIT) of intelligence, 99–102
  predicting IQ from brain images, 130–137
  relationship between intelligence and reasoning, 137–139
  use of templates in brain image analysis, 134
  X-ray imaging, 76
neuromorphic chip technology, 201
neuro-poverty, 212–223
  public policy approach, 219–223
neuro-social economic status, 58, 212–223
neurotransmitters, 61, 69, 171, 194, 204
Newton, Isaac, 4, 10

NMDA (*N*-methyl d-aspartate) receptor, 61, 66
NODDI (neurite orientation dispersion and density imaging), 129
noetron, 200
normal distribution of IQ scores, 14–15
  implications for social policy, 26
NR2B gene, 61
nucleotides (base pairs), 64–65
nutrition in children, influence on intelligence, 169–170

Obama, Barack, 184
obesity research, 45
optogenetic techniques, 196–197

Parieto-Frontal Integration Theory (PFIT) of intelligence, 99–102
  connectivity analyses studies, 120–121
  evidence from brain network studies, 108–122
  evidence from MEG studies, 127–128
  findings related to brain development, 118
  recent neuroimaging evidence, 107
  related to Wechsler IQ scores, 122
  sex differences, 118–119
Parkinson's disease, 179
Pavacinni, Derek, 4, 11
Peek, Kim, 3, 11
PET (positron emission tomography)
  brain activity
    during mathematical reasoning, 85–86
    early PET studies, 76–80
    individual differences, 84–87
    in low-IQ groups, 83–84
    in a nonproblem solving situation, 86–87
  brain efficiency
    and intelligence, 78–80
    and learning, 79–83
  FDG (flurodeoxyglucose) tracer, 77, 107
  multiple areas involved in intelligence, 84–87
  radioactive oxygen tracer, 77
  radioactive tracers, 77, 78

search for a center of intelligence in the brain, 87–89
sex differences in brain activity, 84–85
what early studies revealed, 87–88
phenotype, 45
Pinker, Steven, 49
pleiotropy, 45, 144
Plomin, Robert, 63, 72
polygenic scores (PGSs), 65, 167, 174, 225
predicting intelligence using, 71–72
polygenicity of intelligence, 45
positive manifold of mental ability tests, 7–9, 126
Posthuma, Danielle, 40
Prabhakaran, Vivek, 99
predicting IQ from brain images, 136–137
predictive value of intelligence tests
everyday life functioning, 24–27
job performance, 22–23
learning ability, 21–22
longitudinal studies, 27–35
preschool attendance, effect on IQ, 170
profile analysis, 135
proteins
formation by genes, 64
genes coding for, 43
proteomics, 65
psychometric tests, limitations of, 152–172, 186, 187
psycho-stimulant drugs, 172
public policy
neuro-poverty and the achievement gap, 219–223
relevance of intelligence research, 27

quantitative genetics, 45, 54–61
quantitative trait locus (QTL), 65

radioactive tracers, 77, 78
Rauscher, Francis, 152–156
Raven's Advanced Progressive Matrices (RAPM) test, 16–17, 79–80, 158
Reagan, Ronald, 184
reasoning, relationship to intelligence, 137–139
regression to the mean, 166

relative risk of life events
relationship to IQ, 24
Ritchie, Stuart, 170
RNA, 64
Rosenthal, David, 50

SAT (Scholastic Assessment Test), 17
SAT-Math, 29–31, 85
savant abilities, 2–4, 9–10
schizophrenia, Denmark Adoption Studies, 50
schizophrenia research, 44, 45, 50
Schubert, Anna-Lena, 126
Scottish Mental Survey, 32–33
sex differences
brain activity, 84–85
connectivity patterns, 136
PFIT, 118–119
white matter correlations with IQ, 113
Shaw, Gordon, 156
Shwachman-Diamond syndrome, 114
Simon, Theodore, 12
Single nucleotide polymorphisms (SNPs), 65, 66, 68, 70, 71
Skinner, B.F., 75
small world networks, 108
social economic status (SES)
confounding with intelligence, 58, 212–223
heritability, 58, 215
and heritability of intelligence, 214
neuro-social economic status, 58, 212–223
social policy
relevance of intelligence research, 26
sociobiology, 26
Spearman, Charles, 7, 9
SSGAC (Social Science Genetic Association Consortium), 66
standardized tests, for college admission, 18, 130
Stanford Binet test, 13, 27
Stanley, Julian, 29, 84
statistical issues
effects of restricted range of scores, 35–36

statistical parametric mapping (SPM),
    92–93
Steele, Kenneth, 155
Stern, William, 12
structural connectivity, 128
structural equation modeling, 133
structure of mental abilities, 5–9
Study of Mathematically & Scientifically
    Precocious Youth, 29
Sweden, adoption study, 51
SyNAPSE Program, 202
synesthesia, 3

talent searches, 29–30
Tammet, Daniel, 2–3
Terman, Lewis, 13, 27–28
Tetris, 81, 82
Thompson, Paul, 99, 140, 141
Thurstone, Louis, 35–36
thymine (T), 64
time-limited intelligence tests, 77
traffic accident risk and IQ, 24
transcranial alternating current stimulation
    (tACS), 177–179
transcranial direct current stimulation
    (tDCS), 176–179
transcranial magnetic stimulation (TMS),
    175–179
traumatic brain injury, 67–68, 109
twin studies, 45
    adoption studies, 50–54
    Bouchard's Minnesota study, 54
    common genes for brain structure and
        intelligence, 139–145
    cortical thickness, 95, 141
    dizygotic (fraternal) twins, 50
    Dutch twin study, 55–56
    effect of age at intelligence testing, 55
    genetic contribution to intelligence, 55
    heritability of general intelligence, 55
    heritability of the g-factor, 60–61
    monozygotic (identical) twins, 50
    neuroimaging and molecular genetics
        studies, 145–147

role of genetics in intelligence, 50–54
Sir Cyril Burt, 50–54

UK Biobank Project, adoption study, 51
University of California, 18
US military, IQ cutoff for recruits, 220

Val66Met gene polymorphism, 67, 145
video gaming, relationship to intelligence,
    166–167
voxel-based morphometry (VBM), 92–96
voxels, 90, 110

Watson (IBM computer), 4, 11
Watson, John B., 40
Wax, Amy, 218–219
Wechsler Adult Intelligence Scale
    (WAIS), 13–15
Wechsler Intelligence Scale for Children
    (WISC), 15
Wechsler IQ scores, connectivity
    differences related to, 122
white matter
    correlations with intelligence, 112–113
        sex differences, 113
    evidence of association with working
        memory, 115
    evidence of relationship to g-factor, 113
    fractional anisotropy (FA) studies,
        141–144
    MRI techniques, 97–98
Wilshire, Stephen, 3
Wilson, Edward O., 26
working memory, evidence of association
    with white matter, 115
working memory training programs, effects
    of, 163

X-ray imaging of the brain, 76

Young Adult Human Connectome Project,
    136

Zuckerberg, Mark, 30